Lecture Notes in Electrical Engineering

Volume 1298

The book series *Lecture Notes in Electrical Engineering* (LNEE) publishes the latest developments in Electrical Engineering—quickly, informally and in high quality. While original research reported in proceedings and monographs has traditionally formed the core of LNEE, we also encourage authors to submit books devoted to supporting student education and professional training in the various fields and applications areas of electrical engineering. The series cover classical and emerging topics concerning:

- Communication Engineering, Information Theory and Networks
- Electronics Engineering and Microelectronics
- Signal, Image and Speech Processing
- Wireless and Mobile Communication
- Circuits and Systems
- Energy Systems, Power Electronics and Electrical Machines
- Electro-optical Engineering
- Instrumentation Engineering
- Avionics Engineering
- Control Systems
- Internet-of-Things and Cybersecurity
- Biomedical Devices, MEMS and NEMS

For general information about this book series, comments or suggestions, please contact leontina.dicecco@springer.com.

To submit a proposal or request further information, please contact the Publishing Editor in your country:

China

Jasmine Dou, Editor (jasmine.dou@springer.com)

India, Japan, Rest of Asia

Swati Meherishi, Editorial Director (Swati.Meherishi@springer.com)

Southeast Asia, Australia, New Zealand

Ramesh Nath Premnath, Editor (ramesh.premnath@springernature.com)

USA, Canada

Michael Luby, Senior Editor (michael.luby@springer.com)

All other Countries

Leontina Di Cecco, Senior Editor (leontina.dicecco@springer.com)

**** This series is indexed by EI Compendex and Scopus databases. ****

Shailendra Kumar · Jitendra Kumar ·
Narayana Prasad Padhy · Sukumar Kamalasadan ·
Premalata Jena
Editors

Soft Computing Applications in Modern Power and Energy Systems

Select Proceedings of EPREC 2024, Volume 4

Springer

Editors
Shailendra Kumar
Department of Electrical Engineering
Indian Institute of Technology Bhilai
Bhilai, Chhattisgarh, India

Jitendra Kumar
Department of Electrical Engineering
National Institute of Technology Jamshe
Jamshedpur, Jharkhand, India

Narayana Prasad Padhy
Department of Electrical Engineering
Indian Institute of Technology Roorkee
Roorkee, Uttarakhand, India

Sukumar Kamalasadan
Department of Electrical Engineering
University of North Carolina at Charlotte
Charlotte, NC, USA

Premalata Jena
Department of Electrical Engineering
Indian Institute of Technology Roorkee
Roorkee, Uttarakhand, India

ISSN 1876-1100 ISSN 1876-1119 (electronic)
Lecture Notes in Electrical Engineering
ISBN 978-981-96-0057-1 ISBN 978-981-96-0058-8 (eBook)
https://doi.org/10.1007/978-981-96-0058-8

This Springer imprint is published by the registered company Springer Nature Singapore Pte Ltd.
The registered company address is: 152 Beach Road, #21-01/04 Gateway East, Singapore 189721, Singapore

Preface

The Electric Power and Renewable Energy Conference (EPREC-2024) organized by the Department of Electrical Engineering, National Institute of Technology Jamshedpur, India, from May 24 to 26, 2024. We thank all the contributors for maintaining the high standard and making EPREC-2024 a huge success. This volume, i.e., Soft Computing Applications in Modern Power and Energy Systems-Select Proceedings of EPREC 2024, is one of four volumes to be published by Springer in the book series "Lecture Notes in Electrical Engineering (LNEE)". It contains 25 high-quality papers.

This book entitled "Soft Computing Applications in Modern Power and Energy Systems select proceedings of EPREC-2024" provides the rigorous discussions, case studies, and recent developments in the areas of soft computing and its applications in power systems enabled with power electronics-based equipment, energy systems, and the energy community. The other topics to be covered are optimal planning, analysis, operation, and control related to modern power and energy systems, and applications of various soft computing methodologies. The readers find this book useful for enhancing their knowledge and skills in the domain areas.

We thank all the authors, organizing committee members, technical program committee members, reviewers, and student coordinators for their valuable contributions and volunteer work. We also appreciate the role of session chairs/co-chairs and thank the series editors of LNEE and Ms. Kamiya Katter, Associate Editors, Springer for their help and quick responses during the preparation of the volume.

Toledo, USA
Delhi, India
Oshawa, Canada

Shailendra Kumar
Jitendra Kumar
Narayana Prasad Padhy
Sukumar Kamalasadan
Premalata Jena

Contents

About the Editors

Shailendra Kumar is currently working as an Assistant Professor in the Department of Electrical Engineering Indian Institute of Technology (IIT) Bhilai. He received a B.Tech. Degree in electrical and electronics engineering from Uttar Pradesh Technical University, Lucknow, India, in 2010, and the M.Tech. Degree in power electronics, electrical machine, and drives (PEEMD) from the Indian Institute of Technology, Delhi, India, in 2015. He completed his Ph.D. degree in the Department of Electrical Engineering from the Indian Institute of Technology Delhi, New Delhi, India in May 2019. He has worked as a postdoctoral research fellow at Petroleum Institute of Research Centre Khalifa University from May 2019 to November 2019. He has also worked as an Assistant Professor at Maulana Azad National Institute of Technology, Bhopal from June 2020 to May 2023. His research interests include charging stations for EVs, power quality, grid integration of renewable energy resources, custom power devices, and microgrids. Dr. Kumar was the recipient of the prestigious POSOCO power system award (in Master as well as Doctoral categories) in 2016 and 2019. He is also a recipient of Prof. Som Nath Mahendra's Student Travel Awards for the IEEE PEDES 2018 conference and the IEEE UPCON Best Paper Award in 2016 and 2018. Dr. Kumar is a senior member of IEEE and a member of INAE Societies.

Jitendra Kumar is currently an Assistant Professor at the Department of Electrical Engineering, National Institute of Technology Jamshedpur, India. He received his B.Tech. from IMS Engineering College, Ghaziabad, India, in 2009, his M.Tech from the National Institute of Technology Kurukshetra, Kurukshetra, in 2011, and the Ph.D. degree in electrical engineering from the Indian Institute of Technology Roorkee, Roorkee, India, in 2017. He has over 5 years of teaching experience and taught subjects like Power systems, Advanced Power System, Control System, Control and instrumentation, Power System Operation and Control, Signal and system, Power System Protection, Circuit and Network Theory. He has many papers in reputed journals. His major areas of research interests include power system protection and restructuring, Protection algorithms design in smart grid and microgrid environments, and Design of protection algorithms in FACTS environment. The research area also includes protection strategies during power swings.

Narayana Prasad Padhy (Senior Member, IEEE) received a Ph.D. degree in power systems engineering from Anna University, Chennai, India, in 1997. He is an Institute Chair Professor with the Department of Electrical Engineering, Indian Institute of Technology Roorkee, Roorkee, India. He is heading many national and international projects, such as DSIDES, ID-EDGe, and HEAPD. He is also part of other international projects, namely Indo–US UI-ASSIST and Indo-UK ZED-I. His research interests include power system analysis, optimization, demand-side management, and smart grids. Dr. Padhy is also a Fellow of the Indian National Academy of Engineers, Institution of Engineering and Technology, and Institution of Engineers India.

Sukumar Kamalasadan (Senior Member, IEEE) received the B.Tech. degree in electrical and electronics engineering from the University of Calicut, Malappuram, India, in 1991, the M.Eng. degree in electrical power systems management from the Asian Institute of Technology, Bangkok, Thailand, in 1999, and the Ph.D. degree in electrical engineering from The University of Toledo, Toledo, OH, USA, in 2004. He is currently a Professor with the Department of Electrical and Computer Engineering, University of North Carolina at Charlotte, Charlotte, NC, USA. His research interests include intelligent and autonomous control, power systems dynamics, stability and control, smart grid, microgrid, and real-time optimization and control of the power system. Dr. Kamalasadan was the recipient of several awards, including the NSF CAREER Award and IEEE Best Paper Award.

Premalata Jena is currently an Associate Professor at the Department of Electrical Engineering, Indian Institute of Technology Roorkee, Uttarakhand, India. She received her B.Tech. degree in electrical engineering from the Utkal University, Bhubaneswar, India, in 2001, the M.Tech. degree in Electrical Engineering from the Indian Institute of Technology Kharagpur, Kharagpur, India, in 2006, and the Ph.D. degree from the Indian Institute of Technology Kharagpur, Kharagpur, India, in 2011. She has seven years of teaching experience. She is an IEEE member and INAE, Young Associate. She received many awards such as the Women Excellence Award and Early Carrier Research Award, SERB, DST, Government of India, New Delhi, INAE Young Engineer Award, POSOCO Power System Award, in 2013. She has published many papers in different reputed journals and conferences. Her fields of interest are Smart Grid, Smart grid technology, and protection, Microgrid, Microgrid Protection, Signal processing application to power system relaying, Power system Protection, Protection Issues with FACTS Devices, Protection Scheme Development for a line with FACTS devices, Disturbance localization, Signal processing application for disturbance localization.

Decentralized Energy Transactions in Microgrids Using Blockchain Technology

Debani Prasad Mishra, Sandip Ranjan Behera, Umamani Subudhi, Aditya Ranjan Patro, and Sitanshu Kumar Sahoo

Abstract The increasing integration of decentralized energy sources in microgrids has prompted the need for efficient and secure mechanisms to facilitate energy transactions. Blockchain emerges as most promising solution to address the problems associated with trust, transparency, and coordination in decentralized energy systems. This research paper explores the uses of blockchain in microgrids to enable decentralized energy transactions. The proposed framework leverages smart contracts to automate and validate energy transactions, ensuring a transparent and tamper-resistant record of energy flow. Through the use of a decentralized ledger, participants in the microgrid can securely and autonomously engage in energy transactions, fostering a more resilient and efficient energy ecosystem. The study also addresses key considerations such as scalability, privacy, and consensus mechanisms to enhance the practicality of implementing blockchain in microgrid environments. The findings contribute to the growing body of knowledge on decentralized energy systems, showcasing the potential of blockchain to revolutionize the way energy is produced, consumed, and traded within microgrids.

Keywords Microgrid · Blockchain · Smart contracts · Energy transactions · Distributed ledger · Smart grid · Renewable energy

D. P. Mishra (✉) · S. R. Behera · U. Subudhi · A. R. Patro · S. K. Sahoo
Department of Electrical Engineering, IIIT Bhubaneswar, Bhubaneswar, Odisha, India
e-mail: debani@iiit-bh.ac.in

S. R. Behera
e-mail: b32004@iiit-bh.ac.in

U. Subudhi
e-mail: umamani@iiit-bh.ac.in

A. R. Patro
e-mail: b320004@iiit-bh.ac.in

S. K. Sahoo
e-mail: b320055@iiit-bh.ac.in

S. Kumar et al. (eds.), *Soft Computing Applications in Modern Power and Energy Systems*, Lecture Notes in Electrical Engineering 1298,
https://doi.org/10.1007/978-981-96-0058-8_1

1 Introduction

Microgrids are emerging as an effective solution for promoting energy sustainability and resilience. However, the centralized nature of energy transactions in microgrids poses several challenges, including inefficiency, lack of transparency, and vulnerability to cyber-attacks. To address these issues, decentralization and blockchain technology have gained attention as promising approaches [1, 2]. This research aims to explore the potential of decentralized energy transactions in microgrids using blockchain technology as a means to enhance efficiency, transparency, and security in energy trading [3–5]. By utilizing a decentralized approach, participants in microgrids can engage in peer-to-peer energy trading, enabling greater control and autonomy over their energy transactions. Additionally, the implementation of blockchain technology can ensure the immutability, transparency, and traceability of energy transactions in microgrids [1, 6, 7].

1.1 Background on Microgrids and their Importance in the Energy Sector

Microgrid systems represent compact power networks capable of operating independently or in coordination with the main power grid [8–12]. These systems integrate local generation sources, energy storage, and distribution infrastructure, enabling them to provide reliable and resilient energy supply. In recent years, microgrids have gained significant traction in the energy sector due to their capacity to incorporate renewable energy sources, bolster grid stability, and extend electricity access to remote regions. This research delves into the application of blockchain technology to facilitate decentralized energy transactions within microgrid environments [13–16] (Fig. 1).

1.2 Overview of the Challenges in Energy Transactions Within Microgrids

Implementing decentralized energy systems within microgrids presents various challenges that require careful consideration for successful execution. Firstly, there is a lack of standardized protocols and interoperability among different microgrid systems, hindering efficient energy transactions. Secondly, the fluctuating nature of renewable sources makes it difficult to maintain a balance between supply and demand within microgrids. Thirdly, traditional centralized energy markets have complex regulations and policies that may not be suitable for the decentralized nature of microgrids. Finally, the issue of trust and transparency arises as multiple stakeholders are involved in energy transactions, requiring a reliable and secure system

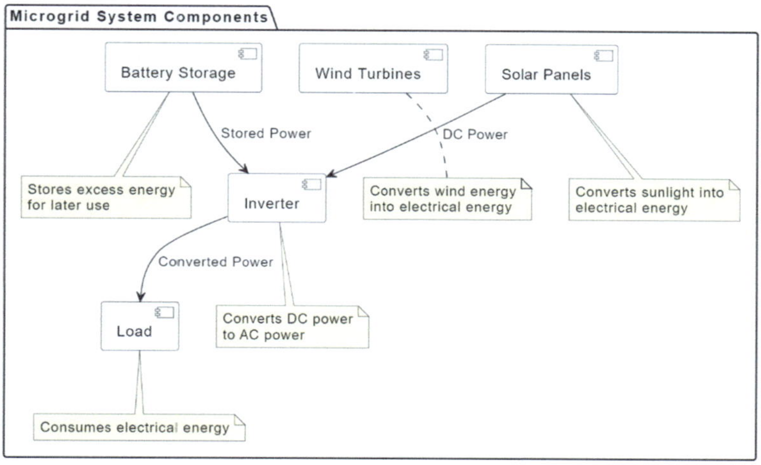

Fig. 1 Microgrid system components

for recording and validating transactions. To overcome these challenges, blockchain technology can offer solutions by offering a blockchain based and secure foundation for transparent and efficient energy transactions within microgrids [17, 18].

1.3 Introduction to Blockchain Technology and its Potential in Decentralized Energy Transactions

Blockchain technology has attracted considerable attention in recent years for its potential to revolutionize various industries, including energy transactions. This technology allows for the creation of a distributed ledger that provides transparency, security, and trust in recording and verifying transactions. In the context of decentralized energy transactions in microgrids, blockchain can offer several benefits such as enabling peer-to-peer energy trading, reducing operational costs, enhancing energy efficiency, and promoting renewable energy integration. With blockchain's ability to eliminate middlemen and provide a decentralized platform, it has the potential to reshape the energy landscape and empower individuals and communities to participate actively in energy markets. Furthermore, blockchain technology can facilitate the development of innovative business models and ensure equitable distribution of renewable energy resources [19, 20].

2 Microgrids and Energy Transactions

Microgrids are compact power grids capable of operating independently or in coordination with the primary grid. They can generate, store, and distribute electrical energy in a localized area. In the context of energy transactions, microgrids are ideal for implementing decentralized systems. These systems enable energy transactions between participants within the microgrid network, such as homes, businesses, or even electric vehicles. By leveraging blockchain technology, the transactions can be securely recorded and verified, ensuring transparency and trust among participants. Figure 6 illustrates the concept of energy transactions within a microgrid using blockchain technology [21, 22].

2.1 *Definition and Characteristics of Microgrids*

Microgrids are small-scale electrical networks capable of operating independently or in conjunction with the main grid. They are distinguished by their capability to generate, store, and distribute electricity locally, often using renewable energy sources. Microgrids have the ability to seamlessly integrate various energy resources and provide reliable power supply to local communities.

They provide enhanced energy efficiency, decreased carbon emissions, and increased power system resilience.

2.2 *Importance of Efficient Energy Transactions Within Microgrids*

Efficient energy transactions within microgrids are of paramount importance for ensuring the seamless operation and optimization of the system. Through the use of blockchain technology, these transactions can be tracked, validated, and settled in a transparent and secure manner, ultimately promoting energy efficiency and sustainability within the grid.

2.3 *Current Challenges in Energy Transactions Within Microgrids*

Current challenges in energy transactions within microgrids include issues such as lack of standardization, complexity in metering and billing systems, and the need for transparent and secure transactions. These challenges hinder the efficient and reliable exchange of energy within microgrids, emphasizing the need for innovative solutions like blockchain technology.

3 Decentralized Energy Transactions Using Blockchain Technology

One of the primary benefits of employing blockchain technology for decentralized energy transactions in microgrids is the elimination of intermediaries, such as utilities or energy suppliers. Blockchain enables peer-to-peer transactions, allowing users to directly sell excess energy to others in the microgrid without the need for a third party. This reduces transaction costs, enhances efficiency, and provides more control and autonomy for the participants.

3.1 Concept of Decentralized Energy Transactions Within Microgrids

In order to facilitate seamless and secure energy transactions within microgrids, the concept of decentralization plays a significant role. Decentralization refers to the distribution of decision-making authority and control among multiple entities within the microgrid. It allows for peer-to-peer energy exchange, enabling the participants to directly engage in transactions without the need for intermediaries. This concept empowers the individual nodes within the microgrid to make autonomous decisions regarding energy production, consumption, and trading.

3.2 Benefits of Utilizing Blockchain Technology for Energy Transactions

One of the primary advantages of employing blockchain for energy transactions is the heightened transparency and security it provides. Blockchain's immutable nature of blockchain guarantees that all transactions are documented and resistant to manipulation. This can help prevent fraud and increase trust among participants in energy transactions. Additionally, blockchain technology allows for real-time monitoring and tracking of energy usage enabling more efficient and accurate billing processes.

3.3 *Case Studies and Examples of Successful Implementation of Blockchain in Microgrid Energy Transactions*

Several case studies and examples have demonstrated the successful implementation of blockchain in microgrid energy transactions. One such example is the Brooklyn Microgrid project in New York City, which enables peer-to-peer energy transactions within a local community. Participants can buy and sell energy using blockchain technology, which ensures transparency, security, and efficiency in the transaction process. Another notable case study is the Power Ledger project in Australia, where blockchain is used to enable solar energy trading among neighbors. This platform allows solar energy producers to sell additional energy directly to nearby customers, thereby optimizing the use of renewable energy sources. The successful implementation of blockchain in these case studies highlights the potential of this technology to revolutionize energy transactions within microgrids, bringing about a more decentralized and sustainable energy system (Fig. 2).

In addition to enabling secure and transparent energy transactions, blockchain technology can also facilitate the decentralized nature of microgrids. Through a peer-to-peer network and consensus protocols, blockchain ensures that all participants in the microgrid have equal access and control over energy transactions, contributing to a resilient and efficient energy system.

4 Implementation of Blockchain in Microgrids

In this section, we delve into the practical aspects of deploying blockchain technology in microgrids, emphasizing the use of Solidity, a programming language specifically designed for writing smart contracts on the Ethereum blockchain. The implementation aims to showcase how smart contracts can facilitate and secure decentralized energy transactions within a microgrid environment.

4.1 *Smart Contract Design*

The foundation of our implementation lies in the design of smart contracts. We employ Solidity to code the essential functions that govern energy transactions. These functions include the following (Fig. 3).

In this simplified example, we create a smart contract named **EnergyTransaction** with functions to transfer energy between participants. The **gridOperator** is initialized during contract deployment, and only the grid operator can execute the **transferEnergy** function.

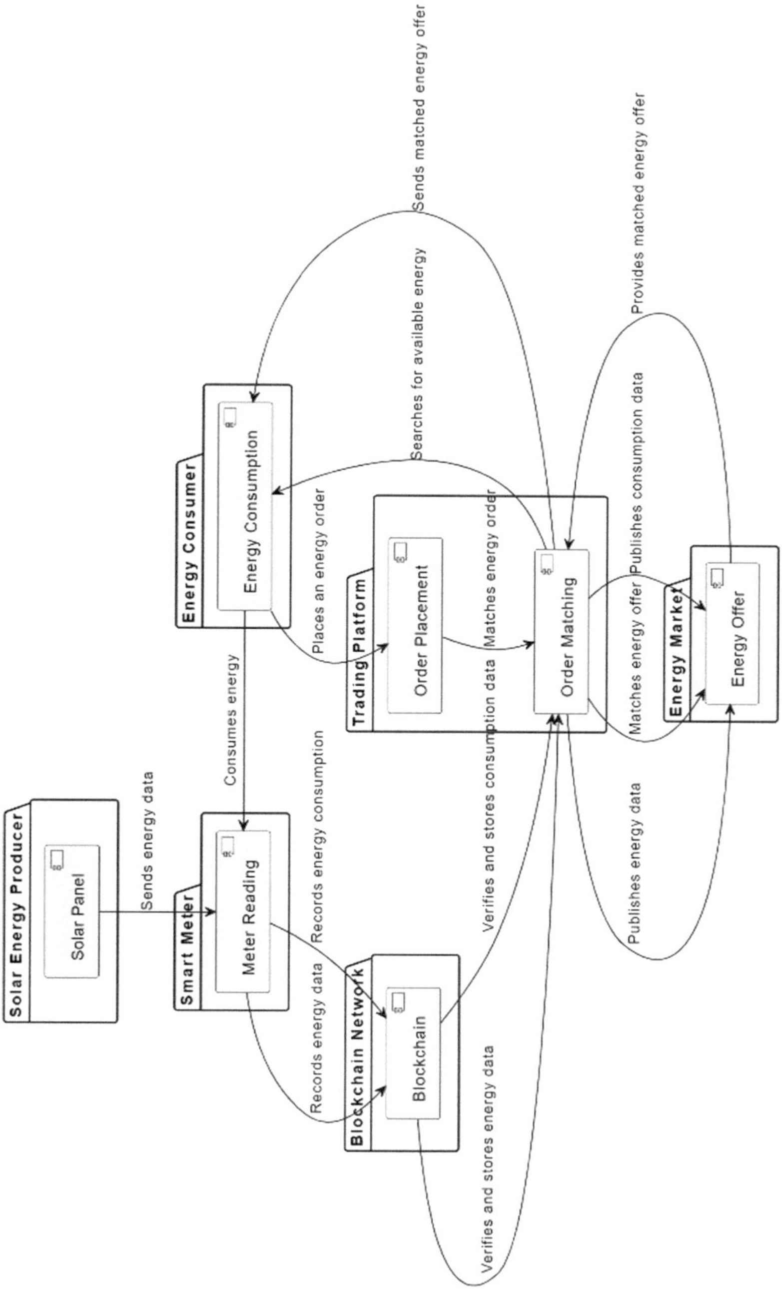

Fig. 2 Solar neighbor blockchain energy trading

```
// Solidity Smart Contract for Energy Transactions
pragma solidity ^0.8.0;

contract EnergyTransaction {
    address public gridOperator;
    mapping(address => uint256) public energyBalances;

    event EnergyTransferred(address indexed from, address indexed to,
    uint256 energyAmount);

    constructor() {
        gridOperator = msg.sender;
    }

    modifier onlyGridOperator() {
        require(msg.sender == gridOperator, "Only the grid operator can
        execute this function");
        _;
    }

    function transferEnergy(address _recipient, uint256 _amount) public
    onlyGridOperator {
        require(energyBalances[msg.sender] >= _amount, "Insufficient energy
        balance");

        energyBalances[msg.sender] -= _amount;
        energyBalances[_recipient] += _amount;

        emit EnergyTransferred(msg.sender, _recipient, _amount);
    }
}
```

Fig. 3 Solidity smart contract for energy transactions

4.2 *Deployment on the Ethereum Blockchain*

The smart contract is deployed on the Ethereum blockchain, providing a decentralized and tamper-resistant platform for microgrid participants to engage in energy transactions. The deployment process involves compiling the Solidity code, generating a bytecode, and then interacting with the Ethereum network through a wallet or a development environment like Remix.

4.3 *Interfacing with the Microgrid*

Microgrid participants, including energy producers and consumers, can interact with the smart contract through user interfaces or automated systems. Transactions are recorded on the blockchain, ensuring transparency and immutability.

4.4 Considerations and Optimizations

To enhance scalability and reduce transaction costs, considerations such as gas optimization and efficient contract design are crucial. Additionally, privacy concerns can be addressed through off-chain solutions or privacy-focused blockchain platforms. This implementation serves as a foundational model for integrating blockchain technology into microgrids, offering a secure and transparent framework for decentralized energy transactions. This section outlines a basic smart contract for energy transactions using Solidity, providing a starting point for further exploration and optimization based on specific microgrid requirements and considerations.

5 Technical Aspects of Blockchain-Based Energy Transactions

One of the crucial technical facets of blockchain-based energy transactions is the consensus mechanism. In a decentralized energy transaction system, a consensus mechanism is needed to validate and record transactions on the blockchain. This ensures that all participants in the network agree on the validity of each transaction, thereby preventing any fraudulent or malicious activities. Various consensus mechanisms such as Proof of Work (PoW), Proof of Stake (PoS), and Byzantine Fault Tolerance (BFT) have been proposed and implemented in blockchain-based energy transaction platforms. These mechanisms determine the process by which new blocks are added to the blockchain, and they have a significant impact role in maintaining the security and efficiency of the system.

5.1 Smart Contracts and their Role in Energy Transactions

Smart contracts play a crucial role in energy transactions within microgrids, as they enable automated and secure execution of agreements. These self-executing contracts are coded on blockchain platforms, ensuring transparency, accuracy, and trustworthiness in managing energy payments, consumption, and distribution. This diagram provides an overview of how smart contracts function in energy transactions in a decentralized microgrid system.

5.2 *Consensus Mechanisms and their Impact on Energy Transactions*

Consensus mechanisms play a crucial role in guaranteeing the validity and security of energy transactions in microgrids using blockchain technology. These mechanisms, like Proof of Work (PoW) and Proof of Stake (PoS), dictate the process by which transactions are authenticated and incorporated into the blockchain. They have a direct impact on the energy consumption and efficiency of the system.

5.3 *Scalability and Interoperability Challenges in Blockchain-Based Energy Transactions*

Scalability and interoperability are key challenges in implementing blockchain technology for energy transactions in microgrids. As the volume of transactions grows, the blockchain network encounters scalability issues related to processing speed and capacity. Additionally, securing compatibility across several blockchain platforms and legacy energy systems is essential for smooth integration and effective implementation of blockchain in the energy industry. Decentralized energy transactions in microgrids using blockchain technology offer a promising solution to optimize energy distribution and mitigate the challenges faced by centralized systems. In this context, a key component of the system is the smart contract, which ensures secure and automated energy transactions between prosumers and consumers (Fig. 4).

6 Security and Privacy Considerations

One of the key concerns when implementing blockchain technology in microgrids for decentralized energy transactions is ensuring the security and privacy of the system. The decentralized nature of blockchain offers some inherent security benefits, such as transparency and immutability. However, it also introduces certain vulnerabilities that must be addressed. For instance, the consensus mechanism used in blockchain can be susceptible to 51% attacks, where a solitary entity commands the most of the network's computational capacity and has the ability to manipulate transactions. To mitigate this risk, a robust consensus mechanism such as proof of stake or proof of authority can be employed. Moreover, privacy is a critical consideration in energy transactions, as sensitive information such as energy consumption patterns and financial transactions need to be protected from unauthorized access. Methods like zero-knowledge proofs and secure computation techniques can be utilized to provide privacy-preserving features in the microgrid blockchain system. Overall, a comprehensive security and privacy framework will be essential to ensure the integrity and confidentiality of decentralized energy transactions in microgrids.

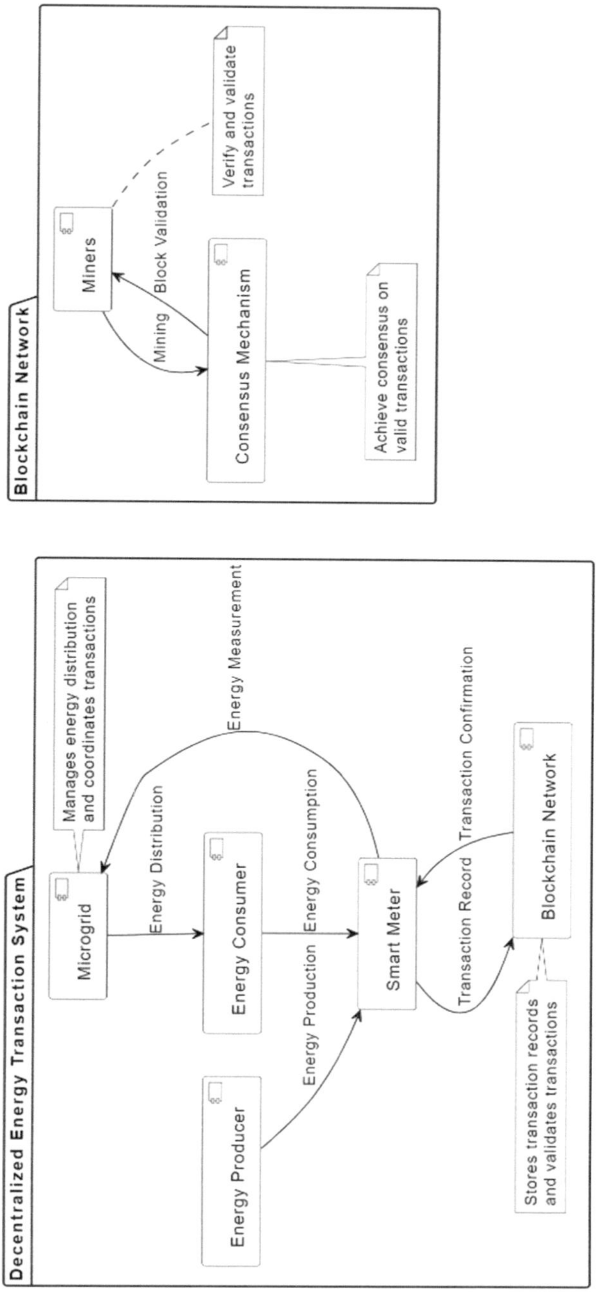

Fig. 4 Architecture of a decentralized energy transaction system in microgrids using blockchain technology

6.1 Overview of Security Risks and Vulnerabilities in Blockchain-Based Energy Transactions

The security risks and vulnerabilities in blockchain-based energy transactions are a crucial aspect to consider. As transactions occur on a decentralized and transparent platform, they are vulnerable to various threats such as double-spending, 51% attack, and smart contract vulnerabilities. These risks highlight the importance of implementing robust security measures to guarantee the confidentiality and integrity of energy transactions in blockchain-based microgrids.

6.2 Privacy Concerns and Potential Solutions in Decentralized Energy Transactions

One major concern in decentralized energy transactions using blockchain technology is privacy. Since blockchain technology stores all transaction data on a public ledger, it raises concerns about the privacy of participants. However, potential solutions such as zero-knowledge proofs and private channels can address these concerns and ensure privacy in decentralized energy transactions.

6.3 Regulatory and Legal Implications of Blockchain in Energy Sector

Regulatory and legal implications of blockchain in the energy sector pose both challenges and opportunities. The adoption of blockchain introduces new complexities related to data privacy, security, and network governance. Additionally, compliance with existing regulations and the development of new regulations specifically tailored to blockchain-based energy transactions are necessary.

Furthermore, the implementation of blockchain technology in microgrids enables decentralized energy transactions. In this system, multiple nodes connected to the microgrid network can participate in energy trading without the need for intermediaries. This decentralized approach ensures transparency, efficiency, and security, as validated transactions are recorded chronologically in a distributed ledger called the blockchain. Transactions within the microgrid are facilitated through smart contracts, which automatically execute predefined conditions once they are met. The diagram below illustrates the concept of decentralized energy transactions in microgrids using blockchain technology.

7 Future Directions and Challenges

As the adoption of blockchain technology in microgrids continues to grow, several future directions and challenges need to be considered. Firstly, there is a need for standardized protocols and frameworks to ensure interoperability and seamless integration of different blockchain systems within microgrids. Secondly, scalability remains a challenge as blockchain networks need to handle a large number of transactions and ensure fast confirmation times. The implementation of sharding techniques and off-chain solutions may help address scalability concerns. Thirdly, privacy and security issues need to be thoroughly addressed to protect sensitive information and prevent any malicious activities within the microgrid ecosystem. Developing efficient consensus mechanisms and secure identification protocols can contribute to enhancing privacy and security. Lastly, regulatory frameworks and policies must be developed to encourage the deployment of blockchain-based microgrid systems, ensuring compliance with existing energy regulations and fostering innovation in the energy sector (Figs. 5, 6 and 7).

This diagram illustrates the need for standardized protocols and frameworks to enable interoperability and seamless integration between different blockchain systems in microgrids. It highlights the importance of standardizing data formats, communication protocols, and consensus mechanisms to facilitate efficient energy transactions within microgrids.

This diagram presents potential solutions for addressing scalability concerns in blockchain-based microgrids. It visualizes the implementation of sharding techniques and off-chain solutions to increase transaction throughput and reduce confirmation times. These techniques help enhance the scalability of blockchain networks within microgrids.

This diagram showcases the privacy and security measures that can be implemented in blockchain-based microgrids to safeguard sensitive information and prevent malicious activities. It depicts the integration of efficient consensus mechanisms and secure identification protocols to enhance the privacy and security of energy transactions within microgrids.

8 Conclusion

In conclusion, decentralized energy transactions in microgrids using blockchain technology offer several advantages. It mitigates the need for intermediaries in energy transactions, increases transparency and trust among participants, enables peer-to-peer energy trading, and promotes the use of renewable energy sources. Furthermore, the use of smart contracts ensures automated and efficient energy transactions. However, challenges such as scalability, regulatory frameworks, and cybersecurity need to be addressed to fully realize the potential of blockchain technology

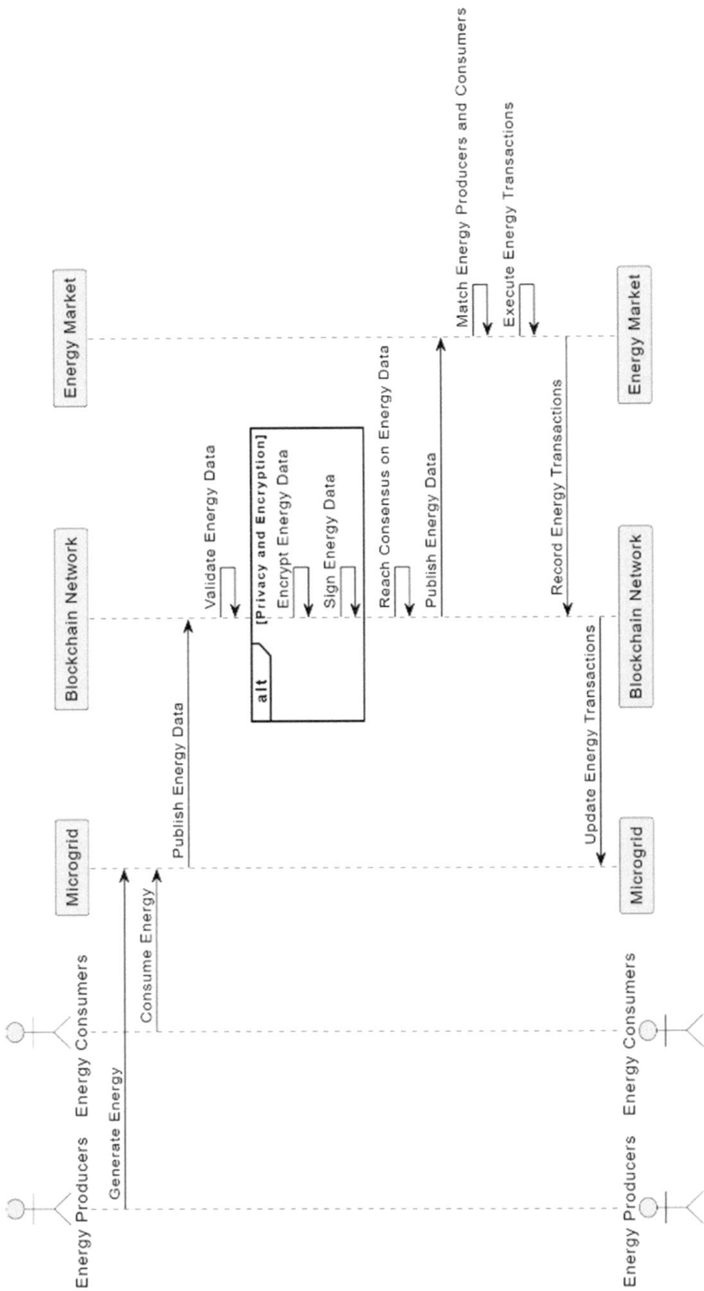

Fig. 5 Standardized protocols and frameworks

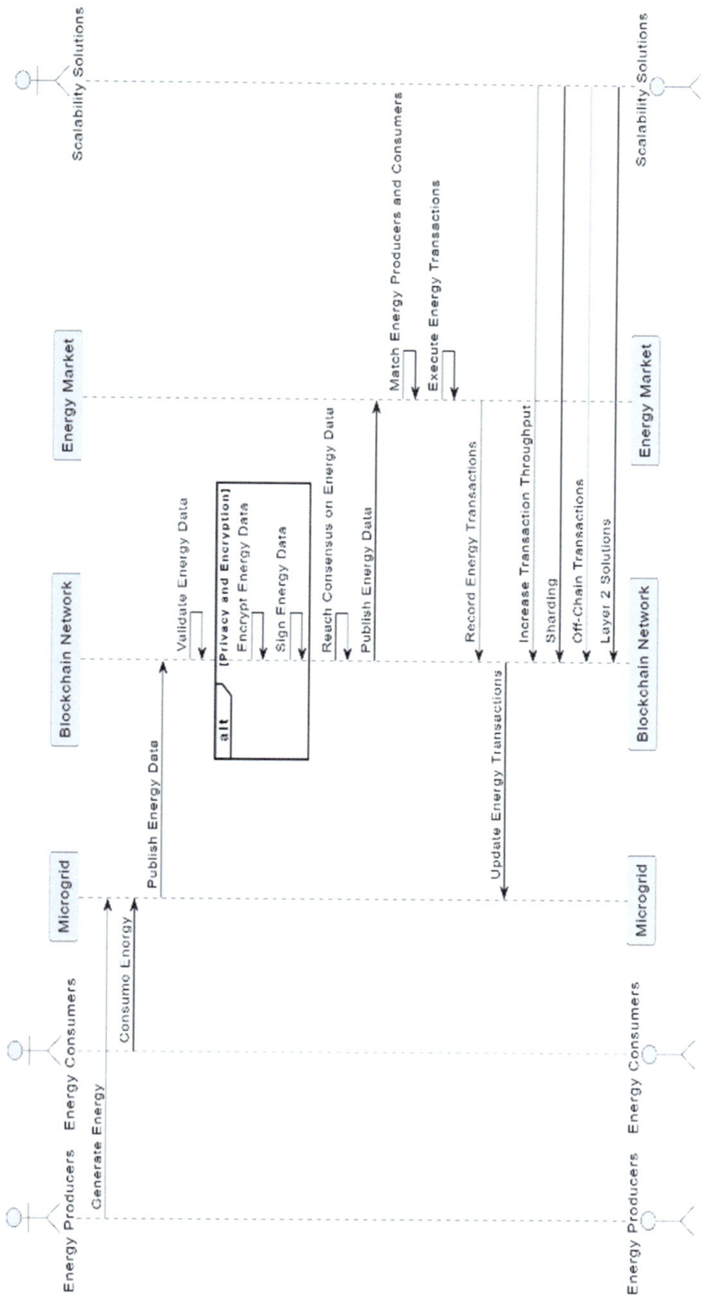

Fig. 6 Scalability solutions

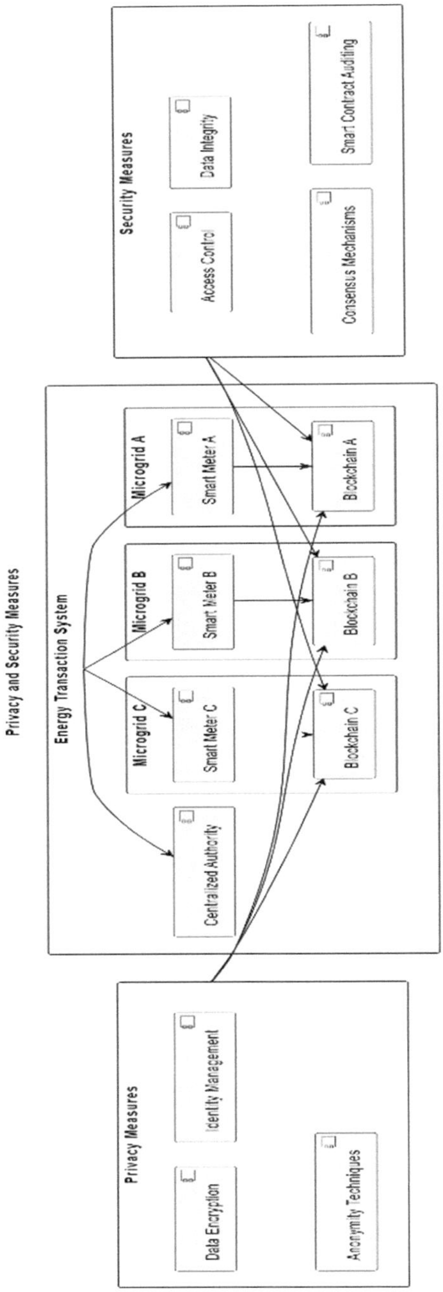

Fig. 7 Privacy and security measures

in microgrid energy transactions. Future research should focus on developing solutions to these challenges and implementing pilot projects to test the feasibility and effectiveness of blockchain-based microgrid systems.

References

1. Ray P, Mishra DP, Dey K, Mishra P (2017) Fault detection and classification of a transmission line using discrete wavelet transform and artificial neural network. In: 2017 international conference on information technology (ICIT), Bhubaneswar, India, pp 178–183. https://doi.org/10.1109/ICIT.2017.24
2. Behera SR, Mishra DP, Parida S, Patro AR (2023) Proposal of user-friendly design Of NFT marketplace. In: 2023 4th international conference on computing and communication systems (I3CS), Shillong, India, pp 1–6. https://doi.org/10.1109/I3CS58314.2023.10127468
3. Ray P, Mishra DP, Lenka RK (2016) Short term load forecasting by artificial neural network. In: 2016 international conference on next generation intelligent systems (ICNGIS), Kottayam, India, pp 1–6. https://doi.org/10.1109/ICNGIS.2016.7854003
4. Ray P, Mishra DP, Mohaptra S (2016) Fault classification of a transmission line using wavelet transform and fuzzy logic. In: 2016 IEEE 1st international conference on power electronics, intelligent control and energy systems (ICPEICES), Delhi, India, pp 1–6. https://doi.org/10.1109/ICPEICES.2016.7853293
5. Sahu JK, Mishra SK, Satapathy S, Behera SR (2023) Comparison of MPPT algorithms using ANN and ANFIS in a standalone photovoltaic system. In: 2023 2nd international conference on automation, computing and renewable systems (ICACRS), Pudukkottai, India, pp 195–200. https://doi.org/10.1109/ICACRS58579.2023.10405330
6. Hota AP, Mishra S, Mishra DP (2022) Power/energy loss allocation in deregulated power distribution system with load factor and load power factor variation. Int J Syst Assur Eng Manag 13:250–266. https://doi.org/10.1007/s13198-021-01227-3
7. Mishra DP, Ray P, Arya SR (2019) Comparative analysis of classification and location of faults in a long transmission line using random forest technique. In: 2019 IEEE 1st international conference on energy, systems and information processing (ICESIP), Chennai, India, pp 1–5. https://doi.org/10.1109/ICESIP46348.2019.8938331
8. Ali SM, Gupta N, Nayak GK, Lenka RK (2016) Big data visualization: tools and challenges. In: 2016 2nd international conference on contemporary computing and informatics (IC3I), Greater Noida, India, pp 656–660. https://doi.org/10.1109/IC3I.2016.7918044
9. Lenka RK, Dey MR, Bhanse P, Barik RK (2018) Performance and load testing: tools and challenges. In: 2018 international conference on recent innovations in electrical, electronics and communication engineering (ICRIEECE), Bhubaneswar, India, pp 2257–2261. https://doi.org/10.1109/ICRIEECE44171.2018.9009338
10. Lenka RK, Satapathy U, Dey M (2018) Comparative analysis on automated testing of web-based application. In: 2018 international conference on advances in computing, communication control and networking (ICACCCN), Greater Noida, India, pp 408–413. https://doi.org/10.1109/ICACCCN.2018.8748374
11. Mallick SR, Goswami V, Lenka RK, Patra S, Kumar V, Barik RK (2023) Performance evaluation of priority queueing assisted IoST-Fog-Blockchain framework in geospatial cloud environment. In: 2023 international conference on microwave, optical, and communication engineering (ICMOCE), Bhubaneswar, India, pp 1–4. https://doi.org/10.1109/ICMOCE57812.2023.10167317
12. Mishra DP, Behera SR, Ojha PJ, Sethy SK, Sahoo AK (2023) Cultivating resilience: advancing indian agriculture through decentralized blockchain crop insurance. In: 2023 2nd international conference on automation, computing and renewable systems (ICACRS), Pudukkottai, India, pp 1064–1070. https://doi.org/10.1109/ICACRS58579.2023.10404961

13. Baral A, Mallick SR, Sahoo A, Sudershana S, Lenka RK, Barik RK (2023) A smart and secure blockchain-enabled framework for healthcare applications. In: 2023 1st international conference on circuits, power and intelligent systems (CCPIS), Bhubaneswar, India, pp 1–6. https://doi.org/10.1109/CCPIS59145.2023.10292016
14. Gaur RK, Lenka RK, Prakash S, Patra SS, Barik RK (2022) FogCrypto: towards cryptographic protocols for securing fog assisted cloud system. In: 2022 IEEE 2nd international symposium on sustainable energy, signal processing and cyber security (ISSSC), Gunupur, Odisha, India, pp 1–5. https://doi.org/10.1109/iSSSC56467.2022.10051284
15. Santhiya K, Vivek Anand M (2023) Agriculture basedfood supply chain traceability using blockchain. In: 2023 2nd international conference on advancements in electrical, electronics, communication, computing and automation (ICAECA), Coimbatore, India, pp 1–6. https://doi.org/10.1109/ICAECA56562.2023.10200445
16. Alam S (2022) Security concerns in smart agriculture and blockchain-based solution. In: 2022 OPJU international technology conference on emerging technologies for sustainable development (OTCON), Raigarh, Chhattisgarh, India, pp 1–6. https://doi.org/10.1109/OTCON56053.2023.10113953
17. Hwerbi K, Amdouni I, Laouiti A, Adjih C, Saidane L (2023) Veterinary drone: blockchain-based system for cattle health monitoring. In: 2023 international wireless communications and mobile computing (IWCMC), Marrakesh, Morocco, pp 1613–1618. https://doi.org/10.1109/IWCMC58020.2023.10182480
18. Surya M, Manohar S (2023) An interpretation of the challenges and solutions for agriculture-based supply chain management using blockchain and IoT. In: 2023 7th international conference on computing methodologies and communication (ICCMC), Erode, India, pp 1199–1205. https://doi.org/10.1109/ICCMC56507.2023.10083747
19. Abraham A, Santosh Kumar MB (2020) A study on using private-permissioned blockchain for securely sharing farmers data. In: 2020 advanced computing and communication technologies for high performance applications (ACCTHPA), Cochin, India, pp 103–106. https://doi.org/10.1109/ACCTHPA49271.2020.9213237
20. Dutta H, Nagesh S, Talluri DJ, Bhaumik DP (2023) A solution to blockchain smart contract based parametric transport and logistics insurance. In: IEEE transactions on services computing. https://doi.org/10.1109/TSC.2023.3281516
21. Nagendra NP, Narayanamurthy G, Moser R, Hartmann E, Sengupta T (2023) Technology assessment using satellite big data analytics for India's Agri-insurance sector. IEEE Trans Eng Manage 70(3):1099–1113. https://doi.org/10.1109/TEM.2022.3159451
22. Uike D, Villavicencio-Mendoza M, Gehlot A, Verma D, Panduro-Ramirez J, Susanto E (20200) Block chain integration with AI for agriculture product supply chain. In: 2022 11th international conference on system modeling and advancement in research trends (SMART), Moradabad, India, pp 774–778. https://doi.org/10.1109/SMART55829.2022.10047358

Synthesizing MPPT Control and Fault Detection Tactics in Autonomous Photovoltaic Systems Employing Residual Incremental Conductance

Debani Prasad Mishra, Jayanta Sahoo, Anubhav Prakash Gaur, and Ishaan Sahu

Abstract Solar photovoltaic (PV) systems are crucial for clean energy generation, but their efficiency and reliability can be hampered by both unpredictable operating conditions and potential faults in components. To address these challenges, this work proposes a comprehensive solution integrating advanced control techniques and fault detection mechanisms. At the heart of this solution lies an enhanced Incremental Conductance (IC) algorithm, which utilizes the mathematical residue theorem for highly precise Maximum Power Point Tracking (MPPT). This ensures the PV system consistently extracts the maximum possible power even under variable conditions. Additionally, a robust Fault Diagnosis (FD) system employs residual formation and threshold comparisons, allowing for the swift detection of faults within sensors, switches, and other critical PV components. This proactive fault detection safeguards the system's reliability. Finally, the integration of a sliding mode compensator enables the autonomous determination of controller gains, rendering the system robust to uncertainties and eliminating the need for precise atmospheric data. Collectively, these innovations substantially improve the energy conversion efficiency, reliability, and overall resilience of solar PV systems.

Keywords Maximum Power Point Tracking (MPPT) · Photovoltaic (PV) Systems · Incremental Conductance (IC) algorithm · Residual-based IC MPPT · Fault Detection (FD) · Fault diagnosis system · Renewable energy · Solar photovoltaic energy · Residue formation approach · Sliding mode compensator ·

D. P. Mishra (✉) · J. Sahoo · A. P. Gaur · I. Sahu
Department of Electrical Engineering, IIIT Bhubaneswar, Bhubaneswar, Odisha, India
e-mail: debani@iiit-bh.ac.in

J. Sahoo
e-mail: jayanta@iiit-bh.ac.in

A. P. Gaur
e-mail: b320010@iiit-bh.ac.in

I. Sahu
e-mail: b320023@iiit-bh.ac.in

S. Kumar et al. (eds.), *Soft Computing Applications in Modern Power and Energy Systems*, Lecture Notes in Electrical Engineering 1298,
https://doi.org/10.1007/978-981-96-0058-8_2

Nonparametric MPPT · Residue Injection Current (RIC) · Model-Residue (MR) based FD · Augmented-Residue (AR) based FD · Observer-based fault detection · Hybrid MPPT algorithm · BLDC motor controller · Energy conversion efficiency · Reliability · Sustainability

1 Introduction

Solar energy is a critical global resource that has attracted significant attention from both researchers and industries. Photovoltaic (PV) energy conversion systems are at the forefront of harnessing this abundant energy source by directly converting sunlight into electrical power in the form of DC output [1]. One of the key techniques for enhancing system efficiency is the Maximum Power Point Tracking (MPPT) algorithm, which aims to maximize energy extraction throughout the day by maintaining the PV system's DC voltage at the Maximum Power Point (MPP) through precise regulation of output power. Notably, nonparametric approaches leverage PV voltage, eliminating the need for control systems to rely on atmospheric conditions and solar irradiance data, which are typical prerequisites for parametric techniques [2]. The latter often require prior knowledge of PV panels, thus limiting their adaptability.

Based on the analysis of existing MPPT methods [1] and the limitations of traditional Perturb and Observe (P&O) techniques [3], a novel incremental conductance (INC) MPPT algorithm has been proposed. This algorithm incorporates adaptive variable step sizes, eliminating the need for a user-defined constant and simplifying PV system design. The proposed method automatically adjusts step sizes through a coefficient, enhancing MPP tracking efficiency. Additionally, a tuning method for initial step sizes, derived from the open-circuit voltage and MPP voltage relationship, has been introduced [4]. This approach results in faster dynamic response and improved steady-state performance, especially under extreme irradiance changes.

In the context of renewable energy challenges and the global commitment to sustainability reflected in the Strategic Development Goals (SDGs) [5], advancements in MPPT algorithms play a crucial role. Sustainable energy solutions are vital for achieving environmental goals and addressing renewable energy challenges effectively. The proposed INC MPPT algorithm, validated through simulations and experimental results, demonstrates its effectiveness in optimizing PV array performance and contributing to sustainable energy practices.

In the domain of PV systems, potential DC-side faults like open-circuit faults, short-circuit faults, hot spots, and partial shading raise significant concerns. Various fault detection methods fall into two categories: Model-Residue (MR) and Augmented-Residue (AR) based Fault Detection (FD) approaches. MR-based FD involves comparing data from multiple sensors to diagnose sensor faults, while AR-based FD incorporates model-based techniques such as parity-space (PS) and temporal redundancy.

Observer-based techniques have gained attention for fault detection in sensors, including current, voltage, and speed sensors [6]. These techniques, ranging from

Kalman-filter-based methods to higher-order sliding mode (HOSM) observers, aim to reduce disruptive chattering and enhance reliability. A novel approach incorporating the generalized super-twisting algorithm (GSTA) within a differential flatness framework efficiently retrieves missing system states, improving fault detection accuracy. This comprehensive fault detection scheme integrates Sliding Mode Observers (SMOs) and customized residual thresholds, offering a robust strategy for ensuring the reliability and longevity of PV energy systems.

On the other hand, while some MPPT techniques are rapid and accurate, they require special design and expertise in subjects like fuzzy logic or neural network methods [2]. MPPT fuzzy logic controllers, known for their performance under varying atmospheric conditions, can outperform traditional methods like P&O control. However, their effectiveness hinges on the engineer's technical knowledge in error computation and rule-based table creation [7]. The success of these techniques is greatly dependent on the designer's skill and experience in arranging the system effectively.

1.1 PV Energy Conversion System

The primary focus of this study is on a compact, battery-free photovoltaic (PV) energy conversion system comprising a PV module and a DC-DC boost converter, designed to power a DC load [3]. This system operates by harnessing solar energy through solar cells within the PV module, capturing both direct and diffused radiations. The PV modules are arranged in series and parallel configurations to form a composite PV array. The total energy output is determined by the area under the curve during peak sunlight hours, which is influenced by the photocurrent (Iph) generated by the solar cell, leading to the production of photovoltaic current (Ipv) and corresponding photovoltaic voltage (Vpv).

Key components in this equation include the electron charge (q), solar irradiance (λ), cell array temperature (T), short-circuit current (Isc), diode leakage current (I0), diode ideality factor (α), Boltzmann constant (k), and Boltzmann's temperature (T).

In practical applications, a photovoltaic (PV) cell array, depicted in Fig. 1, is employed, comprising interconnected PV cells arranged in series and parallel configurations [8]. Parallel connections enhance the photovoltaic voltage (Vpv), while series connections increase the photovoltaic current (Ipv). The mathematical representation of photovoltaic current (Ipv) within a PV array is articulated as follows.

Creating a photovoltaic (PV) and current-voltage (IV) curve entails collecting data points that depict the relationship between voltage and current in a solar cell [6]. Through experimental measurements, the PV curve illustrates the cell's power output under varying conditions, while the IV curve provides insights into its electrical characteristics [9]. Figure 2 outlines the process of systematically adjusting the cell's voltage and recording corresponding current values, enabling a thorough analysis of the cell's performance across different operational parameters [8]. Employing such

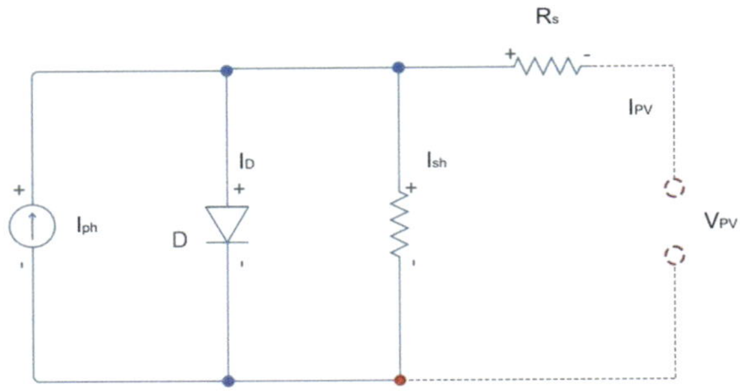

Fig. 1 Energy conservation system

a methodology yields valuable insights for optimizing the efficiency and operational aspects of photovoltaic systems (Fig. 3).

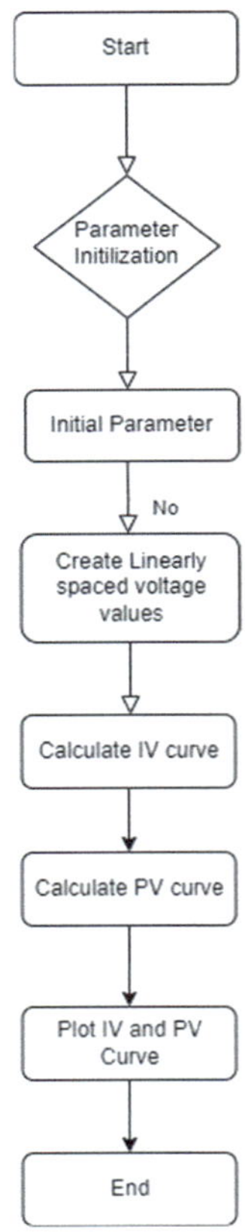

Fig. 2 Flowchart of IV versus PV curve

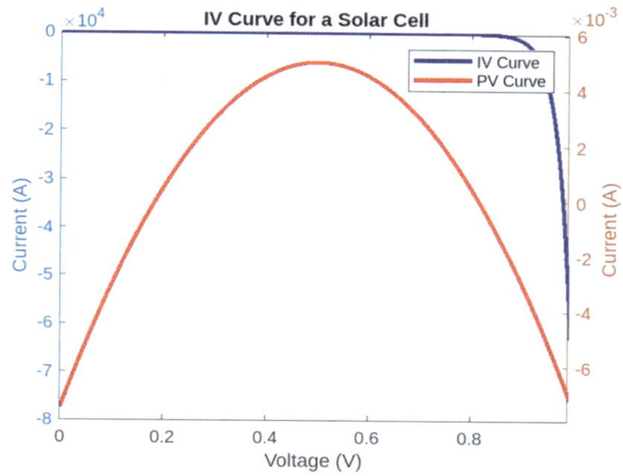

Fig. 3 Graph of IV and PV curve

2 Comparison Analysis

The conventional incremental conductance (IC) algorithm has undergone extensive design and testing, revealing significant discrepancies in actual tracking performance, resulting in longer tracking times and reduced efficiency. In contrast, the proposed residue-based IC MPPT algorithm has been compared with the classical IC MPPT method [10]. The evaluation primarily targets achieving peak power under maximum solar irradiance, as illustrated in the accompanying figure.

The figure prominently showcases the enhanced stability of the proposed residue-based IC MPPT algorithm at the operational point. The presence of a zero residue significantly influences both MPPT attainment and operational point stability [11]. As a result, this advancement leads to improved energy conversion efficiency, potentially increasing PV array ratings. This heightened efficiency serves as a crucial factor in gradually elevating PV array ratings over time (Fig. 4).

Solar irradiance	Classical IC power (W)	Residue IC power (W)
1000	85.14	84.80
800	68.40	68.40
600	51.10	51.25
400	32.10	32.44

It has been proposed that variations in insolation levels contribute to observed shifts between different states [12]. A complex solution has been proposed and validated across various scenarios, resulting in enhanced efficiency and extended duration, albeit with the trade-off of increased algorithmic complexity.

A comparison between the Residue Injection Current (RIC) and Classical Injection Current (CIC) techniques reveals distinct approaches to fault detection in power systems [13]. The Residue IC method utilizes residue signals derived from the

Fig. 4 Comparison of classical IC and residue IC technique

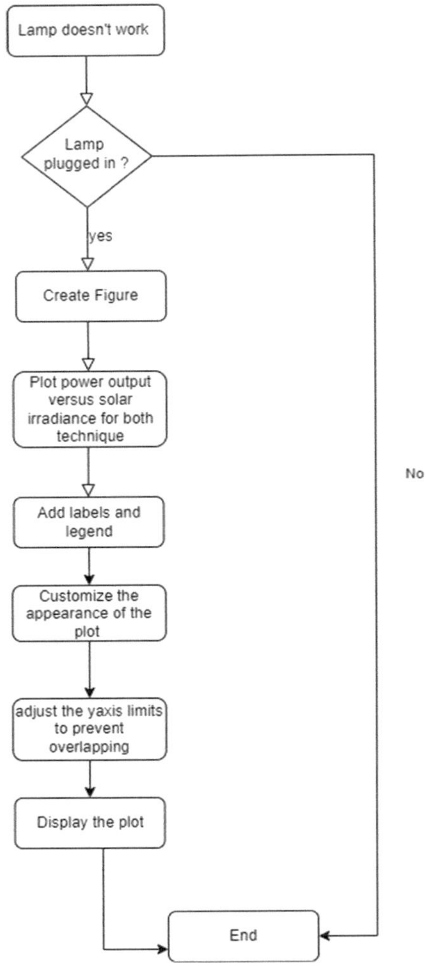

system's response to injected currents, offering improved sensitivity for detecting incipient faults. In contrast, Classical IC techniques rely on predefined injection patterns and require a deep understanding of system behavior. While RIC provides enhanced sensitivity, Classical IC methods are known for their simplicity and ease of implementation. Each technique presents a trade-off between complexity and effectiveness, necessitating the choice based on the specific requirements and intricacies of the power system under consideration [14] (Fig. 5).

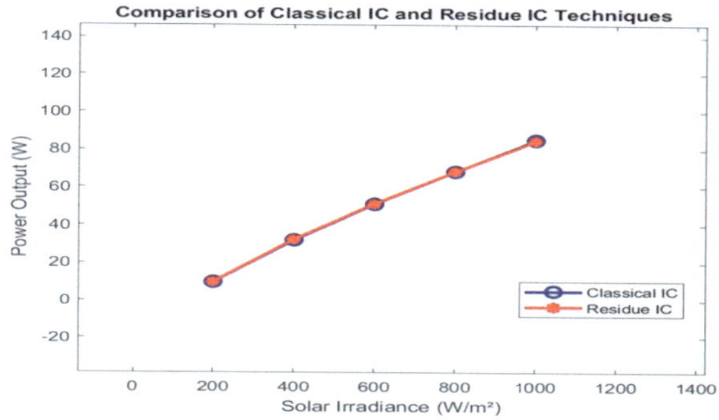

Fig. 5 Graph of power output and solar irradiance

3 Conclusion

Our research has focused on enhancing the efficiency and reliability of photovoltaic energy conversion systems, specifically within the context of PV water pumping systems utilizing BLDC motor controllers. We conducted thorough investigations into various Maximum Power Point Tracking (MPPT) algorithms, resulting in substantial improvements in their convergence characteristics, notably through the implementation of the forward Euler convergence method.

One of the main challenges addressed in our study was the uncertainty arising from the nonparametric nature of the selected MPPT algorithm. To mitigate this uncertainty and ensure robust performance, we introduced a sliding mode compensator. This compensator significantly enhanced the system's ability to handle unpredictable fluctuations, strengthening its stability across diverse operational conditions.

Additionally, the development of our novel hybrid MPPT algorithm represented a pivotal aspect of our research. This hybrid approach integrated advantageous features from the GSS-based MPPT strategy and the Incremental Conductance (INC) MPPT algorithm, enabling seamless adaptation to changing weather conditions and efficient extraction of the Maximum Power Point (MPP) under both stable and dynamic scenarios. The dynamic utilization of the Gradient Search Strategy (GSS) in response to weather variations ensured consistent operation at or near optimal performance levels.

The practical significance of our work lies in its potential to address the persistent issue of oscillations around the reference point inherent in conventional MPPT systems. Through our investigation, we demonstrated that the proposed residue Incremental Conductance (INC) MPPT algorithm effectively mitigates this drawback, paving the way for an efficient and reliable PV water pumping system powered by a BLDC motor controller.

In summary, our study contributes novel insights and innovations to the field of photovoltaic energy conversion systems. The integration of refined convergence methodologies, the introduction of a sliding mode compensator to handle uncertainty, and the development of a hybrid MPPT algorithm represent significant advancements in enhancing the performance and adaptability of PV systems. These advancements have promising implications for more efficient and reliable renewable energy solutions, particularly in remote or off-grid applications such as water pumping systems.

References

1. Alsumiri M (2019) Residual incremental conductance based nonparametric MPPT control for solar photovoltaic energy conversion system. IEEE Access 7:87901–87906
2. Mishra DP, Ray P (2018) Fault detection, location and classification of a transmission line. Neural Comput Appl 30(5):1377–1424
3. Rout KK, Mishra DP, Mishra SK, Patra S, Salkuti SR (2023) Perturb and observe maximum power point tracking approach for microgrid linked photovoltaic system. Indones J Electr Eng Comput Sci 29(2):635–643
4. Chendi L, Yuanrui C, Dongbao Z, Junfeng L, Jun Z (2016) A high-performance adaptive incremental conductance MPPT algorithm for photovoltaic systems. Energies 9:288
5. Deshmukh NR (2015) Comparison of perturb and observe and incremental conductance MPPT based solar tracking system. Int J Eng Res 4(5):222–223
6. Alam Z, Khan MA, Khan ZA, Ahmad W, Khan I, Khan Q, Irfan M, Nowakowski G (2023) Fault diagnosis strategy for a standalone photovoltaic system: a residual formation approach. Electronics 12(2):282
7. Safari A, Mekhilef S (2011) Simulation and hardware implementation of incremental conductance MPPT with direct control method using cuk converter. IEEE Trans Ind Electron 58:1154–1161
8. Shang L, Guo H, Zhu W (2020) An improved MPPT control strategy based on incremental conductance algorithm. Prot Control Mod Power Syst 5(1)
9. Soon T, Mekhilef S (2015) A fast-converging MPPT technique for photovoltaic system under fast-varying solar irradiation and load resistance. IEEE Trans Ind Inf 11:176–186
10. López-Santos O, García G, Martínez-Salamero L, Giral R, Vidal-Idiarte E, Merchán-Riveros M, Moreno-Guzmán Y (2019) Analysis, design, and implementation of a static conductance-based MPPT method. IEEE Trans Power Electron 34:1960–1979
11. Abdulkadir M, Yatim A, Yusuf S (2014) An improved PSO-based MPPT control strategy for photovoltaic systems. Int J Photoenergy 2014:1–11
12. Mishra DP, Jena S, Senapati RN, Panigrahi A, Salkuti SR (2023) Global solar radiation forecast using an ensemble learning approach. Int J Power Electron Drive Syst 14(1):496–505
13. Jhunjhunwalla D, Mishra DP, Hembram D, Salkuti SR (2024) Revolutionizing domestic solar power systems with IoT-enabled blockchain technology. Int J Appl Power Eng 13(1):255–262
14. Tousi S, Moradi M, Basir N, Nemati M (2016) A function-based maximum power point tracking method for photovoltaic systems. IEEE Trans Power Electron 31:2120–2128

Solar Module Modeling and Parameter Extraction: A Novel Approach Using Crayfish Optimization Algorithm for Enhanced Accuracy

Shikhar Gupta, Tanishq Prasad, Debashis Kundu, and Saurav Raj

Abstract Over the past few decades, there has been remarkable surge in the utilization of Solar Energy. This significant growth can be attributed to pressing concern like the depletion of conventional energy sources, the threat of global warming, and the economic downturn brought by the COVID-19 pandemic. Open circuit, short circuit, and maximum power points are typically provided by manufacturers and are used to calculate PV system parameters using data based on I-V features. This study presents a novel method for accurately modeling solar modules, with particular emphasis on the photovoltaic single-diode model (SDM) that has five unknown parameters: I_{light}, I_{sat}, a, R_{sh}, and R_s. This study deviates from conventional experimental methods by putting forth a novel approach to parameter extraction. In order to effectively ascertain the unknown parameters of the SDM, it offers a new meta-heuristic algorithm, Crayfish Optimization Algorithm (COA). Rather than depending solely on lengthy tests, it creates a unique cost function based on datasheet values. Two of the parameters are estimated using standard test conditions (STC) data, while the remaining three are optimized by the COA method. The COA is used to extract parameters from commercial PV panel, the KC200GT, in order to verify the efficacy of this method. The I-V curves for these PV modules are then produced for five simulation runs under STC.

Keywords Crayfish optimization algorithm · Single diode model · Parameter estimation

---

S. Gupta · T. Prasad · D. Kundu · S. Raj (✉)
Institute of Chemical Technology, Mumbai, Jalna, India
e-mail: s.raj@marj.ictmumbai.edu.in

D. Kundu
e-mail: d.kundu@marj.ictmumbai.edu.in

S. Kumar et al. (eds.), *Soft Computing Applications in Modern Power and Energy Systems*, Lecture Notes in Electrical Engineering 1298,
https://doi.org/10.1007/978-981-96-0058-8_3

1 Introduction

Photovoltaic (PV) cells, which collect solar energy, have become a key component in the drive for environmentally benign and sustainable power generation. The reliance on traditional fossil fuels has created significant environmental is-sues as the demand for power continues to rise around the globe. The world has focused on using solar energy to address these problems and pave the way for a cleaner, more sustainable future. Photovoltaic cells, or PV cells, are the corner-stone of this amazing solar energy revolution. These cells are in charge of converting solar energy into electricity, providing a reliable and renewable substitute for conventional energy sources. Complex mathematical frameworks such as different models, which offer vital insights into their performance, are used to describe the complex behavior of PV cells.

A novel fitting method for precisely determining the single-diode model's parameters as well as the temperature and irradiance of photovoltaic modules' operating conditions, allowing for efficient monitoring of module deterioration and aging using internal electrical current-voltage curve measurements is explained in [1]. A combined analytical and iterative approach for precisely extracting the values of diode model used in the evaluation of PV device performance, with the parameter Rs obtained iteratively to reduce the error between theoretical and investigational data and demonstrate its dependability and resilience in contrast to other methods, is being presented in [2]. An improved HGWOSCA for precisely extracting parameters of different solar PV module types. Devarapalli et al. [3] shows how effective it is at minimizing errors in both single and double diode models and validates its performance through comparative study with alternative approaches found in the literature. In order to precisely determine unknown parameters in various PV models, the Adolescent Identity Search Algorithm (AISA), which was inspired by adolescent identity formation behavior, is used in [4], achieving lower root mean square errors (RMSE) than competing algorithms and closely matches simulated I-V curves with experimental data. In order to accurately determine the parameters of the single-diode model and the operational circumstances (temperature and irradiance) of PV modules, Kalliojärvi-Viljakainen et al. [1] discusses the significance of monitoring the condition of PV modules and proposes a novel fitting procedure based on internal electrical current-voltage curve measurements.

A sophisticated algorithm for adaptive differential evolution, namely the L-SHADE algorithm, based on scant data from the manufacturer's datasheet, and achieving accurate parameter optimization for both single-diode and double-diode PV models discussed in [5] without the need for pre-assumed parameter values. In [6], a modified probabilistic fractal search algorithm is introduced for accurately estimating parameters in different diode models of solar cells and PV modules. The equilibrium optimizer algorithm is applied to accurately estimate parameters in one and two-diode models of a photovoltaic solar cell, validating its effectiveness for simulating solar photovoltaic systems are described in [7]. The Whippy Harris Hawks Optimization (WHHO) algorithm is introduced in [8] as a highly effective

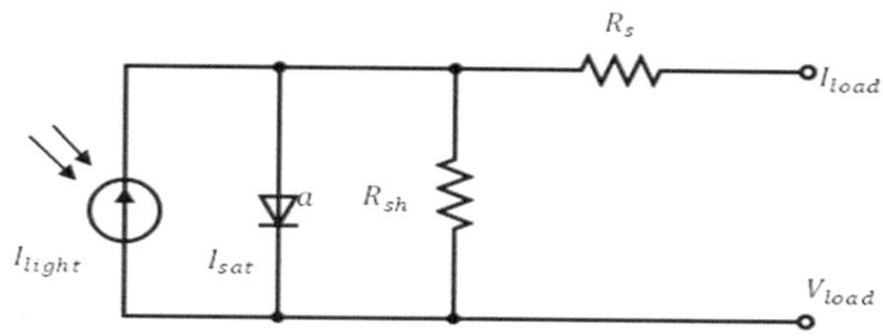

Fig. 1 Circuit diagram of single-diode model

technique for precisely estimating parameters in solar cell and photovoltaic module models. Its superiority to other optimization techniques is shown, and its practical efficacy for parameter estimation in various conditions and commercial modules is confirmed.

2 Modeling of Single-Diode Photovoltaic Cells

The photogenerated current (I_{light}) produced by solar irradiance is given to the load after passing through the diode which is changed by the diode current in this concept. The model may be expanded to account for other parameters for instance the cumulative resistance of contacts, metal grids, and p & n layers in series with the PV cell (R_s), as well as the influence of p-n junction leakage current with (R_{sh}). These changes serve to improve the model's accuracy in reflecting real-world settings and its capacity to function as a visualization of a PV cell's electrical activity (Fig. 1).

Applying Kirchoff's current law on load terminals to find the current I_{load}, we get

$$I_{\text{load}} = I_{\text{light}} - I_d - I_{sh} \tag{1}$$

After the substitution of parameters and constants in the above equation, we get

$$I_{\text{load}} = I_{\text{light}} - I_{\text{sat}} \cdot \left(\exp\left(\frac{q(V_{\text{load}} + R_s I_{\text{load}})}{akT} \right) - 1 \right) - \frac{V_{\text{load}} + R_s I_{\text{load}}}{R_{sh}} \tag{2}$$

3 Problem Formulation

Authors have concentrated on different types of test in a PV system in the context of solving the issue. By applying analytical techniques, authors hope to predict two parameters—I_{sat} and I_{light}—while employing the COA algorithm to optimize three additional parameters—R_s, R_{sh}, and a—for the PV module.

Effective formulation and solution of an objective function is crucial to ensuring that the PV panel operates in line with the I-V relationship and minimizing differences in its operating points. This calls for the application of a meta-heuristic optimization method.

According to the available research, solving the PV model equation's nonlinearity may be done using either numerical calculations or the use of linear equations. Therefore, while extracting parameters for PV models, the use of a trustworthy and accurate approach is crucial. It is important to remember that Eq. (2) has five unknown parameters, respectively, and are nonlinearity implicit transcendental equations.

After taking into account all the information, the model's objective function is expressed as follows:

$$obj_1 = I_{sc} + I_{sat}.\left(exp\left(\frac{q(R_s I_{SC})}{akTN_c}\right) - 1\right) + \frac{R_s I_{SC}}{R_{sh}} - I_{light} \tag{3}$$

$$obj_2 = I_{sat}.\left(exp\left(\frac{q(V_{OC})}{akTN_c}\right) - 1\right) - \frac{V_{OC}}{R_{sh}} - I_{light} \tag{4}$$

$$obj_3 = I_{load} = I_{light} - I_{MPP} - I_{sat}.\left(exp\left(\frac{q(V_{MPP} + R_s I_{MPP})}{akTN_c}\right) - 1\right) - \frac{V_{MPP} + R_s I_{MPP}}{R_{sh}} \tag{5}$$

The major objective function is expressed as

$$obj_f = obj_1^2 + obj_2^2 + obj_3^2 \tag{6}$$

4 Crayfish Optimization Algorithm

The behavioral pattern of crayfish like summer vacation resort, competition, and foraging, has inspired the COA. There are two stages in COA, such as exploration and exploitation. The exploration stage is summer resort vacation, whereas the exploitation stages are foraging and competition.

At certain temperature ranges, crayfish search for caves to reside in as a summer vacation. The availability of caves depends on the size of colony and the positions of crayfishes in it. In this event, the goal of crayfish is to approach the cave directly. Therefore, the summer vacation stage of the crayfish colony is the exploration part of the COA.

When the availability of caves does not measure up with the size of the crayfish colony, there happens to be competition between crayfishes for the residing in the cave by adjusting their position with competitors of the cave. Even though the happening of this event is random. This is one of the exploitation stages.

Crayfish judge the size of intake from the size of food after searching the location of food. If the size of food is bigger than their intake size, they tear it to their intake size. If they find it in the proper intake size, they will directly eat. Therefore, the foraging stage is also one of the exploitation stages (Fig. 2).

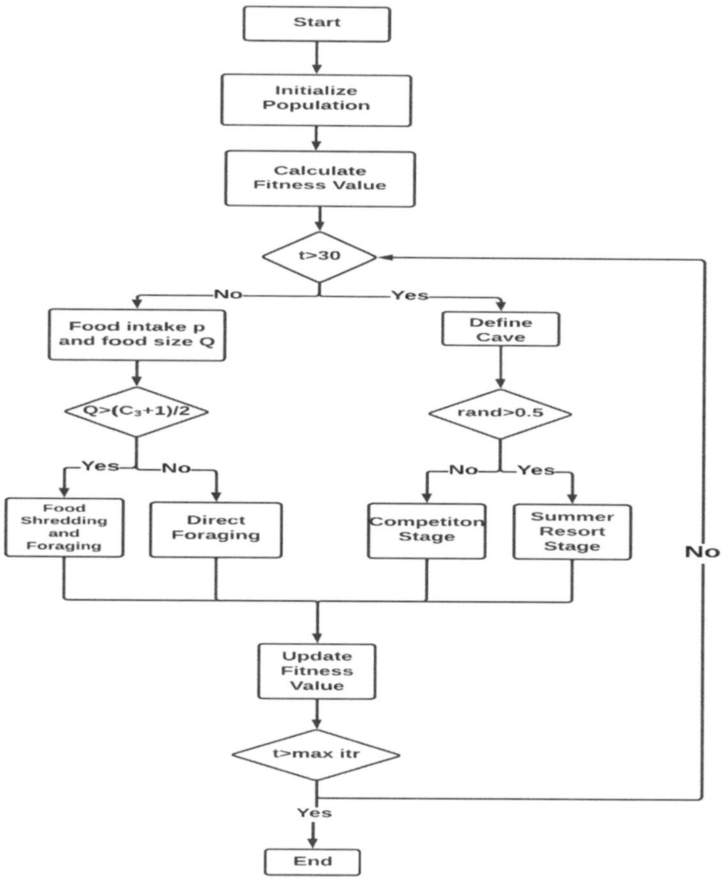

Fig. 2 Crayfish optimization algorithm flowchart

5 Result and Discussion

In this proposed work, real experimental data is obtained from [9] to validate and analyze the performance of the single-diode models. The characterization of PV module under standard test conditions and the number of I-V data permits at different environmental condition is taken from [9]. Simulation result for single-diode polycrystalline cell, KC200GT, by implementing Crayfish Optimization Algorithm is reviewed and discussed in this section. When the error function value is less than 10^{-12}, it is considered to be zero. In fact, the parameters vary between every run. For the single-diode model of the KC200GT, the thirty I-V are created and plotted in Fig. 3 in accordance with the thirty optimal parameter sets shown in Table 1. The values of the (R_s), (R_{sh}), and (a) are some of the variables that can explain variations in the I-V characteristics of a PV cell. In a study using 30 different combinations of these parameters, it was discovered that each combination could generate an I-V curve that precisely matched the open circuit, short circuit, and maximum power points on the I-V characteristic curve (Fig. 4).

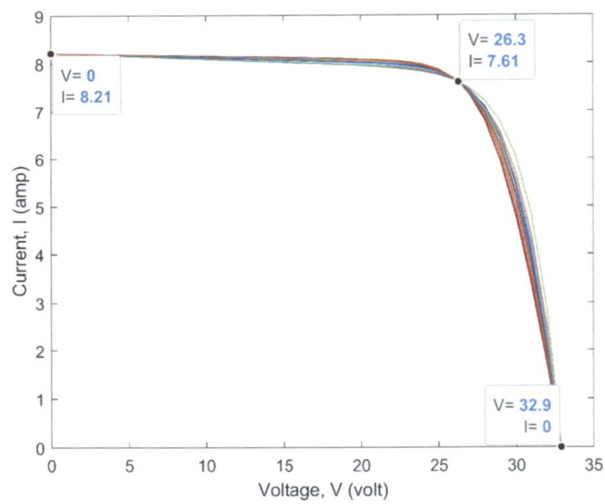

Fig. 3 I-V characteristics for SDM

Table 1 Optimal parameters for SDM

No. of execution	Optimized values			Estimated values		Error
	a	R_s	R_{sh}	I_{sat}	I_{light}	
1	1.2197	0.1666	100.0104	2.84e-08	8.2237	0
2	1.1910	0.2255	142.0625	1.80e-08	8.2230	0
3	0.8660	0.3320	77.5674	1.00e-11	8.2451	0
4	1.4284	0.1145	169.7111	4.95e-07	8.2155	0
5	1.3125	0.1295	110.5619	1.13e-07	8.2196	0
6	0.9999	0.2684	85.3972	3.94e-10	8.2358	0
7	1.3053	0.1789	168.8303	1.03e-07	8.2187	0
8	1.0964	0.2698	131.7465	3.23e-09	8.2268	0
9	1.1415	0.2394	124.3114	7.57e-09	8.2258	0
10	0.9491	0.3654	164.4750	1.13e-10	8.2282	0
11	1.4901	0.0909	189.4922	9.86e-07	8.2139	0
12	1.2570	0.1972	154.0947	5.13e-08	8.2205	0
13	1.1787	0.2127	117.7847	1.45e-08	8.2248	0
14	1.4447	0.0967	153.5702	5.95e-07	8.2152	0
15	1.4102	0.0811	115.4876	3.95e-07	8.2158	0
16	1.5432	0.0489	160.9354	1.70e-06	8.2125	0
17	0.9023	0.3700	122.8228	3.07e-11	8.2347	0
18	0.9459	0.3575	141.7675	1.04e-10	8.2307	0
19	1.4414	0.0964	150.3509	5.73e-07	8.2153	0
20	1.2619	0.1484	103.7753	5.46e-08	8.2217	0
21	1.0418	0.3150	161.2627	1.04e-09	8.2260	0
22	1.2898	0.1841	162.4076	8.31e-08	8.2193	0
23	1.1474	0.2568	156.4854	8.48e-09	8.2235	0
24	1.3149	0.1579	141.0425	1.17e-07	8.2192	0
25	1.0460	0.2166	76.3531	1.11e-09	8.2333	0
26	1.3773	0.0162	77.9348	2.59e-07	8.2117	0
27	1.4657	0.0903	161.3909	7.54e-07	8.2146	0
28	0.9095	0.3516	105.8934	3.77e-11	8.2373	0
29	1.1475	0.2242	112.2321	8.41e-09	8.2264	0
30	1.0835	0.2836	142.3614	2.50e-09	8.2264	0

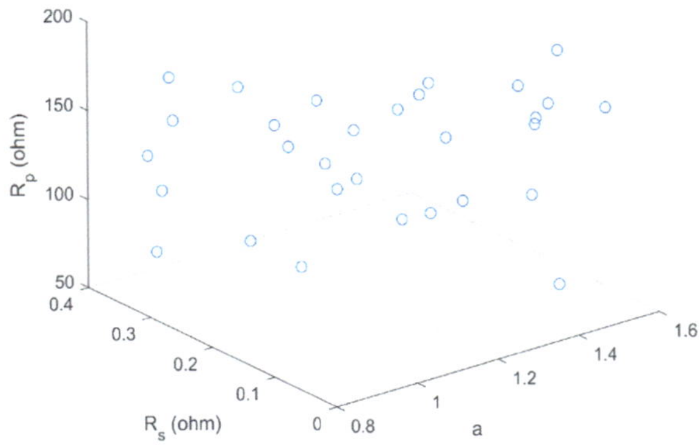

Fig. 4 Sprinkle of optimal solutions

6 Conclusion

This study uses the Crayfish Optimization system, an evolutionary system, to investigate parameter estimates of single-diode polycrystalline PV cells. The estimate is based on datasheet information at the highest power, short circuit, and open circuit points of the I-V characteristic curve. Using the single-diode model, the study entails a complete circuit analysis. Parameter extraction is accomplished by combining analytical techniques with optimization algorithms.

The analysis and simulation show that an exclusive I-V characteristic cannot be obtained by depending just on the restricted datasheet information available at the three primary sites. Designers can select from a range of optimal solutions. The research highlights that more than just the three key points' worth of experimental data are needed to obtain an accurate and distinct I-V characteristic. Even though the research is done using PV cells under standard test settings (STC), the findings remain valid at different temperatures and irradiance levels. Regardless of the circumstances, many trial runs can produce multiple sets of ideal parameters that meet the I-V relationships at the three main locations.

References

1. Kalliojärvi-Viljakainen H, Lappalainen K, Valkealahti S (2022) A novel procedure for identifying the parameters of the single-diode model and the operating conditions of a photovoltaic module from measured current–voltage curves. Energy Rep 8:4633–4640
2. Choulli I, Elyaqouti M, Saadaoui D, Lidaighbi S, Elhammoudy A, Obukhov S, Ibrahim A (2023) A novel hybrid analytical/iterative method to extract the single-diode model's parameters using Lambert's W-function. Energy Convers Manag: X 18:100362

3. Devarapalli R, Rao BV, Al-Durra A (2022) Optimal parameter assessment of solar photovoltaic module equivalent circuit using a novel enhanced hybrid GWO-SCA algorithm. Energy Rep 8:12282–12301
4. Lekouaghet B, Khelifa MA, Boukabou A (2022) Adolescent identity search algorithm for parameter extraction in photovoltaic solar cells and modules. J Comput Electron 21(4):859–881
5. Biswas PP, Suganthan PN, Wu G, Amaratunga GAJ (2019) Parameter estimation of solar cells using datasheet information with the application of an adaptive differential evolution algorithm. Renew Energy 132:425–438
6. Jia H, Rao H, Wen C, Mirjalili S (2023) Crayfish optimization algorithm. Artif Intell Rev 1–61
7. Xu S, Qiu H (2022) A modified stochastic fractal search algorithm for parameter estimation of solar cells and PV modules. Energy Rep 8:1853–1866
8. Naeijian M, Rahimnejad A, Ebrahimi SM, Pourmousa N, Gadsden SA (2021) Parameter estimation of PV solar cells and modules using Whippy Harris Hawks optimization algorithm. Energy Rep 7:4047–4063
9. Ndi FE, Perabi SN, Ndjakomo SE, Abessolo GO, Mengata GM (2021) Estimation of single-diode and two diode solar cell parameters by equilibrium optimizer method. Energy Rep 7:4761–4768

Fuzzy-Based Control Techniques for Voltage Regulation in DC Microgrids with Nonlinear Load Dynamics

Ila Rai

Abstract Renewable energy technology has led to the widespread adoption of DC microgrids, particularly for their environmental and resilience. DC microgrids offer higher efficiency in applications like electric vehicles and spacecraft due to reduced interface converters and the absence of reactive current. However, they face challenges regulating DC link voltage and ensuring proper current sharing among diverse sources and loads. One major obstacle to these microgrids' ability to regulate voltage is the existence of nonlinear constant power loads (CPLs). DC microgrid performance can degrade because of constant power load (CPL) dynamics. By using fuzzy-based control strategies, this research suggests a novel way to deal with this problem. The research identifies voltage fluctuations, describes the nonlinear behavior of CPLs in DC microgrids, and develops a fuzzy logic controller to control system voltage under various load scenarios. By utilizing fuzzy sets and linguistic rules, the fuzzy controller efficiently reduces overshoot and responds more quickly than conventional techniques to the voltage fluctuations caused by CPLs. In this paper, a fuzzy controller is suggested to deal with the effects of CPLs during transients. The suggested approach is simulated with a DC microgrid feeding two CPLs. The simulation is done in MATLAB/SIMULINK and the results program that the suggested fuzzy-based controller can stabilize a DC microgrid faster in comparison with the conventional PI controller during transients.

Keywords DC microgrid · Fuzzy control · Constant power load · Converters · Renewable energy sources

I. Rai (✉)
Department of Electrical and Electronics Engineering, Alliance School of Applied Engineering, Alliance University, Bengaluru, India
e-mail: Ila.rai@alliance.edu.in

S. Kumar et al. (eds.), *Soft Computing Applications in Modern Power and Energy Systems*, Lecture Notes in Electrical Engineering 1298,
https://doi.org/10.1007/978-981-96-0058-8_4

1 Introduction

The concept of a microgrid is now extensively used in power systems thanks to the quick advancement of renewable energy technology. These days, much attention is paid to DC microgrids because of their benefits, including the simplicity with which energy storage may be integrated and the low system loss rate [1]. Traditional power networks have met the increasing electricity demand, but to address environmental concerns and improve the resilience of the energy infrastructure, there is a growing emphasis on moving toward more sustainable and decentralized models that incorporate energy storage, innovative grid technologies, and renewable energy sources [2].

Aside from that, combining various energy sources seeks to offer maximum electricity generating capacity at the lowest cost for places served by traditional power networks while ensuring sustainable and balanced electricity in remote areas. With cutting-edge technologies and renewable energy sources, microgrids provide a decentralized and resilient method of supplying electricity in various environments [3]. Also, the challenges posed by the distribution network can be answered by the idea of microgrids, in particular when distributed generation sources, which include solar and wind energy or modern loads such as intelligent buildings and electric vehicles, are present [3, 4].

Microgrid power systems, sometimes called hybrid renewable energy systems or multiple energy resource systems, are a new generation of electrical systems that produce balanced and clean electricity using renewable sources, including solar and wind energy. Integrating renewable energy generation, energy storage systems (ESS), and loads to a microgrid is one of the smartest approaches for improving power balance in distribution networks and reducing the transmission lines losses [5].

Microgrids can be categorized as AC and DC. The most common form is an AC microgrid, which offers a simple, direct way to incorporate distribution generation units into the existing utility grid. Even though most distribution networks use AC, the significant integration of DC-based distribution generation, ESS, and loads, besides other factors, make DC-controlled distribution networks more viable. Their key benefit is increased efficiency because of less no. of interface converters and no network-wide reactive current circulation. Additionally, unlike AC microgrids, DC microgrids do not experience challenges with synchronization, reactive power management, harmonics, or frequency management. Nevertheless, controlling the DC bus voltage, including the power-sharing, becomes more complex by adding numerous distributed generators to the shared DC bus, including solar, fuel cells, wind, various loads, and energy storage devices. More and more applications, such as electric automobiles, spacecraft, and aircraft, are adopting DC microgrids [6–8].

The efficient functioning of a DC microgrid requires proper DC link voltage regulation and current sharing among numerous sources and loads. The main challenges include maintaining stability while operating because it is connected to multiple intermittent renewable sources and different loads. DC microgrids are coupled to

DC-DC power converters to improve flexibility and dependability. However, these tightly controlled points of load (POL) converters act as constant power loads (CPLs). The CPLs identified here have negative incremental impedance and are nonlinear. These properties degrade the system's damping and introduce instability in the DC bus voltage connected to the DC microgrid [9, 10].

The regulation of the DC-bus voltage is essential since the stability of DC microgrids is dependent on its regulation, which is seen as a sign of power balance in these kinds of systems. Nonlinear CPL poses significant issues in DC microgrid design since it can cause oscillation, failure, and stress or damage system equipment. Researcher interest in this problem led to the introduction of numerous ways to counteract and offset the instability brought on by CPLs [11–13].

Different linear control strategies have been put forth in [14, 15] for the DC microgrid's output voltage control. Due to its simplicity, the linear control technique, a standard approach in literature and industry, cannot address the control issues caused by nonlinear loads coupled to DC microgrids.

The linear controller has some practical drawbacks, including the inability to handle the power system's nonlinearities and tuning of gains that shift the operating point of the converter toward instability when system parameters vary. Also, CPL utilization is rising quickly due to the growing number of DC microgrids and smart grids [16, 17].

This paper suggests a fuzzy-based control technique to address the nonlinearity and unpredictability of the CPL in a DC microgrid. Control designs based on fuzzy logic resist parameter changes over time and don't require a precise mathematical model.

The primary goals of this research study are:

1. Regulating the output DC bus voltage at the correct level
2. Verifying the suggested controller's real-time viability using MATLAB/SIMULINK.

In this paper, sect. 2 discusses DC microgrid dynamics and problem formulation for CPL. Section 3 presents the overview of the fuzzy-based control algorithm; Sect. 4 explains the simulation outcomes performed in MATLAB/SIMULINK; Sect. 5 concludes the proposed work.

2 The DC Microgrid Dynamics and Problem Formulation

DC microgrids are localized, direct current electricity distribution systems promoting local energy production, storage, and consumption, offering increased efficiency, reduced losses, and resilience during power outages, aligning with decentralized energy systems. The DC microgrid components include:

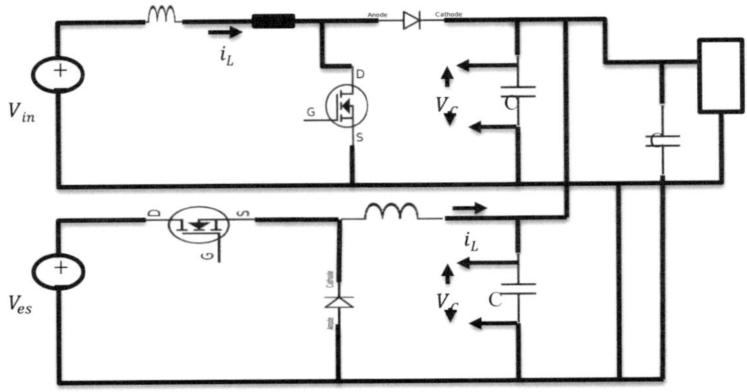

Fig. 1 A solar-based DC microgrid's circuit diagram with energy storage

1. Generation Sources
   (a) Renewable Energy Sources (RES)
   (b) Energy Storage Systems (ESS)
   (c) Backup Generators.
2. DC Bus
3. DC-DC Converters
4. DC/AC Loads.

Figure 1 displays the DC microgrid schematic that was the subject of this investigation. The components to store energy significantly influence the DC Microgrid's overall power management. They supply high-quality power, guarantee a secure network, and keep the voltage of the common DC grid constant [18].

The system has many associated DC/DC converters and inverter-supplied loads. The power output of the DC/DC converter will stay constant, leading to practically constant input power, if the output voltage of the converter that supplies a resistive load is closely regulated. Also, when the speed of an motor drive system using inverter is closely controlled, and the torque is stable for a specific working duration, the inverter's input and output power are constant. These loads tightly regulated by the converter are CPLs.

The CPL's negative impedance features introduce nonlinearity to the system, harming its performance. It can also lead to instability if the CPL is coupled in cascade to a source converter, reducing the damping of the system [19–21]. Figure 2 displays a simplified equivalent circuit of a DC microgrid. A DC/DC converter links the ESS and the PV array to the DC grid. Here, Vc is the capacitor voltage, and the i_L is the inductor current, R is the equivalent resistance of the loads linked to the DC microgrid, and M is the controllable switch. V_{in} denotes the DC source's input voltage.

The circuit equation can be expressed as follows by applying the state average method:

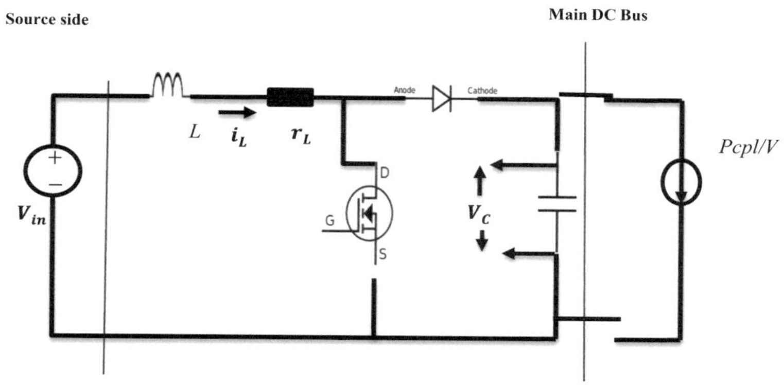

Fig. 2 A basic DC microgrid schematic connected to a single source and CPL

$$\begin{cases} L\frac{di_L}{dt} = V_{\text{in}} - (1-u)V_c - r_L i_L \\ C\frac{dV_c}{dt} = (1-u)i_L - \frac{V_c}{R} - \frac{P_{\text{cpl}}}{V_c} \end{cases} \tag{1}$$

Here u is the converter's duty cycle. The switch state is defined as an ON state and an OFF state. When switch S is in the ON position, $u = 1$, and when it is in the OFF position, $u = 0$. The equivalent resistive load, denoted by R, is the sum of all the resistive loads in parallel. Here R can be defined as

$$\frac{1}{R} = \left(\frac{1}{R1} + \frac{1}{R2} + \ldots + \frac{1}{Rn}\right) \tag{2}$$

$\frac{P_{\text{cpl}}}{V_c}$ represents the CPL current i_{cpl} and P_{cpl} represents the CPL power.

Figure 2 represents a basic circuit diagram of a DC microgrid connected to a single load and single source. The connected load is a CPL and is shown as a current source. The system presented in (1) must first be transformed into a standard form to develop the suggested TS-Fuzzy-based controller. Given that the total stored energy equals the stored energy's rate of change, the dynamic system's overall stored energy can be represented as follows:

$$x_1 = \frac{1}{2}Li_L^2 + \frac{1}{2}Cv_c^2 \tag{3}$$

Taking the derivative of (2) results in

$$\dot{x}_1 = Li_L\dot{i}_L + Cv_c\dot{v_c} = V_{in}i_L - \frac{v_c^2}{R} - P_{cpl} \tag{4}$$

An uncertain term w_1 and a new state x_2 drawing from (3) can be defined as follows:

$$\begin{cases} x_2 = V_{in} i_L - \frac{v_c^2}{R} \\ w_1 = -P_{cpl} - \frac{v_c^2}{R_0} - \frac{v_c^2}{R} \end{cases} \tag{5}$$

where R_0 is the resistive load's nominal resistance.

Taking the derivative of (4) results in

$$\dot{x}_2 = \frac{V_{in}^2}{L} + \frac{2v_c^2}{R_0^2 C} - \left(\frac{V_{in} v_c}{L} + \frac{2i_L v_c}{R_0 C}\right)(1-u) + \frac{2}{R_0 C}\left(P_{cpl} - \frac{v_c^2}{R_0} + \frac{v_c^2}{R}\right) \tag{6}$$

Therefore, using the relevant work, the simplified system (6) can be changed into the subsequent canonical form [22]:

$$\begin{cases} \dot{x}_1 = x_2 + w_1 \\ \dot{x}_2 = x_2 + w_1 \end{cases} \tag{7}$$

Here, x_1, x_2 are state variables and u_1 is the intermediate control signal, while w_1, w_2 represent uncertainties in the system. Now derived from (5), the uncertain term w_2 can be expressed as

$$w_2 = \frac{2}{R_0 C}\left(P_{\text{cpl}} - \frac{v_c^2}{R_0} + \frac{v_c^2}{R}\right) \tag{8}$$

System (1) has been converted to Brunovsky's canonical form (7) for fuzzy control methods [23]. The primary goal of a DC/DC converter in real-world engineering applications is to get the voltage v_c to track its initial value using the appropriate control law. To put it another way, the new state x_1 must quickly track toward its reference value z_r, which is given by

$$z_r = x_1^+ == \frac{1}{2} L i_{\text{Lref.}}^2 + \frac{1}{2} C v_{cref.}^2 = \frac{1}{2} L \left(\frac{P_{\text{ref.}}}{v_{\text{in}}}\right)^2 + \frac{1}{2} C v_{\text{cref.}}^2 \tag{9}$$

where $P_{\text{ref.}} = P_{\text{cpl}} + \frac{v_{\text{cref.}}^2}{R}$ provides the total load power. The control term u is defined as follows:

$$u = 1 - \frac{V_{\text{in}}^2}{L} + \frac{2v_c^2}{R_0^2 C} - u_1 / \left(\frac{V_{\text{in}} v_c}{L} + \frac{2i_L v_c}{R_0 C}\right) \tag{10}$$

3 Propose Control Architecture Using the Fuzzy-Logic-Based Algorithm

This section discusses the control algorithm using fuzzy logic. The purpose is to adjust DC bus voltage by injecting the required current into the system. The fuzzy logic control algorithm adjusts the converter's control signal to provide an appropriate current to the DC bus. The inputs chosen for the fuzzy control are errors in load power and the error change rate. The fuzzy control output is the control signal for the converter. Figure 3 shows a simple circuit diagram for a fuzzy-based control algorithm.

Figure 4 illustrates the basic architecture of a fuzzy logic controller. The fuzzy logic circuit (FLC) consists of a fuzzifier with a knowledge base, a fuzzy rule-based inference engine, and a defuzzifier. The design process involves identifying variables, dividing data into fuzzy subsets, obtaining membership functions, formulating a rule base and merging fuzzy outputs, and finally defuzzifying the data to produce a precise output. This ensures all variables are considered in the fuzzy system. In the fuzzy control algorithm, an output of the physical system under control is regulated with the help of an error signal as displayed in Fig. 4. The error signal is obtained by calculating the deviation of the actual response of the plant (output) from the desired response (reference). The final output is our desired control signal to control the plant.

The foremost duty of the fuzzy logic control algorithm is to construct rules based on analysis of the language input variables and system behavior inside the system's framework. Before delivering the output, the inputs fed into the fuzzy logic control must pass via the three essential steps of fuzzification, decision-making, and defuzzification. Using preset membership functions (MFs), the input variable is transformed into a linguistic variable throughout the fuzzification level. Following the established guidelines, the fuzzified output is subsequently produced using the fuzzification stage's output. Ultimately, the output (fuzzified) is transformed into the required output for control of the system at the defuzzification level.

The following principles govern the structure of fuzzy rules:

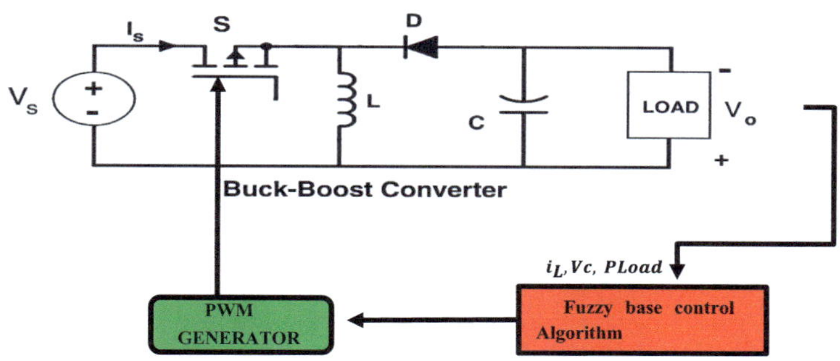

Fig. 3 A simple circuit diagram for the proposed fuzzy control algorithm

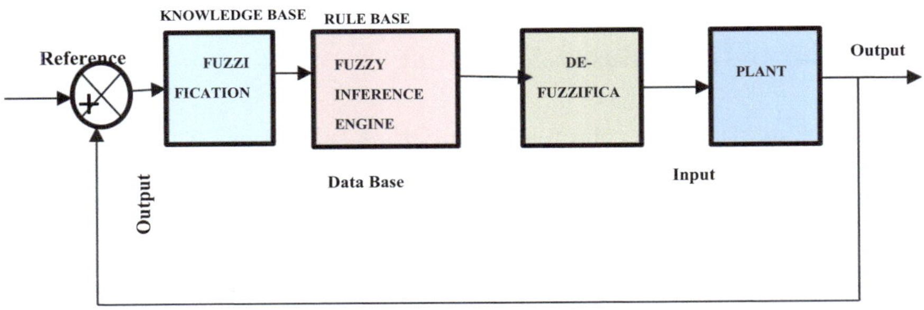

Fig. 4 Block diagram of fuzzy logic architecture

- The control signal change should be according to the variation in load power
- The difference needs to be small if the required value is near the set point
- The control signal must be adjusted so that output will become nearly constant near the set point
- The sign of the control signal will vary according to the set value required.

The output control signal can be calculated as

$$u_k = u_{k-1} + \partial.\nabla u_k \quad (11)$$

where ∂ is the gain factor and ∇u_k is the change in the control signal.

4 Simulation Outcomes

The outcomes of the suggested fuzzy-based control design are presented in this section. The system framework shown in Fig. 3 is simulated using MATLAB/Simulink to confirm the efficacy of the suggested approach. The constraints utilized to simulate the suggested control design technique are listed in Table 1.

A DC microgrid with renewable energy as a source must consider high-power transients to ensure stability. Non-linearity from CPLs can cause system instability during transients. The controller design is crucial to maintain stability and ensure reliable operation during transients. Similarly, the integration of CPL to DC microgrid can potentially delay reaching the stable point under large load variations.

Table 1 System parameters for proposed fuzzy control design

Vdc = 200 V	P1 = 900 W	P2 = 600 W
Rs = 0.4 Ω	r1 = 0.8 Ω	r2 = 0.42 Ω
Ls = 17.3 mH	L1 = 40 mH	L2 = 19.5 mH
Cs = 1.05 mF	C1 = 1 mF	C2 = 1.05

Therefore, two scenarios are illustrated using simulations in MATLAB/SIMULINK. Initial transient performance is tested and compared in scenario 1 with the present PI-based control architecture.

In scenario 2, the effect of load variation is investigated. The initial load was connected at 900 W, and an additional 600 W was supplied to test the consequences of load change. Simulation under load fluctuation conditions is performed and compared for the fuzzy-based control design technique and the standard PI-based control design.

The simulation outcomes are exhibited in Fig. 5a–d, which compares the performance of the bus voltage for the traditional PI control and suggested fuzzy control.

The comparison outcomes exhibit that the fuzzy-based control method works better than the traditional output feedback controller. As shown in Fig. 5a–d, the recommended controller enhances the transient response and quickly stabilizes during significant load changes.

Fig. 5 **a** Transient response at load1 (Voltage) **b** Transient response at load2 (Voltage) **c** Load change response at load1 (Voltage) **d** Load change response at load2 (Voltage)

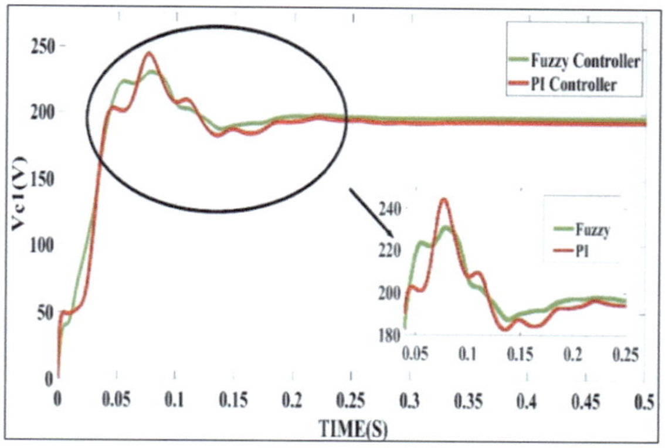

(a)

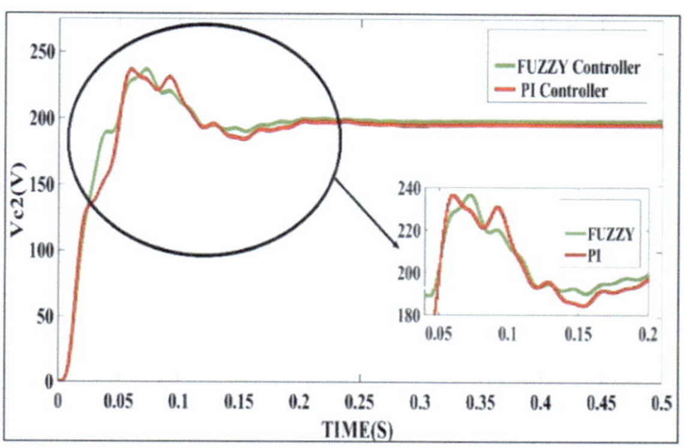

(b)

Fig. 5 (continued)

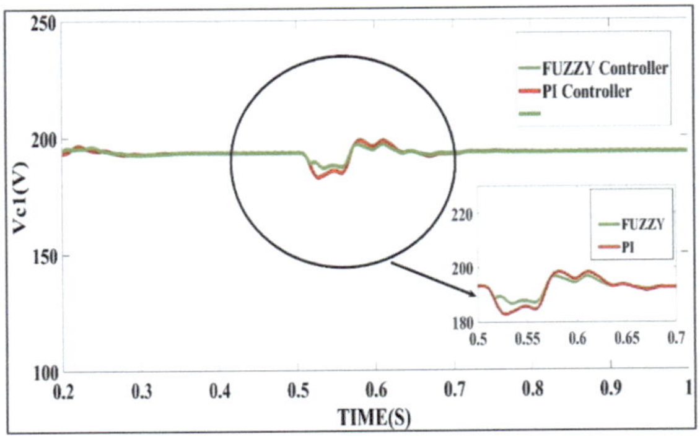

(c)

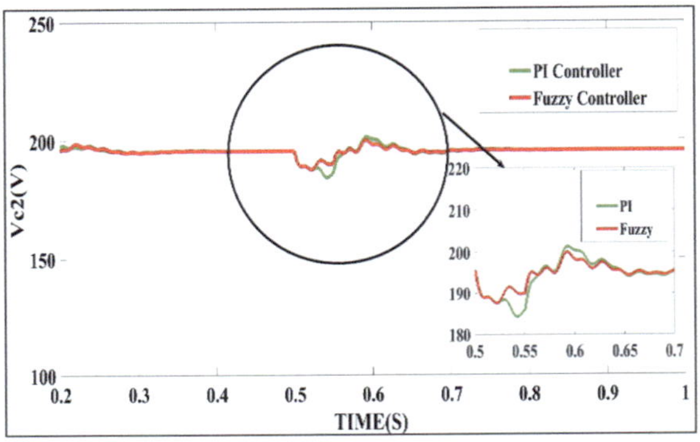

(d)

5 Conclusions

This paper presents a fuzzy-based control design approach for stabilizing the voltage caused by a CPL linked to a DC microgrid. The converter uses fuzzy logic control to maintain voltage stability in the DC bus. Fuzzy logic in control design ensures stability even under extreme load changes and transients. When comparing the output voltage to traditional PI control, it reaches the equilibrium point faster. MATLAB/Simulink is utilized for simulation and outcome verification of the suggested design methodology. Also, a comparison is made between the outcome and the traditional PI control design methodology. The simulation outcomes justified the effectiveness of the suggested control design methodology. The results of this work support the development of voltage control techniques for DC microgrids, increasing their reliability and robustness against nonlinear constant power loads. The proposed fuzzy control methods provide a viable alternative to enhance the functioning of the DC microgrid system and open the door to more flexible and sustainable power distribution networks.

The utilization of fuzzy control methods for voltage regulation in DC microgrids with nonlinear constant power demand is a promising and advantageous approach. Fuzzy logic controllers improve voltage regulation efficiency by adapting to dynamic, non-linear, balanced power loads. This research highlights the benefits of fuzzy-based control techniques in mitigating voltage fluctuations and maintaining stability in DC microgrids, ensuring reliable operation. Also, it offers a user-friendly framework for designing control strategies in microgrid systems, addressing voltage regulation challenges. As the demand for renewable energy grows, this research improves system performance. Future work may explore optimizations, multi-objective control strategies, and advanced technologies to enhance DC microgrid voltage regulation efficiency.

References

1. Nazir R, Laksono HD, Waldi EP, Ekaputra E, Coveria P (2014) Renewable energy sources optimization: a micro-grid model design. Energy Procedia 52:316–327. ISSN 1876-6102. https://doi.org/10.1016/j.egypro.2014.07.083
2. Rehman AU, Zeb S, Khan HU, Shah SSU, Ullah A (2017) Design and operation of microgrid with renewable energy sources and energy storage system: a case study, pp 1–6. https://doi.org/10.1109/ICETSS.2017.8324151
3. Uddin M, Mo H, Dong D, Elsawah S, Zhu J, Guerrero JM (2023) Microgrids: a review, outstanding issues, and future trends. Energy Strat Rev 49:101127. ISSN 2211-467X. https://doi.org/10.1016/j.esr.2023.101127
4. Zhou X, Guo T, Ma Y (2015) An overview on microgrid technology, pp 76–81. https://doi.org/10.1109/ICMA.2015.7237460
5. Abbasi M, Abbasi E, Li RP, Aguilera DL, Wang F (2023) Review on the microgrid concept, structures, components, communication systems, and control methods. Energies 16(1):484. https://doi.org/10.3390/en16010484
6. Shahgholian G (2021) A brief review on microgrids: operation, applications, modeling, and control. Int Trans Electr Energ Syst 31:e12885. https://doi.org/10.1002/2050-7038.12885
7. Hossain E, Kabalcı E, Bayindir R, Perez R (2014) A comprehensive study on microgrid technology. Int J Renew Energy Res 4:1094–1104
8. Planas E, Andreu J, Gárate JI, de Alegria IM, Edorta E (2015) AC and DC technology in microgrids: a review. Renew Sustain Energy Rev 43:726–749. https://doi.org/10.1016/j.rser.2014.11.067
9. Singh S, Gautam A, Fulwani D (2017) Constant power loads and their effects in DC distributed power systems: a review. Renew Sustain Energy Rev 72:407–421. https://doi.org/10.1016/j.rser.2017.01.027
10. Dragicevic T, Vasquez JC, Guerrero JM, Skrlec D (2014) Advanced LVDC electrical power architectures and microgrids: a step toward a new generation of power distribution networks. Electrification Mag. IEEE, 2:54–65. https://doi.org/10.1109/MELE.2013.2297033
11. Whaite S, Grainger B, Kwasinski A (2015) Power quality in DC power distribution systems and microgrids. Energies 8(5):4378–4399. https://doi.org/10.3390/en8054378
12. Dragicevic T, Lu X, Quintero JCV, Guerrero JM (2016) DC microgrids—part I: a review of control strategies and stabilization techniques. IEEE Trans Power Electron 31(7):4876–4891. https://doi.org/10.1109/TPEL.2015.2478859
13. Herrera L, Zhang W, Wang J (2015) Stability analysis and controller design of DC microgrids with constant power loads. IEEE Trans Smart Grid 2015:1–1. https://doi.org/10.1109/TSG.2015.2457909

14. Nasirian V, Moayedi S, Davoudi A, Lewis FL (2015) Distributed cooperative control of DC microgrids. IEEE Trans Power Electron 30(4):2288–2303. https://doi.org/10.1109/TPEL.2014.2324579
15. Al-Ismail F (2021) DC microgrid planning, operation, and control: a comprehensive review. IEEE Access 9:36154–36172. https://doi.org/10.1109/ACCESS.2021.3062840
16. Rai I, Ravishankar S, Anand R (2019) Review of DC microgrid system with various power quality issues in real time operation of DC microgrid connected system. Majlesi J Mechatron Syst 8(3):35–44
17. Cui J, Tian G, Liu G, Wu Y, Guo Z (2022) Research on coordinated control strategy of isolated DC microgrid with PV/hybrid energy storage. Energy Rep 8(10):893–899. ISSN 2352-4847. https://doi.org/10.1016/j.egyr.2022.05.274
18. Meng L et al (2017) Review on control of DC microgrids and multiple microgrid clusters. IEEE J Emerg Sel Top Power Electron 5(3):928–948. https://doi.org/10.1109/JESTPE.2017.2690219
19. Xu Q, Zhang C, Wen C, Wang P (2017) A novel composite nonlinear controller for stabilization of constant power load in DC microgrid. IEEE Trans Smart Grid 1(1). https://doi.org/10.1109/TSG.2017.2751755
20. Rai I, Rajendran A, Lashab A, Guerrero J (2021) Hardy space nonlinear controller design for DC microgrid with constant power loads. Int J Electr Power Energy Syst 133:107300
21. Siad SB, Malkawi A, Damm G, Lopes L, Dol LG (2019) Nonlinear control of a DC microGrid for the integration of distributed generation based on different time scales. Int J Electr Power Energy Syst 111:93–100. ISSN 0142-0615. https://doi.org/10.1016/j.ijepes.2019.03.073
22. Iovine A, Carrizosa MJ, Damm G, Alou P (2018) Nonlinear control for DC microgrids enabling efficient renewable power integration and ancillary services for AC grids. IEEE Trans Power Syst 1(1). https://doi.org/10.1109/TPWRS.2018.2871369
23. Vafamand N, Asemani MH, Khayatiyan A, Khooban MH, Dragicevic T (2017) TS fuzzy model-based controller design for a class of nonlinear systems including nonsmooth functions. IEEE Trans Syst Man Cybern: Syst 1–12. https://doi.org/10.1109/TSMC.2017.2773664

GOOSE-Driven Distance Interlocking: Implementation Strategies and Lessons Learned in Uganda

Pankaj Kumar Jha, Ashwani Devgune, and M. S. Hada

Abstract In 2023, Uganda achieved a significant milestone with the commissioning of four 132/33 kV substations as part of the Grid Expansion and Renovation Project. One of the challenges during the energization of a 132 kV substation arose due to the presence of a double circuit line connecting two substations, Substation-A and Substation-B. Substation-C was situated between these two substations, and the existing line, referred to as line AB, needed to undergo a Loop In Loop Out (LILO) at Substation-C. The LILO operation would transform the existing line AB into two separate lines, AC and BC. The LILO operation at Substation-C was crucial for energizing this substation. However, complications emerged as the communication equipment at Substation-C was not ready, affecting the proper functioning of the line differential protection for lines AC and BC. Additionally, some SCADA-related work at Substation-C remained incomplete, leading to a lack of visibility from the national control center. To address this, a decision was made to connect line AB in such a way that the bus bar at Substation-C would become part of line AB. This solution ensured that there would be no change in the communication link, and the line could continue to be monitored from Substation-A and Substation-B. The double circuit line AB would pass through two different bus bars at Substation-C, with the bus coupler remaining open. Both ends of the existing line AB already had one distance and one-line differential protection. The line differential and distance protection for line AB remained unchanged. However, a new challenge arose when it was decided to commission the 40 MVA 132/33 kV transformer at Substation-C. The energization of the transformer acted like a tap load in line AB, creating a risk of the line tripping due to the operation of line differential protection. Considering the critical role of line AB in power evacuation and supplying major cities in Uganda, tripping

P. K. Jha (✉)
Power Grid Corporation of India Ltd, Vadodara, Gujarat, India
e-mail: pankaj.jha@powergrid.in

A. Devgune
Power Grid Corporation of India Ltd, Gurugram, India

M. S. Hada
IndiGrid, Noida, India

S. Kumar et al. (eds.), *Soft Computing Applications in Modern Power and Energy Systems*, Lecture Notes in Electrical Engineering 1298,
https://doi.org/10.1007/978-981-96-0058-8_5

during the transformer energization was not acceptable. To address this, a scheme was implemented at both Substation-A and Substation-C, where the line differential protection was interlocked with line distance protection. To enhance the scheme's robustness, the inherent distance protection functions of both the line differential relay and the distance protection relay in the second relay were employed to formulate logic for interlocking line differential protection with line distance protection. All line distance and line differential protections were IEC 61,850 compatible, facilitating the creation of interlocked logic through GOOSE communication between the relays. This paper provides a detailed discussion of the implementation of this distance-interlocked line differential protection and the challenges faced during the logic implementation. We aim for this paper to serve as a reference guide for utility engineers seeking to apply similar logic at their substations.

Keywords Distance protection · Differential protection · Power system protection · Generic Object-Oriented Substation Events (GOOSE)

1 Uganda's Energy Evolution: Navigating Challenges in Grid Expansion and Protection

In 2023, Uganda achieved a significant milestone with the commissioning of four 132/33 kV substations under the Grid Expansion and Renovation Project (GERP) [1]. Please refer to Fig. 1 for the connectivity of these substations with the existing network in Uganda. In Fig. 1, Substation-A, Substation-B, and Substation-G are the existing substations, while Substation-C, Substation-D, Substation-E, and Substation-F are the four substations commissioned under GERP. The double bold lines connecting any two substations denote the 132 kV double circuit transmission lines between them. Both existing and new substations are equipped with distance and line differential protection to safeguard the transmission lines connected to the substations.

The GERP project held significant prestige and bore immense socio-political importance for the stakeholders. Consequently, a decision was made to energize 132 kV Substation-C by connecting it to the existing transmission line AB, which runs between Substation-A and Substation-B. To facilitate this, it was necessary for the existing line AB to undergo a Loop In Loop Out (LILO) at Substation-C. However, post LILO, the primary challenges associated with the energization of lines AC and BC at Substation-C were related to the protection and control of the new lines. Transmission lines AC and BC are equipped with one distance and one differential protection at both ends. Further details regarding the communication equipment at Substation-C, as well as the protection and control equipment for the lines connecting Substation-C, are depicted in Fig. 2.

The differential protection relies on healthy communication links between the ends for its proper functioning. Similarly, the line distance protection requires a

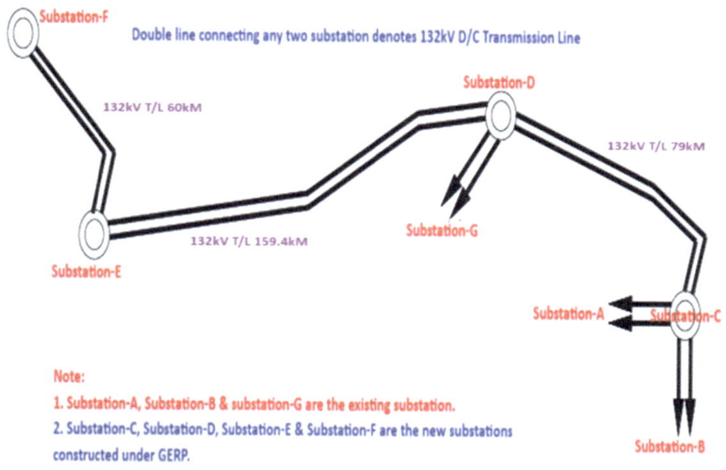

Fig. 1 Connectivity of the substations in Uganda

robust communication link to ensure the protection of 100% of the transmission line. Otherwise, faults on the 20%-line length (considering Zone-1 as 80%) from both ends will be cleared with a time delay equal to Zone-2 time. Therefore, to achieve complete 100% protection of the transmission line without intentional time delays, maintaining healthy communication links is crucial.

During the commissioning of this substation, the communication link from Substation-C to Substation-A and Substation-B was not ready. Consequently, the line differential protection for lines AC and BC could not function. As discussed earlier, the absence of healthy communication links made it impossible to achieve 100% instantaneous fault clearance with distance protection alone. Moreover, due to the non-readiness of communication links, monitoring and control of Substation-C from the national control centre (NCC) were not possible.

Given the political importance of Substation-C, a decision was made to energize line AB in such a way that the bus bar at Substation-C would become part of line AB, making line ACB. Since Substation-A and Substation-B were already monitored from the NCC, there were no immediate requirements for monitoring line AB at Substation-C. Existing protection at Substation-A and Substation-B remained intact, but the transmission line protection at Substation-C was disabled to avoid inadvertent tripping and complicated protection coordination.

The bus switching scheme at Substation-C utilized one main and one reserve bus type. Since the existing line AB was a double-circuit line and energized at Substation-C with the bus bar as part of the transmission line, the two lines were connected to different bus bars at Substation-C with the bus coupler in an open condition. To instantly clear any bus fault at Substation-C, the decision was made to keep the bus bar protection at Substation-C in service.

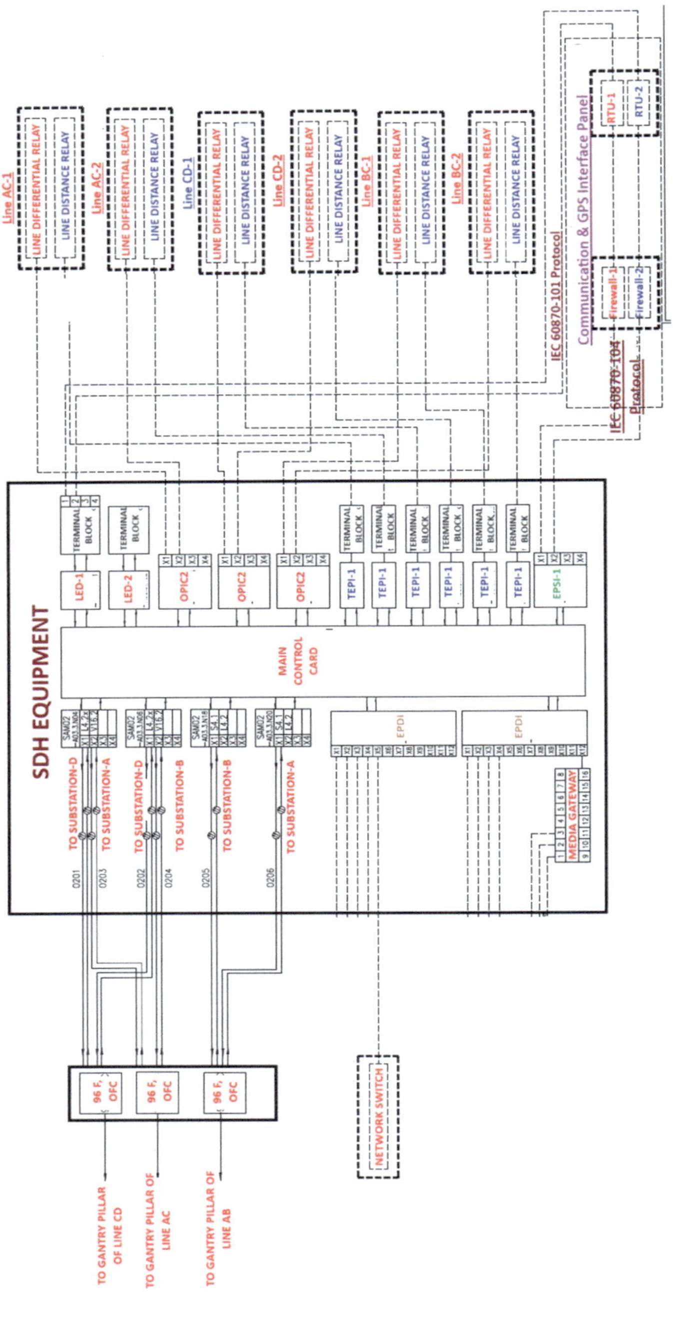

Fig. 2 Communication arrangement at Substation-C

Although it was possible for the line differential protection at the ends of line ACB to clear a bus fault at Substation-C even without active bus bar protection, the concern arose in case the line differential protection was not in service. In such a scenario, the bus fault at Substation-C, without bus bar protection, might not be cleared instantaneously, relying on the distance zone pickup at Substation-A and Substation-B.

Considering these arrangements, the transmission line AB was eventually energized with the Substation-C bus bar as part of the existing transmission line. Details of the active protection functions at Substation-A, Substation-B, and Substation-C at the time of energization of the transmission line AB at Substation-C are detailed in Table 1.

After the energization of line AB at Substation-C, ensuring a reliable auxiliary power system at the substation became a prime requirement. Without reliable auxiliary power, there was a risk of non-operation of switchgears at Substation-C when needed. To ensure reliable auxiliary power at Substation-C, it was deemed prudent to energize the 40 MVA 132/33 kV transformer at Substation-C. Subsequently, the auxiliary transformer of 100 kVA could be energized from the tertiary of this transformer to supply auxiliary power to Substation-C.

However, a new challenge emerged when the decision was made to commission the 40 MVA 132/33 kV transformer at Substation-C. The energization of the transformer acted as a tap load in line AB, posing a risk of line tripping due to the operation of line differential protection at Substation-A and Substation-B. Given the critical role of line AB in power evacuation and supplying major cities in Uganda, tripping during the transformer energization was deemed unacceptable.

To prevent the tripping of the transmission line ACB due to the operation of line differential protection at Substation-A and Substation-B during the energization of transformers at Substation-C, the differential protection at Substation-A and Substation-B was interlocked with the line distance protection at these substations. According to the existing utility practice, the distance protection of the transmission line remains inactive as long as the line differential protection is active. Consequently,

Table 1 Details of active protection function at the time of energization of Substation-C

	Substation-A	Substation-B	Substation-C
Line distance protection for line AB, LILO at Substation-C	All zones active but the settings were same as before the LILO of line AB at Substation-C	All zones active but the settings were same as before the LILO of line AB at Substation-C	All zones inactive
Line differential protection for line AB, LILO at Substation-C	Active	Active	Inactive
Bus bar protection	Active	Active	Active

in the event of a healthy communication link between Substation-A and Substation-B, the line differential protection remains active, and the distance protection becomes inactive.

Since the line distance protection remains inactive, implementing the distance interlocked differential protection using the distance function of the same relay was not possible. The second relay at these substations, being a distance protection relay, keeps its protection function active when the differential protection in the other relay is active. Therefore, it was decided to implement the interlocked scheme using GOOSE from the distance to the differential relay.

To address the limitation posed by VT fuse failure at Substation-A or Substation-B, a non-directional over-current protection function was activated in the distance protection relay. This non-directional over-current protection becomes active only when the relay detects VT fuse failure. The pickup current settings for this non-directional over-current protection were set higher than the inrush current of the transformers at Substation-C. The differential protection was interlocked with this non-directional over-current protection in such a way that the line differential protection would not operate if the non-directional over-current protection pickup in the distance relay. The time of operation of the non-directional over-current protection function was kept at a definite time with a duration longer than the zone-3 time of the distance relay.

The subsequent sections will provide a detailed discussion of the implementation of the distance interlocked differential protection scheme. The contingency condition involving VT fuse failure and the method of implementing the interlock through non-directional over-current stages will be examined. Additionally, a brief overview of the concept of GOOSE and the procedure for implementing the distance interlocked differential protection scheme using GOOSE will be covered.

2 Implementation of Distance Interlocked Differential Protection

In this section, we will discuss in detail the implementation of the distance interlocked differential scheme. Biased low-impedance line-differential protection was used at both ends of the existing line AB. Figure 3 illustrates the scheme of biased low-impedance line-differential protection.

The scheme operates based on a pickup setting and slope for the differential protection. Notably, the deployed differential protection at both substations lacks a harmonic blocking feature. Consequently, there was a significant risk that, during the energization of the transformer at Substation-C, the differential protection might trip if the inrush current of the transformer exceeded the set pickup values at Substation-A and Substation-B.

As per the existing settings, the differential current threshold was configured at 160 A primary, and the slope of the low-impedance differential protection was set at

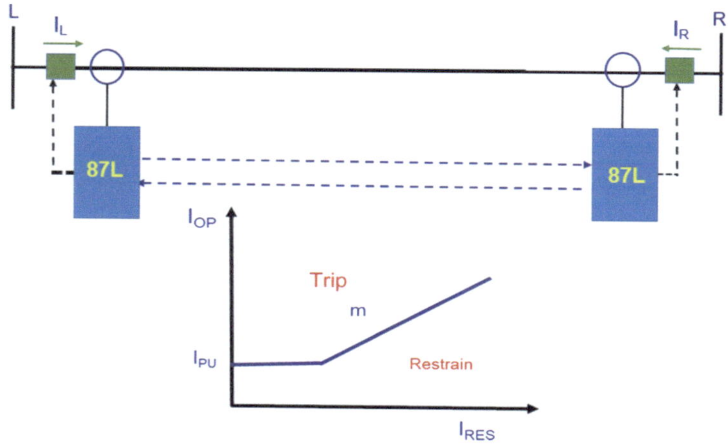

Fig. 3 Biased low impedance line differential protection

0.4. Considering the 40 MVA 132/33 kV transformer being energized, with a full load current of 175 A, the initial inrush current was expected to be around 900–1000 A. Without harmonic blocking, the differential relay could issue a trip command as the inrush current exceeded the set threshold.

Considering the transformer's per-unit impedance of 16% on a 132 kV base (approximately 70 Ω) and a fault current of around 1088 A in the event of a terminal fault, a safety margin of 30% required setting the pickup of the differential protection at Substation-A and Substation-B to 1414 A. However, setting the pickup value this high was not advisable due to the potential non-operation of the differential relay in the case of a high-resistance fault on the transmission line.

During transformer energization, the maximum inrush current was anticipated to be between 700 and 1050 A. The differential relay at Substation-A and Substation-B measured fundamental frequency differential current. Assuming 50% of inrush current as fundamental frequency components, the expected differential current ranged from 350 to 525 A. To prevent transmission line ACB tripping during transformer energization at Substation-C, the differential current threshold needed to be increased to around 550 A. However, this increased pickup value was likely to compromise the dependability of the differential protection.

With a total length of 80 km and a line impedance of 0.26 Ω/km for line ACB, the total impedance was 20.8 Ω. The existing practice set zone-1 and zone-2 of the distance protection at 80 and 120% of the line length, with time delays of 0 ms and 500 ms, respectively. Due to the transformer's impedance and expected fault impedance, there was no chance of inadvertent tripping of distance protection zone-1 and zone-2 at Substation-A and Substation-B. While both distance and differential protection would operate during a fault on the transmission line, only the differential protection would operate during transformer energization and a fault on the transformer terminal.

Considering these facts, interlocking the differential protection with distance protection appeared crucial to avoid misoperation of the differential protection during transformer energization at Substation-C. However, due to a limitation in the utility's existing practice, where the distance protection function becomes active only when the line differential protection is in a blocked condition, the distance protection function of the same relay could not serve as an interlock for line differential protection. Therefore, GOOSE from the separate distance relay was employed to the line differential relay to create the interlocked protection logic.

The reliability of GOOSE was deemed critical for the correct operation of the line differential relay. An additional logic was implemented in the line differential relay to monitor the quality of GOOSE. In case the line differential relay received a bad GOOSE from the distance relay, a non-directional over-current stage would be enabled in the line differential relay to provide the necessary interlock. The pickup of this non-directional over-current stage was set higher than the expected fundamental component of the transformer inrush current. If the overcurrent pickup occurred, it would block the line differential protection; otherwise, it would not. A similar non-directional stage was also enabled in the distance protection, and this non-directional over-current protection would activate when the distance protection was blocked due to VT fuse failure. The pickup of this non-directional over-current stage was also transmitted to the line differential relay through GOOSE, with a time delay of 20 ms to prevent any race-around conditions.

3 Overview of GOOSE and Implementation of Scheme

This section provides a detailed discussion of the logics for implementing the interlocked scheme. Additionally, we will briefly explore the concepts of GOOSE.

3.1 Overview of the GOOSE in IEC61850

The IEC 61,850 standard integrates contemporary technologies derived from computer science and the communication industry, incorporating OSI (Open Systems Interconnection) [2], the layer model [3], Object-Oriented Design [4], UML (Unified Modeling Language), the publish-subscribe model [5], and XML (Extensible Markup Language) [6]. These technologies are employed to achieve long-term stability, interoperability among different vendors, and a reduction in system engineering time. In contrast to traditional communication protocols, IEC 61,850 establishes a semantic hierarchical object data model and communication services based on this model.

Within this data model, each IED functions as a container for logic devices. Each logic device constitutes a group of logic nodes with specified functions outlined in IEC 61850-7-4 [6]. A logic node contains multiple data objects for storing parameters, data values, and control outputs. Each data object serves as an instance of a data class defined by IEC 61850-7-3 [7]. This approach ensures a standardized and structured framework that enhances stability, interoperability, and efficiency in system engineering processes.

GOOSE (Generic Object-Oriented Substation Event) serves as a communication service in the IEC 61850 standard, replacing traditional hardwired connections between IEDs. Utilizing a Publish-Subscribe service model, GOOSE achieves high-speed and high-availability communication.

An example of the distance protection zone-1 operated status, no longer presented as a simple binary input, would be in this format: "LD0.PDIS1.Op.general." This format signifies that an IED has a logic device named LD0, which includes all protection functions. Logic device Prot contains a PDIS1 logic node. Logic node PDIS1 has a single point status data object [6] Op. The attribute general of this data object Op stores the distance protection zone-1 operated value.

A GOOSE data set consists of a comprehensive list of detailed data attributes associated with data points, following the IEC 61850 data model. This is illustrated in Fig. 4 for an IED from vendor-A. Users have the capability to generate a GOOSE data set, including the addition or removal of data attributes from the list.

The GOOSE Control Block (GCB) outlines the characteristics and actions of the GOOSE message, defining the specific data set it manages. The Destination MAC address and App ID serve as a unique message identifier. While multiple GOOSE messages are broadcasted to all devices on the network, the GOOSE receiver utilizes this information to verify if the received GOOSE message matches the expected one.

The Max time is set to control the frequency of GOOSE heartbeats in milliseconds. In this configuration, the IED sends a GOOSE message every 10,000 ms when there is no new event. The GOOSE receiving IED waits twice as long, 20,000 ms. If it does not receive a GOOSE message within 20 s, the receiving IED concludes that communication with the GOOSE-sending IED is disrupted.

The VLAN ID is employed to restrict GOOSE broadcasting messages to a contained network, preventing unnecessary flooding to other devices. The Configure revision is a non-configurable parameter, changing whenever the GOOSE data set is updated. These parameters provide crucial information for troubleshooting GOOSE communication.

Communication in IEC 61850 GOOSE relies on a publish-subscribe mechanism, where IEDs broadcast GOOSE messages to all connected devices on the network. To process these messages, the receiving IED must identify which GOOSE messages it is set to receive. In IEC 61850 configuration tools, check boxes are commonly used to enable subscription from IEDs. These check boxes are typically found in the column corresponding to GOOSE control blocks in the configuration tool's rows (refer to Fig. 4).

When GOOSE subscription is enabled, information from the GOOSE data set and control block is copied from the sending IED to the receiving IED.

Upon receiving a GOOSE message, the information is compared with the GOOSE control block configuration in the receiving IED. If there is a match, the GOOSE communication is processed. However, if the GOOSE control block information, including destination MAC address, App ID, and configuration revision, does not

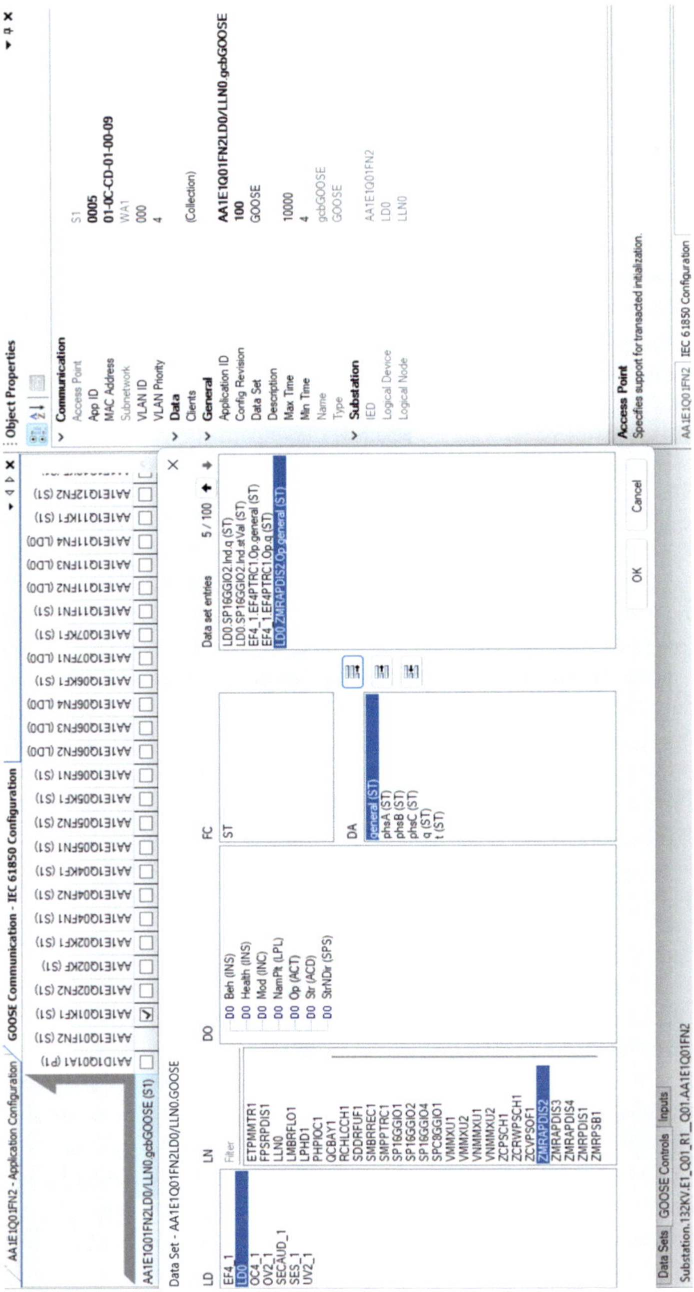

Fig. 4 Configuration of GOOSE data set in vendor-A IED

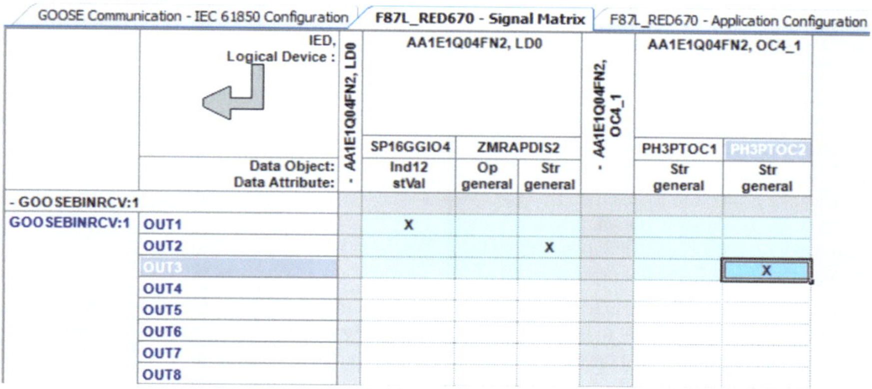

Fig. 5 GOOSE subscription in vendor-A IED

align with the configuration, the receiving IED will reject the GOOSE message. The configuration of GOOSE subscription in an IED from vendor-A is illustrated in Fig. 5.

3.2 Implementation of the Scheme

The SPS logics to be implemented at this substation are as follows:

Condition	Action
1. Communication Channel Healthy: Differential protection is in service, GOOSE communication is healthy between distance and differential protection, and distance protection is active. If Zone-2 of the distance protection picks up, and the line differential protection also picks up	Initiate a trip command to the line-controlling circuit breaker (CB)
2. Communication Channel Healthy: Differential protection is in service, GOOSE communication between distance and differential protection is healthy, and distance protection is active. If Zone-2 of the distance protection does not pick up and only the line differential protection picks up	Do not issue a trip command to the line-controlling CB. This is not a fault condition; it may be transformer energization or a fault beyond the transformer

(continued)

(continued)

Condition	Action
3. Communication Channel Healthy: Differential protection is in service, GOOSE communication between distance and differential protection is healthy, and distance protection is inactive. In this scenario, first enable non-directional overcurrent protection in the distance relay. If only the line differential protection picks up	Issue a trip command to the line-controlling CB only if the non-directional relay does not pick up; otherwise, do not issue the trip command
4. Communication Channel Healthy: Differential protection is in service, but GOOSE communication between distance and differential protection is unhealthy, and distance protection is inactive. In this case, first enable non-directional overcurrent protection in the differential relay. If the line differential protection picks up	Issue a trip command to the line-controlling CB only if the non-directional relay does not pick up; otherwise, do not issue the trip command
5. If communication channel is not healthy, enable distance protection in the line differential protection. If any one out of the two relay detects fault	Issue a trip command to the line-controlling CB

The detailed implemented logic is shown in Fig. 6.

4 Conclusion

The deployment of distance-interlocked line differential protection through GOOSE was introduced as a temporary solution, strategically preventing the mis-operation of line differential protection during the energization of a transformer at an intermediate substation. This approach is particularly advantageous for transmission lines with tapped loads. The decision to employ GOOSE was influenced by the simultaneous inactivity of line differential and line distance protection in the same relay. Adhering to the utility's established practice, line distance protection remains inactive until line differential protection becomes active.

If the distance protection function in the line differential relay remains consistently active, regardless of the healthiness of the line differential communication channel, implementing the scheme without GOOSE becomes viable due to the availability of both functions in the same relay. While integrating distance-interlocked differential protection in the same relay reduces the risk of scheme failure in the event of GOOSE failure, the viability of this decision is contingent upon utility practices. It is crucial to acknowledge that the concurrent activation of both distance and differential protection functions in the same relay carries the potential for relay mis-operation during line faults. Therefore, the scheme's implementation strategy must align with utility practices to ensure its effectiveness.

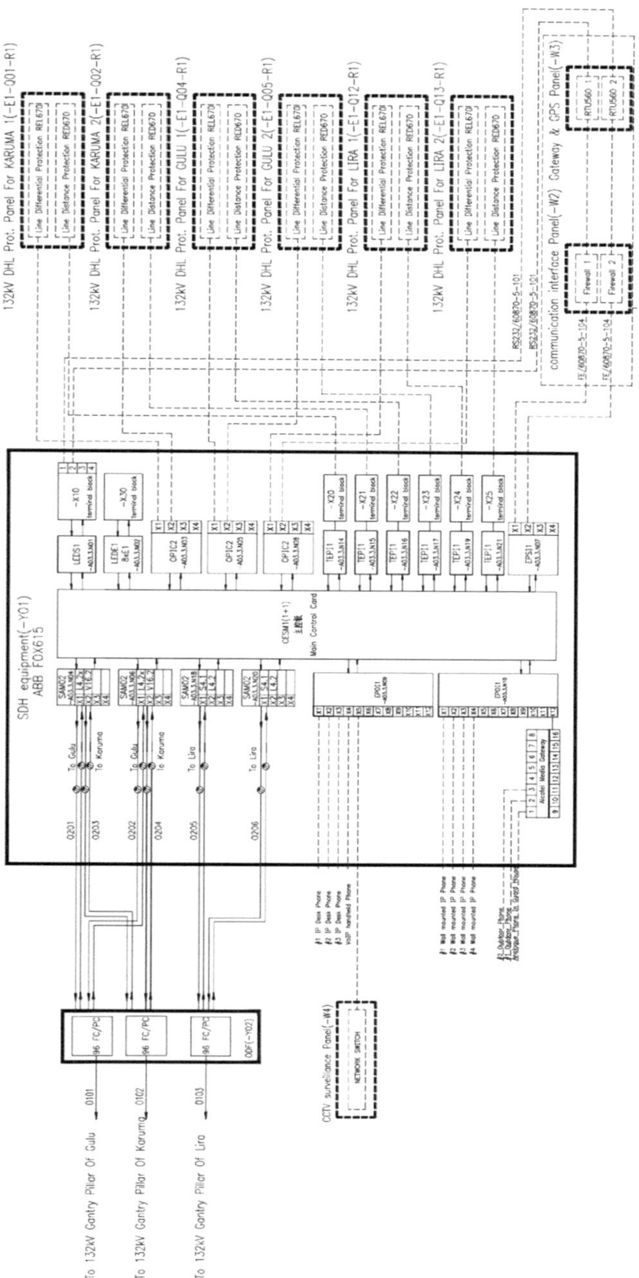

Fig. 6 Detailed implemented logic

In conclusion, this paper has undertaken a comprehensive exploration of the implementation of distance-interlocked differential protection in Uganda, presenting valuable insights for utilities contemplating the deployment of a similar scheme in their substations. By delving into crucial concepts such as GOOSE, the paper has provided a detailed understanding of the intricacies involved in GOOSE logics, publication, and subscription across different numerical microprocessor-based IEDs. The discussion has extended to include the meticulous calculations of diverse settings, elucidating the methodology employed to derive the final scheme logic.

The authors anticipate that the knowledge disseminated in this paper will serve as an invaluable resource for both academia and industry, fostering a deeper understanding of the practical challenges associated with line differential and distance protection in real-world applications. It is our hope that readers will find practical guidance within these pages, enabling them to implement analogous schemes tailored to their specific contexts.

References

1. https://documents1.worldbank.org/curated/en/790251611238166469/pdf/Disclosable-Version-of-the-ISR-Uganda-Grid-Expansion-and-Reinforcement-Project-GERP-P133305-Sequence-No-09.pdf
2. ISO/IEC 7498-1 (1994) Information technology—open systems interconnection—basic reference model: the basic model. IEC international standard, 2nd edn
3. Gamma E, Helm R, Johnson R, Vlissides J (1950) Design patterns: elements of reusable object-oriented software. Addison-Wiley. ISBN 0-201-63361-2
4. W3C Extensible Markup Language (XML) 1.0. W3C recommendation, 5th edn
5. Birman K, Joseph T (1987) Exploiting virtual synchrony in distributed systems. In: Proceeding of the eleventh ACM symposium on operating systems principles, pp 123–138. ISBN: 0-89791-242-X
6. IEC 61850-7-4 (2010) Communication networks and systems for power utility automation—basic communication structure—compatible logic node classes and data object classes. IEC international standard edn 2.0
7. IEC 61850-7-3 (2010) Communication networks and systems for power utility automation—basic communication structure—common data classes. IEC international standard edn 2.0

Comparative Study of Maximum Power Point Tracking of DC Microgrid Using Evolutionary Optimization Techniques

Gaurav Kumar and Santosh Kumar Verma

Abstract Renewable Energy Resources are viable and dependable solutions for achieving universal electricity access rapidly. Among these, Solar Photovoltaic (PV) stands out, despite its variable output power caused by fluctuations in solar radiation, cell performance, and ambient temperatures. Additionally, the modules utilized often exhibit lower conversion efficiency. Consequently, the implementation of maximum power point tracking becomes imperative to optimize energy harvesting from the sun and enhance the overall efficiency of photovoltaic systems. A microgrid has been designed with a Solar PV Cell and battery system in MATLAB, and a comparative study of maximum power point tracking using Particle Swarm Optimization (PSO), Flying Squirrel Search Optimization (FSSO), and Horse Herd Optimization (HHO) techniques is presented in this work. The Comparison of output power with different changing input parameters like irradiation and temperature has been carried out in this work. The comparative study of the said algorithm is also depicted by figures and tables.

Keywords Renewable Energy (RE) · PV system · Maximum Power Point Tracking (MPPT) · PSO Algorithm · FSSO Algorithm · HHO Algorithm

1 Introduction

The inherent correlation between energy access and socioeconomic development is widely recognized. India, which has the highest proportion of the world's population without access to electricity, faces the challenge of providing this essential service to approximately 237 million people, according to the International Energy Agency.

G. Kumar (✉) · S. K. Verma
Assam Energy Institute, Sivasagar, a Center of Rajiv Gandhi Institute of Petroleum Technology, Jais Amethi, India
e-mail: gauravk22ea@rgipt.ac.in

S. K. Verma
e-mail: skverma@rgipt.ac.in

S. Kumar et al. (eds.), *Soft Computing Applications in Modern Power and Energy Systems*, Lecture Notes in Electrical Engineering 1298,
https://doi.org/10.1007/978-981-96-0058-8_6

Ironically, even among Indian households that have access to electricity, many experience inadequate and intermittent power supply. A recent study conducted by the Council for Energy, Environment, and Water (CEEW) in Bihar, Jharkhand, Madhya Pradesh, Uttar Pradesh, West Bengal, and Odisha revealed that nearly 50% of households in these states lacked reliable electricity access despite being connected to the power grid. This underscores the urgent need to focus not only on expanding the grid but also on enhancing the quality, affordability, and reliability of electricity supply.

The conventional power system depends on large-scale fossil fuel generators, which deliver power through extensive transmission lines to consumers. This approach leads to significant environmental issues, including air, water, and soil pollution. The associated health impacts are wide-ranging, encompassing respiratory issues, lung diseases, hepatitis, typhoid, cancer, neurological disorders, and other conditions, ultimately resulting in fatalities [1–6].

Moreover, a significant portion of our thermal power stations rely heavily on coal. Given that India's coal reserves account for only 10% of the global total and are depleting rapidly, it is crucial to transition to renewable energy sources. The dwindling coal reserves highlight the urgent need to adopt renewable alternatives to sustainably meet our energy demands in the near future. Additionally, the power shortage in India contributes to a substantial number of homes being off the grid and those connected experiencing prolonged periods of load shedding. The country generates insufficient power relative to its increasing consumption needs. Furthermore, even in villages with grid connectivity, the power infrastructure often lacks the necessary extensions to reach every individual home.

Rooftop solar panels are widely promoted as a viable alternative for power generation. In many regions of India, as few as fifteen 413W solar panels can generate a significant portion of the required power. This is particularly advantageous as it eliminates transmission and distribution losses associated with conventional power sources [7].

In a Photovoltaic (PV) system, the power output reaches its maximum at a dynamic position known as the Maximum Power Point (MPP). This point continually shifts due to variations in solar radiation and temperature. As the location of the MPP changes, it directly affects the power output of the PV system, which is predetermined during the system's design and installation phase [8–12]. To ensure the effective utilization of a PV generator's peak power, it is essential to track the MPP using a Maximum Power Point Tracking (MPPT) algorithm. This algorithm continuously adjusts the operating point to align with the MPP on the PV curve, representing the optimal power output a PV module can generate at any given moment. Consequently, a reliable MPPT algorithm is crucial for consistently maximizing the efficiency of a PV system [13].

Maximum Power Point Tracking (MPPT) is currently a prominent area of research [14], with numerous algorithms developed by researchers and engineers to optimize the power output from PV arrays. Recently, several intelligent MPPT techniques have been introduced for solar module arrays, designed to precisely track the Maximum

Power Points (MPPs) and improve both dynamic and steady-state tracking performance. However, these techniques are typically applicable to MPPT in solar module arrays that are unaffected by shading [15].

A Particle Swarm Optimization (PSO)-based MPP tracker has been proposed for solar module arrays [16]. While this tracker can monitor global MPPs with multiple peak characteristic curves, its performance was not robust due to the use of fixed values for weighting in the algorithm, leading to poor success rates in tracking global MPPs. Although the MPPs were effectively monitored, the dynamic response was suboptimal. To address these issues, this research employed Particle Swarm Optimization (PSO) with enhancements to avoid getting stuck in local MPPs, thus allowing the algorithm to quickly and reliably locate global MPPs on the complex, multi-peak characteristic curves of photovoltaic module arrays. In 2021, Jalal et al. conducted a comprehensive analysis of various optimization techniques for maximizing MPPT parameters in photovoltaic (PV) systems [13].

In this paper, three algorithms—Particle Swarm Optimization (PSO), Flying Squirrel Search Optimization (FSSO), and Horse Herd Optimization (HHO)—are implemented for Maximum Power Point Tracking (MPPT) in a microgrid system with a PV system and battery. A comparative study is then conducted to determine which algorithm yields the most optimal results. The selection of these three algorithms is based on their simplicity of implementation and rapid convergence, which are crucial attributes for effective MPPT tracking.

2 Photovoltaic System

A photovoltaic (PV) system, or solar power system, is an electric power system designed to supply usable solar power. It consists of several photocells connected in series and parallel combinations. When photons penetrate the depletion region of a diode, they impart high energy to atoms, facilitating the absorption of energy by electrons, and leading to their release. This process results in the formation of free electrons and holes, thereby initiating a flow of current. The generation of electron–hole pairs occurs when photons with energy exceeding 1.1eV strike the diode. The built-in electric field within the depletion region plays a crucial role in this mechanism. Under the influence of this electric field, the electron–hole pairs move away from the junction. Consequently, the holes migrate toward the anode, while electrons move toward the cathode, orchestrating the production of photocurrent. Multiple cells are interconnected electrically in series or parallel within a PV module to meet specific voltage and current requirements. Typically, cells are arranged in series to attain higher voltage at the terminals, while parallel connections result in increased current at the PV panel's output.

The solar cell shares similarities with a p–n junction diode in terms of characteristics but operates in a regenerative mode due to the photon energy of light. Consequently, the solar cell can be effectively modeled as a p–n junction diode, incorporating additional parameters to accurately replicate the I–V characteristics

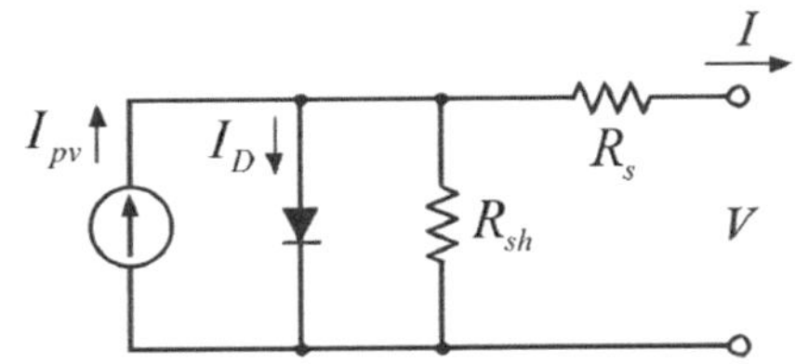

Fig. 1 Circuit diagram of PV cell

of the physical cell. The resulting circuit, comprising a current source and a diode, forms the electrical equivalent circuit of a solar cell, as depicted in Fig. 1 [17–19].

where.

I_{pv} : Is the current generated by the diode in response to incident light.

I_D : Is the current flowing through the diode.

I : Output current of PV cell.

R_{sh} : It shows the resistance of the flow of electrons from the cathode to the anode.

R_s : It shows the resistance of contact of wire bonds and semiconductor material.

If I_{pv} increases, due to an increase in solar radiation, V_{oc} will increase logarithmically whereas I_{sc} will increase linearly. The closer rectangle area1 (made by point V_{mp}) to the rectangle area (made by point I_{sc}, V_{oc}), the better the cell is. As temperature increases, I_{sc} increases but V_{oc} & P will decrease.

The quality of PV cells is measured by the Fill Factor (FF). FF should be near to 1 for better PV Cell (FF $\leq$ 1).

$$Fill\ Factor = \frac{V_m * I_m}{V_{oc} * I_{oc}} \tag{1}$$

3 Maximum Power Point Tracking:

The output current and voltage of PV modules undergo changes due to variations in irradiation and temperature. Consequently, the efficiency of power generation from PV modules fluctuates with external climate conditions. MPPT technique allows us to continuously monitor the Maximum Power Point and optimize the utilization of the PV module's maximum power output under diverse environmental conditions [14]. The Maximum Power Point or peak power voltage represents the maximum power output of a PV module, as depicted in Fig. 2. At the MPP, there exists a singular combination of current (I_{mp}) and voltage (V_{mp}) values that define the conditions under which the maximum power (P_{mp}) is achieved.

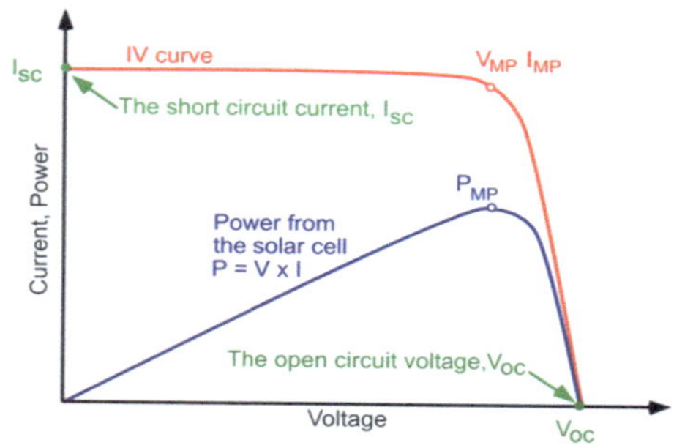

Fig. 2 A typical I-V and P–V characteristics curve of a PV cell

Among various existing algorithms implemented for Maximum Power Point Tracking (MPPT) three most common and effective algorithms are implemented in this work. The discussion will specifically delve into the following algorithms:

1. Particle swarm optimization algorithm
2. Flying squirrel search optimization algorithm
3. Horse herd optimization algorithm.

3.1 Particle Swarm Optimization Algorithm

The Particle Swarm Optimization (PSO) algorithm was introduced by Eberhart and Kennedy in 1995. It is a stochastic optimization technique inspired by swarm behavior observed in nature, such as in insects, herds, birds, and fishes as shown in Fig. 3 [20].

In PSO, swarms emulate the cooperative nature of these animal groups as they collectively search for food. Each member of the swarm continuously adapts its search pattern based on its individual learning experiences and interactions with other swarm members [21].

Fig. 3 Particle swarm optimization

The methodology originates from the empirical study of avian and fish aggregations during flight, in which individual organisms acquire knowledge from their counterparts. The members take turns exchanging data with one another, which allows them to change their spot as well as velocity. The creation of clusters assists to reduce the amount of work that is required to hunt for sources of food and water on the ground. PSO is characterized by its rapid processing rate and lower memory requirements as a consequence of its development [22]. A collection of particles, each of which is characterized by a proposed solution, are the components that make up the PSO algorithm. Each and every particle is affected by the best solution of all particles in the entire solution space, which is denoted by the symbol G_best, as well as the best solution of the particle (P_{best}) surrounding it [21–23]. The searching equation is expressed as.

Updated position of particle i is

$$x_i^{k+1} = x_i^k + V_i^{k+1} \tag{2}$$

Additionally, the step size is indicated by the letter V, which stands for the velocity of the particle. The velocity factor is determined using the following formula:

$$V_i^{k+1} = W_i V_i^k + C_1 r_1^k \left(\mathrm{P_{best}}_i^k - x_i^k\right) + C_2 r_2^k \left(\mathrm{G_{best}}_i^k - x_i^k\right) \tag{3}$$

where

${P_{best}}_i^k$: Best location of the particle I at iteration k.
${G_{best}}_i^k$: Global best location of the group.
x_i^k : Current location of particle i.
W_i : Inertia factor of particle I (lie between 0.9 to 0.4).
$C_1 \& C_2$: : Constant value (lie between 0.5 to 2).
V_i^{k+1} : Updated velocity at k + 1 iteration.
V_i^k : Updated current velocity at iteration k of particle i.
$r_1^k \& r_2^k$: Random number (lie between 0 to 1).

G_best represents the optimal position in the overall population, while P_best, i denotes the best spot for each particle. The PSO algorithm consists of five primary processes in determining the optimal parameters for the controller. The process consists of the following steps: Step-1, initiation; Step-2, fitness evaluation; Step-3, updating the best data; Step-4, updating velocity and location; Step-5, determining convergence. In order to initiate the algorithm, specific parameters must be initialized, and the details of these parameters may be found in Table 1.

3.2 Flying Squirrel Search Optimization Algorithm

The Flying Squirrels Optimizer is an emerging metaheuristic swarm-based optimization algorithm designed for solving mathematical and engineering problems. Geographic range of flying squirrels (Glaucomys Sabrinus) includes the tree line

Table 1 Constraints of the PSO

Description	Constraint
Maximum iteration	50
Population size	20
Acceleration factor C_1	2
Acceleration factor C_2	2

in Canada and Alaska, as well as Northern California and Colorado in the Western hemisphere.

In the central and eastern parts of the continent, they extend from Central Michigan and Wisconsin to northern North Carolina and Tennessee. Despite being somewhat clumsy on the ground, flying squirrels exhibit graceful gliding abilities between trees. Their ground movement is modeled as a (a) random walk with steps following a normal distribution, while their flying phase is simulated based on the principles of (b) Lévy flight theory as shown in Fig. 4 [24].

Notably, flying squirrels emit a soft, low chirp, serving as a means of communication, particularly when distressed. They rely on scent and touch as preferred modes of communication with each other.

A random walk is a stochastic process characterized by the taking of a series of consecutive random steps. Random walks find applications in various fields such as physics, economics, statistics, computer sciences, environmental science, and engineering. Mathematically let S_n denote the sum of each consecutive random step X_i, then Sn forms a random walk in N dimensions:

$$S_n = \sum_{i=1}^{N} X_i = X_1 + X_2 + X_3 \ldots\ldots\ldots\ldots\ldots\ldots + X_N \tag{4}$$

where X_i represents a stochastic increment sampled from a probability distribution. This relationship can alternatively be expressed as a recursive formula:

(a) (b)

Fig. 4 Two motion phases of flying squirrels to find food **a** random walking, and **b** Levy Flying

$$\Rightarrow S_n = \sum_{i}^{N-1} X_i + X_N = S_{N-1} + X_N \tag{5}$$

This implies that the subsequent state (S_n) will solely be contingent on the current state (S_{N-1}) and the motion or transition (X_N) from the current state to the next state. The magnitude of the step in a random walk can either remain constant or change. Furthermore, a constant step length is not inherently necessary. Yes, the magnitude of the step can vary according to a predetermined distribution. When the step length follows a Gaussian distribution, the random walk transitions into Brownian motion.

Levy flights have been observed in the foraging patterns of various species, including albatrosses, fruit flies, and spider monkeys. Even human hunter-gatherers have demonstrated the ability to trace paths with Levy-flight patterns. Moreover, Levy flights find widespread applications, extending to physical phenomena such as the diffusion of fluorescent molecules, cooling behavior, and noise, which can exhibit Levy-flight characteristics under specific conditions. This concept can be applied to model the flying phase of flying squirrels' motion. It has been demonstrated that levy flights are more effective than Brownian random walks for examining uncharted large-scale search regions.

3.3 Horse Herd Optimization Algorithm

The Horse Herd Optimization (HHO) algorithm is also a nature-inspired optimization approach that draws inspiration from the collective behavior of horse herds and was introduced by Farid Miar in 2021 [25, 26]. This algorithm replicates various social aspects of horses across different ages, encompassing six key features: grazing, hierarchy, sociability, imitation, defense mechanism, and roam. HHO has demonstrated excellent performance in tackling complex problems within high-dimensional spaces, leveraging a plethora of control parameters derived from the behavioral traits of horses at various life stages. Notably, the HHO algorithm surpasses other widely recognized optimization algorithms in terms of accuracy and efficiency, achieving this with the added advantages of lower computational costs and reduced complexity. Despite having a large number of control parameters, this algorithm has a major advantage in that Complex high-dimensional optimization problems can be solved easily by this technique.

The HHO technique is categorized into four groups based on age and population as shown in Fig. 5.

This study is grounded in the observation of behavior patterns exhibited by horses in their natural environment. These behavior patterns, which encompass Grazing (G), Hierarchy (H), Sociability (S), Imitation (I), Defence Mechanism (D), and Roaming (R) [25–27], serve as the inspiration for the algorithm. Therefore, the algorithm draws influence from these six general behaviors displayed by horses at various ages.

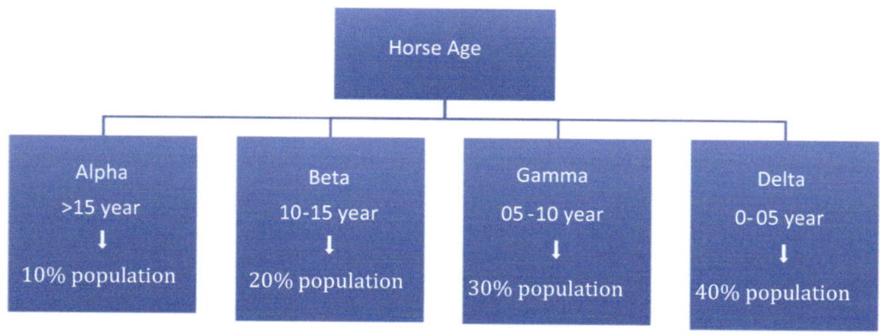

Fig. 5 Classification of horses in the HHO technique

(A) Grazing (G)

Horses, being grazing animals, consume plants, grasses, and forages. They typically spend 16 to 20 h a day grazing in pastures, with short resting periods. This continuous eating behavior is integral to their routine, often observed in pastures where mares graze alongside their foals [28].

(B) Hierarchy (H)

Horses exhibit a social structure that follows a leader, often designated by human influence. In the wild, adult stallions or mares take on the responsibility of leadership, adhering to the law of hierarchy within horse herds [23].

(C) Sociability (S)

Horses thrive in social settings and often coexist with other animals, a behavior developed for safety against predators. The herd life ensures their security, as the presence of multiple individuals increases survival chances and facilitates escape. While occasional conflicts arise due to their social nature, horses generally prefer companionship over solitude [28].

(D) Imitation (I)

Horses demonstrate the ability to imitate one another, learning both positive and negative habits, such as identifying suitable pasture locations.

(E) Defence Mechanism (D)

Horses' reactions are shaped by their history as prey animals. They exhibit the fight-or-flight response as a defense mechanism, with a natural inclination to escape perceived threats. Additionally, they may resort to bucking when feeling trapped. In their quest for resources, horses may engage in competitive behaviors to secure food and water, instinctively avoiding hazardous environments where predators might be present [25, 28].

(F) Roaming (R)

Horses engage in roaming behaviors, moving from pasture to pasture in search of food. Even when kept in stables, horses retain their roaming instincts. They may unexpectedly move to different locations to graze, driven by their curiosity and eagerness to explore new pastures and familiarize themselves with their surroundings. The design of stable side walls often accommodates their curiosity, allowing horses to see each other and satisfying their inquisitive nature [25].

The movement applied to horses at each iteration is represented as

$$X_m^{iter.age} = V_m^{iter.age} + X_m^{(iter-1).age} \tag{6}$$

where $m = 1,2,3 \ldots.$ pop size

$$iter = current\ iteration$$

$$age = \alpha, \beta, \gamma\ and\ \delta$$

The steps to simulate the six behaviors of the horses are mathematically implemented to detect the velocity vector. Horse Velocity Vectors are as.

for α;

$$V_m^{iter.\alpha} = G_m^{iter.\alpha} + D_m^{iter.\alpha} \tag{7}$$

for β;

$$V_m^{iter.\beta} = G_m^{iter.\beta} + D_m^{iter.\beta} + S_m^{iter.\beta} + H_m^{iter.\beta} \tag{8}$$

for γ;

$$V_m^{iter.\gamma} = G_m^{iter.\gamma} + D_m^{iter.\gamma} + S_m^{iter.\gamma} + H_m^{iter.\gamma} + I_m^{iter.\gamma} + R_m^{iter.\gamma} \tag{9}$$

for δ;

$$V_m^{iter.\delta} = G_m^{iter.\delta} + I_m^{iter.\delta} + R_m^{iter.\delta}. \tag{10}$$

Table 2 Parameters of single PV module

Single PV module	
Open circuit voltage (V_{oc})	90.5 V
Short circuit current (I_{sc})	6.21 A
Cell per module (N_{cell})	128
Series resistance (R_s)	0.548 Ω
Shunt resistance (R_{sh})	508.288 Ω
Maximum power (P_{mp})	444.86 W
Voltage at maximum power point (V_{mp})	76.7 V
Current at maximum power point (I_{mp})	5.8 A
Changes are done in the following data of the PV array block for tracking and achieving the required voltage and power	
Number of parallel strings	5
Series connected modules per string	3
Overall voltage = 90.5*3	271.5 V
Overall current = 6.21*5	31.05 A
Overall maximum power = 5*3*444.86	6.673 KW

4 Simulink Models of Microgrid

A Simulink model of a microgrid having 6KW capacity is designed using MATLAB with the help of the "Sunpower SPR-445-WHT-D" solar module with its given parameter in Table 2.

The Simulink model of the whole system is designed with a PV array connected to a battery system with an MPPT algorithm as shown in Fig. 6.

5 Results and Discussion

In this section, the output power of the PV system with different algorithms and different variations in solar radiation and temperature has been analyzed with the help of output response. The performance of algorithms has been analyzed for the output power of both the PV array and load.

5.1 Output Power of PV Array

The modeling of the entire system involves employing three different techniques to track the maximum power from the PV array. These algorithms play a crucial role in maximizing the power output from the PV system. In Table 3, it is demonstrated that

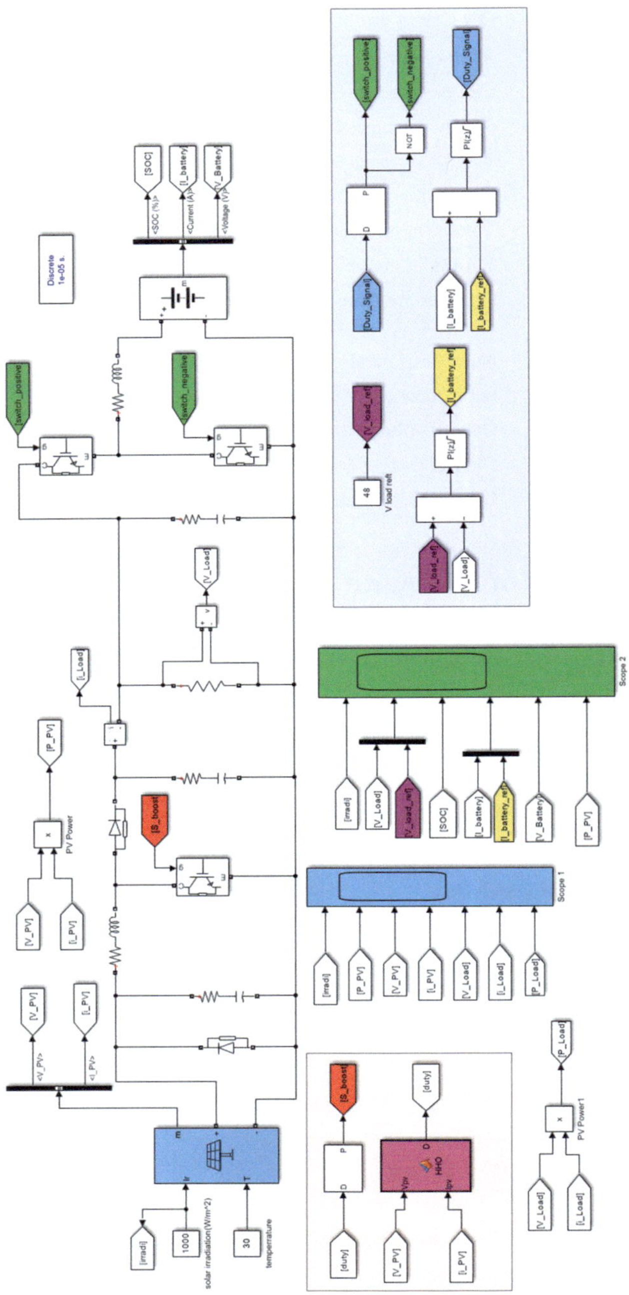

Fig. 6 Simulink model of the whole system

Table 3 Comparison of output power of PV array of all three algorithms at different irradiation

When $R_L = 1\Omega$, temperature 25 °C is fixed, then PV output Power P_{PV} (in KW) is			
i_{rr}	HHO	PSO	FSSO
800	5.328	5.251	5.070
900	5.725	5.590	5.275
1000	5.830	5.805	5.520
1100	6.115	5.963	5.560

at a fixed temperature (t = 25 °C), as irradiation increases, PV power also exhibits an increase across all algorithm cases. Specifically, at an irradiation level of 1000, the HHO algorithm yields the highest output power for the PV array (5.830 kW). In comparison, the PSO algorithm produces 5.805 kW, and the FSSO algorithm results in an output power of 5.520 kW for the PV array.

Figure 7 illustrates the output power of the PV array under 800 irradiation and a temperature of 25 °C where y-axis is Output power of PV array and x-axis is time. As depicted in the figure, HHO outperforms both PSO and FSSO, yielding a higher output power of 5328W for the PV array. Additionally, HHO achieves a steady state in just 0.5 s, demonstrating a faster convergence compared to PSO and FSSO, which require more than 1 s to stabilize. This indicates that HHO not only delivers superior results but also exhibits faster convergence in comparison to PSO and FSSO.

In Fig. 8, the waveform of the output power of the PV array is depicted under 900 irradiation and a temperature of 25 °C. It is evident from the figure that HHO outperforms both PSO and FSSO, delivering a superior output power of 5725W for the PV array. Furthermore, HHO achieves a steady state in only 0.75 s, showcasing a faster convergence compared to PSO (more than 1 s) and FSSO (0.9 s). This underscores that HHO not only provides better results but also demonstrates quicker convergence when compared to PSO and FSSO.

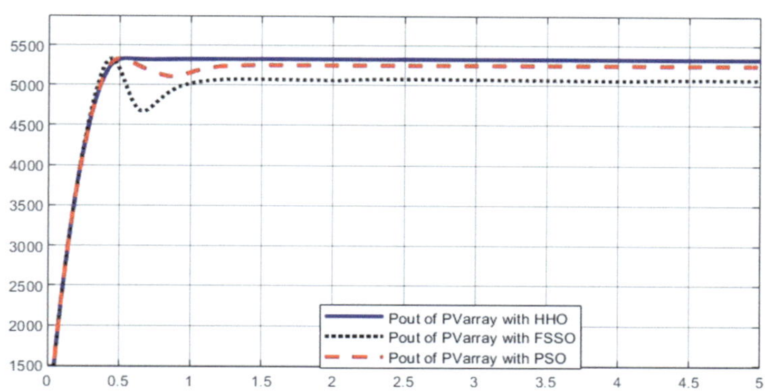

Fig. 7 Output power of PV array at 800 irradiation

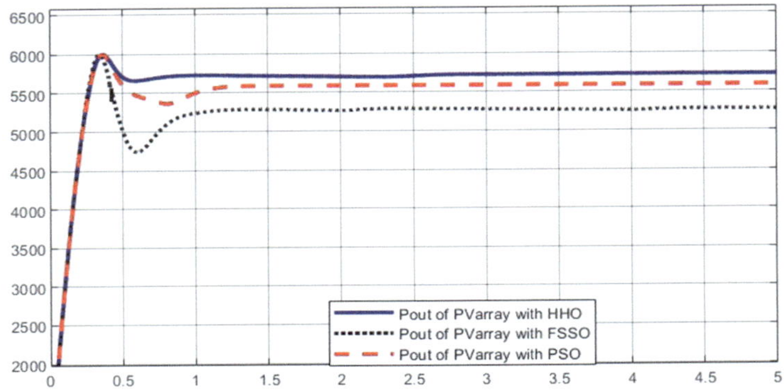

Fig. 8 Output power of PV array at 900 irradiation

In Fig. 9, the waveform illustrates the output power of the PV array under 1000 irradiation and a temperature of 25 °C. Discernible from the figure, HHO outperforms both PSO and FSSO, yielding a superior output power of 5830W for the PV array.

In Fig. 10, the waveform illustrates the output power of the PV array under 1100 irradiation and a temperature of 25 °C. It is evident from the figure that HHO outperforms both PSO and FSSO, delivering a superior output power of 6115W for the PV array.

Similarly, Table 4 illustrates that at a fixed irradiation level ($i_{rr} = 1000$), as the temperature increases, the output power of the PV array also increases for each algorithm case. Specifically, at a temperature of 20 °C, the HHO algorithm produces the highest output power of the PV array (6.116 kW) compared to the other two algorithms (PSO and FSSO). In comparison, the PSO algorithm yields 5.977 kW, while the FSSO algorithm results in an output PV power of 5.665 kW.

The output power of the PV array at temperatures of 20 °C, 25 °C, 30 °C, and 35 °C are also validated with waveforms shown below in Figs. 11, 12, 13, and 14

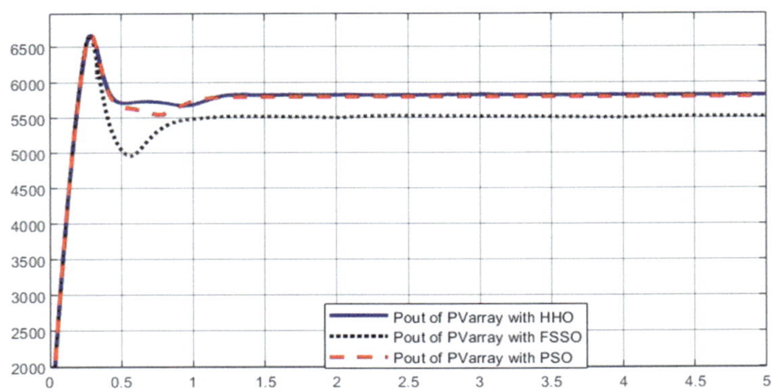

Fig. 9 Output power of PV array at 1000 irradiation

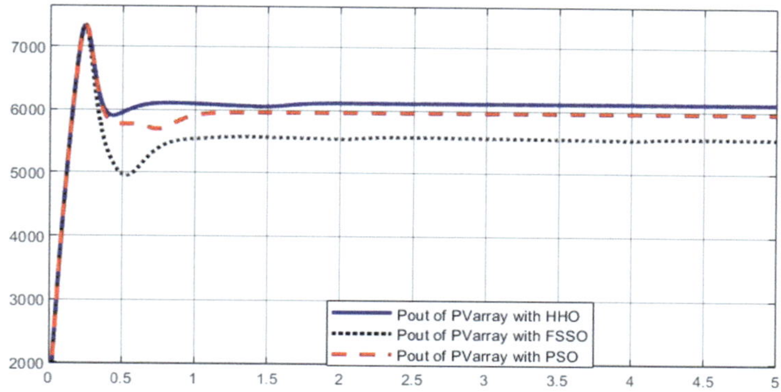

Fig. 10 Output power of PV array at 1100 irradiation

Table 4 Comparison of output power of PV array of all three algorithms at different temperatures

When $R_L = 1\Omega$, irradiation $i_{rr} = 1000$ is fixed, then PV output Power P_{PV} (in KW) is			
Temperature	HHO	PSO	FSSO
20^0C	6.116	5.977	5.665
25^0C	5.830	5.805	5.520
30^0C	5.755	5.636	5.315
35^0C	5.585	5.469	5.195

respectively, with fixed irradiation at 1000 where y-axis is Output power of PV array and x-axis is time.

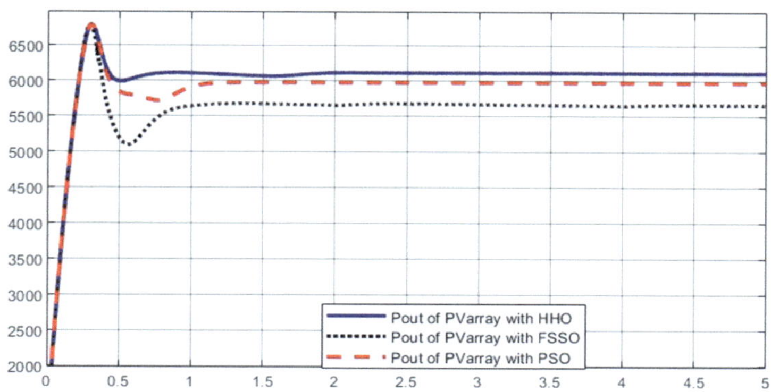

Fig. 11 Output power of PV array at 20 °C temperature

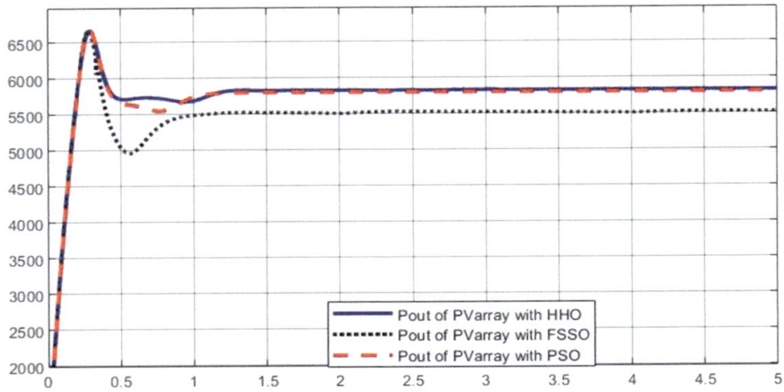

Fig. 12 Output power of PV array at 25 °C temperature

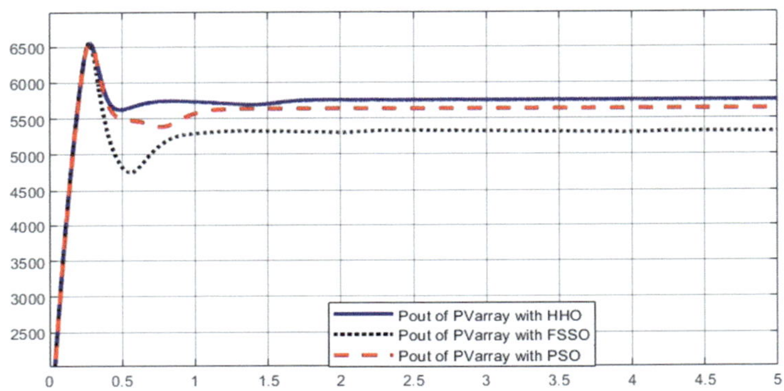

Fig. 13 Output power of PV array at 30 °C temperature

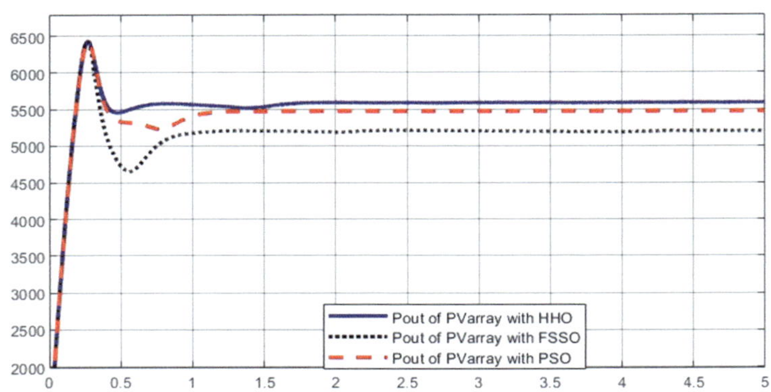

Fig. 14 Output power of PV array at 35 °C temperature

Table 5 Comparison of output power of Load of all three algorithms at different irradiation

When $R_L = 1\Omega$, temperature $t = 25^0C$ is fixed, then load Power P_L (in W) is			
i_{rr}	FSSO	HHO	PSO
800	352	…	178
900	365	300	180
1000	380	320	188

Table 6 Comparison of output power of Load of all three algorithms at different temperatures

When $R_L = 1\Omega$, irradiation $i_{rr} = 1000$ is fixed, then load Power P_L (in W) is			
Temperature	FSSO	HHO	PSO
20^0C	404	350	192
25^0C	380	320	188
30^0C	357	300	200

5.2 *Output Power of Load*

The modeling of the entire system focuses on tracking the maximum load power from the system. Table 5 illustrates that at a fixed temperature (t = 25 °C), the load power increases as the irradiation levels increase for each algorithm. Specifically, at an irradiation level of 1000, the FSSO algorithm yields the highest load power (380W). In comparison, the HHO algorithm produces 320W, while the PSO algorithm results in a load power of 188W.

Similarly, Table 6 demonstrates that, at fixed irradiation ($i_{rr} = 1000$), when temperature increases, load power also increases in each case of the algorithm. At temperature t = 25 °C, the FSSO algorithm gives the highest load power (380W) than the other two algorithms (HHO & PSO). Whereas HHO gives 320W and PSO gives 188W load power.

6 Conclusion

In this work, a microgrid has been designed with a PV array and Battery storage system in MATLAB. Additionally, the three most recent and effective algorithms namely Particle Swarm Optimization, Flying Squirrel Search Optimization, and Horse Herd Optimization have been implemented for MMP tracking. The performance of each algorithm has been analyzed for various irradiation levels (800, 900, 1000, and 1100) and temperatures (20, 25 and 30 °C). The validation of results has been shown with tables and figures. The results indicate that, concerning irradiation and temperature, the HHO algorithm yields the most favorable outcomes for output PV power as compared to PSO and FSSO. Conversely, in scenarios where irradiation varies at a fixed load and temperature, the FSSO algorithm demonstrates superior performance for load power when contrasted with the other algorithms. Additionally, in situations where temperature fluctuates at a fixed load and irradiation, the HHO algorithm proves to be the most effective for output PV power, while the FSSO algorithm excels in optimizing load power.

References

1. Xu Hanqing, Yang Jia, Zhendong Sun, Qunfang Zhou (2022) Environmental pollution, a hidden culprit for health issues, Eco-Environment & Health 1(1)
2. World Health Statistics (2019) Monitoring health for the sdgs, sustainable development goals. World Health Organization, Geneva, p 2019
3. Prüss-Üstün A, Corvalan CF (2006) Preventing disease through healthy environments: towards an estimate of the environmental burden of disease. World Health Organization, Geneva
4. Laborde A, Tomasina F, Bianchi F, Brune MN, Buka I, Comba P et al (2015) Children's health in Latin America: the influence of environmental exposures. Environ Health Perspect 123:201–209
5. Brimblecombe P (2019) Air pollution episodes, J. Nriagu (Ed). Encycl Environ Health, Elsevier, Norwich, pp 41–48
6. M.L. Bell, D.L. Davis, Reassessment of the lethal London fog of (1952) Novel indicators of acute and chronic consequences of acute exposure to air pollution. Environ Health Perspect 109(2001):389–394
7. Tomislav Dragičevic, Xiaonan Lu, Juan C Vasquez, Josep M Guerrero (2016) DC Microgrids-Part II: a review of power architectures, applications, and standardization issues, IEEE Trans Power Electron
8. Marannano G, Puma GL, Giallanza (2018) Porretto 7. The case of Sicily illustrates a methodology for determining the size of standalone hybrid photovoltaic-wind battery systems. J Clean Prod. 199, 817–830
9. Murphy F, McDonnell K (2017) A case study for the Dublin area: a feasibility assessment of photovoltaic power systems in Ireland. Sustain 9, 302
10. Khatib T, Kazem HA (2013) A novel numerical algorithm that utilizes the loss of load probability index to optimize the scaling of a photovoltaic/wind/diesel generator/battery microgrid. Annu J Photoenergy, 1–8
11. Gebremeskel AG, Mondal KF, Ringler C, Gebrehiwot A (2019) Optimisation and cost-benefit analysis of hybrid power systems for off-grid rural electrification in Ethiopia. Energy 177, 234–246
12. Murugaperumal K, Srinivasn S, Prasad GS (2019) Optimal design of a hybrid renewable energy system using various operational methods and load forecasting for rural electrification. Energy Technol Assess Sustain. 37:100613
13. Jalal D, Mehdi N (2021) Optimization methods of MPPT parameters for PV systems: Review, classification, and comparison. J Mod Power Syst Clean Energy 9:225–236
14. Jazayeri M, Uysal S, Jazayeri K (2014) Evaluation of maximum power point tracking techniques. In PV systems using MATLAB/Simulink, in sixth annual IEEE green technologies conference, corpus christi, TX, pp 54–60
15. Mao M, Cui L, Zhang Q, Guo K, Zhou L, Huang H (2020) Classification and summarization of solar photovoltaic MPPT techniques: A review based on traditional and intelligent control strategies, Energy Rep,—Elsevier
16. da Rocha MV, Sampaio LP (2019) Comparative analysis of ABC, Bat, GWO and PSO algorithms for MPPT in PV systems In: 8th international conference on renewable energy research and application (ICRERA 2019)

17. Haq IU, Khan Q, Ullah S, Khan SA, Akmeliawati R, Khan MA, Iqbal J (2022) Neural network-based adaptive global sliding mode MPPT controller design for stand-alone photovoltaic systems. PLoS ONE 17:e0260480
18. Bellia H, Youcef R, Fatima M (2014) A detailed modeling of photovoltaic module using MATLAB. NRIAG J Astron Geophys 3:53–61
19. Baghzouz Y (2015) Basic photovoltaic theory, in Handbook of Clean Energy Systems. University of Nevada, Las Vegas, USA
20. Stuart Reid, Andries Engelbrecht, Ms. Katherine Malan. Portfolio Optimization Using Particle Swarm Optimization, turning finance
21. Deng Wu, Zhao H, Yang X, Xiong J, Sun M (2017) Li Bo Study on an improved adaptive PSO algorithm for solving multi-objective gate assignment. Appl Soft Comput 59:288–302
22. Marini F (2015) Walczak B Particle swarm optimization (PSO). A Tutorial. Chemom Intell Lab Syst 149:153–165
23. Lekouaghet B, Boukabou A, Lourci N (2018) Bedrine K Control of PV grid-connected systems using MPC technique and different inverter configuration models. Electr Power Syst Res 154:287–298
24. Golamreza Azizyan, Farid Miarnaeimi, Mohsen Rashki, Naser Shabakht (2019) Flying Squirrel Optimizer (FSO): A novel si-based optimization algorithm for engineering problems. Iran J Optim. 11: 2, 177–205 Research Paper
25. Waring GH (1983) Horse behavior, in The Behavioral Traits and Adaptations of Domestic and Wild Horses. Noyes Publications, Mill Road, Including Ponies
26. McDonnell SM (2003) The equid ethogram: a practical field guide to horse behavior, Eclipse Press
27. Levine MA (2005) Domestication and early history of the horse, in The Domestic Horse: The Origins, Dev Manag Its Behav, pp 5–22
28. Krueger K, Heinze J (2008) Horse sense: Social status of horses (Equus Caballus) affects their likelihood of copying other horses' behavior. Anim Cogn 11(3):431–439

Power System State Estimation A Critical Review

Mohd Hussain and Neeraj Gupta

Abstract This research explores the critical domain of power system state estimation(PSSE), aiming to enhance the accuracy and reliability of real-time monitoring. Dynamic load, communication channel errors, and measurement errors affect the generation and power flow in a system, this can cause current and power flow to cross the limits which will eventually damage the system. To estimate the states accurately various mathematical models and algorithms were devised that are explained comprehensively in this paper. Recent developments in power system state estimation are discussed. Additionally, the paper discusses challenges faced by existing methodologies and outlines potential directions for future research in the field. This paper also discusses the algorithms developed for the detection of bad data that occur due to malfunctioning of some meters. Error analysis of estimated states is discussed to check for the authenticity of estimated states …

Keywords State estimation · State estimation algorithm · Bad data · Error analysis

1 Introduction

State estimation (SE) of power systems is a crucial process used in various industries and sectors to ensure the reliable and efficient operation of electrical grids and networks. It is an essential tool in electrical engineering used to determine the most appropriate operating conditions using measurements. Measurements involved are voltage at buses, injected real/reactive power, and real/reactive power flow in the system branches. Some measurements dependent on weather conditions because of the integration of distributed energy resources (DERs) are taken from forecasted values hence called pseudo-measurements [1]. Because of the dynamic nature of the load,

M. Hussain (✉) · N. Gupta
National Institute of Technology, Srinagar, India
e-mail: mh380372@gmail.com

S. Kumar et al. (eds.), *Soft Computing Applications in Modern Power and Energy Systems*, Lecture Notes in Electrical Engineering 1298,
https://doi.org/10.1007/978-981-96-0058-8_7

generation varies. This shift in generation has the potential that current and power flow exceed maximum power and current flow restrictions, which could harm the system. Hence it is necessary to monitor the system continuously. Also, the measuring instruments and communication channels are associated with the errors [2]. Estimation of power system states should be such that errors are mitigated. Measurements Power system state estimation is estimating the value of unknown quantities in the power system to minimize the effect of measurement errors. Various methods are devised to estimate the states some of which are explained in this paper.

Error analysis and bad data detection are the essential steps to ensure the accuracy of estimated states and finding out the malfunctioning meters used to take measurements (Fig. 1).

2 Power System States and Measurements

2.1 The Power System States

They are the parameters used to determine all other parameters of the power system. They are denoted using a vector x known as a state vector. State vector x two parameters node voltage and phase angles, all other measurements like injected real/reactive power, current flow, and real/reactive power flow can be calculated using parameters of the state vector.

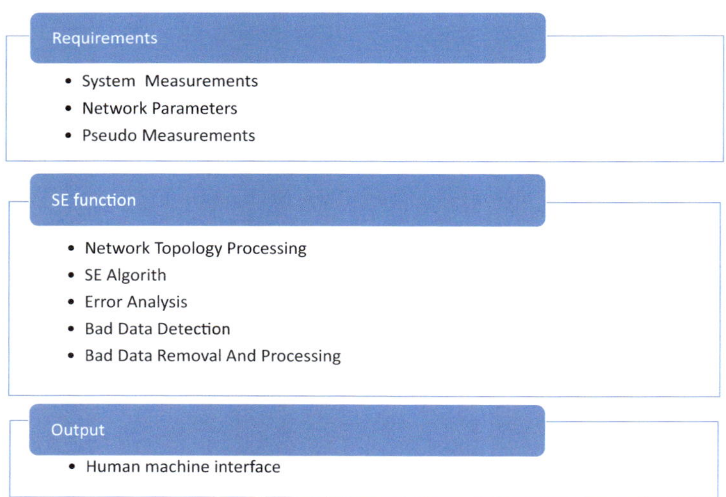

Fig. 1 Overview of steps in PSSE

2.2 *Power System Measurements*

The stability and dependability of the electrical grid depend heavily on the measurements. Power system measurements denoted z comprise measurements of injected power/current at buses, magnitude of voltage at the system buses, measurements of the flow of active and reactive power in branches, pseudo-measurements of the quantities not possible to measure, or any combination of the above. To estimate the present state of the system, these observations are usually combined with a power system model and SE algorithms like the Kalman filter or the weighted least squares approach. Furthermore, contemporary state estimate methods take advantage of the high-resolution and time-synchronized data from PMUs to increase the estimating process's precision and efficiency, particularly with the growing usage of synchrophasor measurements from these devices.

2.3 *Pseudo-Measurements*

Pseudo-measurements are artificially created or developed measurements not derived directly from actual measurements or sensors. To increase the precision and dependability of the predicted state, these pseudo-measurements are incorporated into the state estimation procedure. pseudo-measurements can be generated in various ways environmental constraints, Feature Matching, and Derived Relationships.

Pseudo-measurements can be added to state estimation algorithms to produce more dependable and precise state estimates, particularly when there are noisy, insufficient, or unreliable sensor data available. To prevent adding biases or errors into the estimating process, care should be taken in the design and incorporation of pseudo-measurements. It must be remembered that the use of pseudo-measurements varies based on the features of the system being estimated as well as the situation at hand (Fig. 2).

3 State Estimation Methods

Various SE methods are devised to tackle the SE problem. The choice of method depends on the size of the system, linearity, and existence of measurement errors. Estimation methods are discussed for transmission and distribution systems are explained separately (Fig. 3).

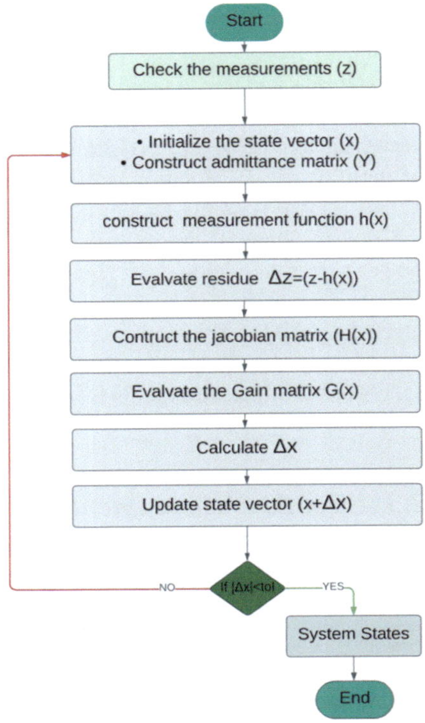

Fig. 2 Flow chart representing various steps in power system state estimation using WLS method

3.1 Transmission System State Estimation (TSSE) Methods

(a) **Weighted Least Square (WLS)** It is the most widely used method of state estimation. In this method weighted square error is made least for estimating the state vector. This method gives more weightage to more accurate meters and less weightage to less accurate meters. Weight is chosen as

$$W_i = \frac{1}{\sigma^2} \tag{1}$$

where σ is the variance of the meter. The estimated state vector is given by the equation

$$\hat{\bar{x}} = \bar{x} + [G^{-1}][H]^T[W][\bar{e}] \tag{2}$$

G is a gain matrix given by the equation

$$G = [H^T][W][H] \tag{3}$$

Fig. 3 Various algorithms for PSSE

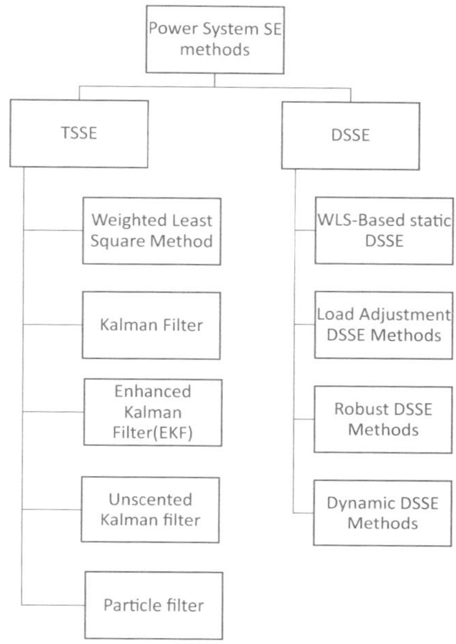

(b) **Kalman Filter (KF)** The Kalman filter is an optimal recursive algorithm that estimates the state of a linear dynamic system. It is named after Rodulf Kalman who published his famous recursive method of estimating dynamic states [3]. It is used in power system state estimation by considering the power system as a dynamic system. It handles both state and parameter estimation and can adapt to changes in the system. It can manage both the presence of measurement noise and the dynamic character of the power system, the KF is especially helpful in power system state estimation. Taking into account the available data and the underlying system dynamics, it offers a statistically optimal estimate of the system state.

(c) **Enhanced Kalman Filter (EKF)** This algorithm is the enhancement of the classical Kalman filter. Kalman filter deals with linear dynamic systems. when dealing with systems that have nonlinear dynamics, the EKF is employed to provide reasonably accurate state estimates. In this method, estimated states are given by the equation

$$\hat{x}_{k|k+1} = f(\hat{x}_{k-1|k-1}, u_{K-1}) + w_{k-1} \tag{4}$$

where f is a state transition function, w_{k-1} is a process error and v_k is the measurement error. This method is essential for state estimation when linear models are insufficient to represent the true dynamics of the system.

(d) **Unscented Kalman filter** An essential nonlinear transformation technique is utilized by the unscented Kalman filter. Since applying a transformation to a

group of individual variables is not difficult, it performs a point-by-point function of a state vector. metamorphosis as opposed to assuming the entire role. This method of estimating the probability density is far simpler.
Although UKF is suitable for highly nonlinear instances since it does not assume that the state or measurement model is nonlinear, it also has the issue of numerical instability.

(e) **Particle filter** It is a method based on probability, a fully nonlinear estimator, but its implementation was impeded by the requirement for significant processing capacity. It does not linearize the extremely non-linear system and does not assume anything about the kind of noise that is present [4].
The challenge in putting this technique into practice is that the quantity of randomly created particles needs to be carefully selected as a
A high particle count will result in a higher computing load, while a low particle count will reduce estimation accuracy.

3.2 Algorithms for Distributed System State Estimation (DSSE)

Due to the incorporation of DERs and demand side management, there is a change in load profile and configuration. The realtime measurements in the case of a distribution system are limited and system observability cannot be achieved hence pseudo measurements are used. Tuning of weights is a difficult task in the case of DSSE.To address these issues methods developed for TSSE are to be adapted to apply them for DSSE. The many paradigms of extant DSSE approaches can be distinguished by the way in which load and bad data are handled, the state variables selected.

(a) **WLS-Based static DSSE** In this state estimation method, we have two choices of state variables node voltage and branch current-based.Weights associated with the actual measurements are proportional to the accuracy of the measurements and non-zero mutual terms in the measurement weight matrix can be included if correlations of measurements are considered.

(b) **Load Adjustment DSSE Methods** This SE methods Generally, modify the load values (power or bus current injection) such that they match the measurement data [5].
A hybrid particle swarm optimization-based method was presented in [6] taking into account the nonlinear properties of distribution system equipment.

(c) **Robust DSSE Methods** Finding and identifying faulty data is essential to getting precise SE results. Generally speaking, a robust estimator lowers the weights given to the suspected poor data points to attenuate the impact of bad data during the solution process. A limited number of redundant measurements without significant variations from the estimated states indicate statistical robustness for the relevant estimator. Machine learning algorithms can generate robust DSSE techniques and load estimates. The influence function idea from the robust estimation theory serves as the foundation for the weight function used by the robust DSSE algorithms presented in [7].

(d) **Dynamic DSSE Methods** The recursive estimate technique known as Dynamic State estimate (DySE), is based on many measurement snapshots taken throughout a time sequence. When dealing with a big system where computing requirements are an issue and measurements come in at varying frequencies, it is possible to forecast state variations between full DSSE executions by processing the recently received measurements in conjunction with the available priority estimate or projected [8].

4 Error Analysis

In the case of state estimation, The measurements are taken at regular intervals and estimate the states of the network at regular intervals. To find out whether the estimated state is accurate error analysis is done. In WLS method of Estimation of errors is done using the estimated value of state vectors as

$$\hat{\bar{e}} = (I - HG^{-1}H^T W)\bar{e} \tag{5}$$

Errors are random and follow Gaussian distribution. Since the Gaussian distribution has zero mean which means the expected value of error should be zero.

Errors should be independent of each other that is

$$E(e_i e_j) = 0 \tag{6}$$

however when we calculate $E(\hat{\bar{e}}\hat{\bar{e}}^T)$ it does not come out to be zero which means that errors are dependent. Therefore any estimated state depends on all measurements.

5 Bad Data Detection

Bad data refers to erroneous or misleading measurements that can adversely affect the precision of the SE results. Some meters can develop a problem such that error due to that gets exceptionally large, hence the estimated results will be very wrong. An expanded version of the WLS method for SE is used as the basis for an approach to detect and identify erroneous data in measurements and network parameters, which is described in [9].

6 Recent Developments in PSSE

As scientists and engineers strive to handle the changing opportunities and problems in the sector, state estimation of power systems is still advancing. The following are a few recent developments and trends in power system state estimation:

(a) **Synchrophasor Technology** Phasor Measurement Units (PMUs) offer high-resolution, real-time data on voltage and current phasors through synchronized phasor measurements. To improve state estimation's precision and speed, these measurements are incorporated into the algorithms.
(b) **PMU-Based Robust State Estimation** Researchers are developing state estimation techniques that can handle a large number of synchrophasor measurements, improving observability, fault detection, and dynamic state estimation capabilities.
(c) **Big Data Analytics**Modern power systems create enormous amounts of data, which are being processed using sophisticated data analytics and machine learning approaches. In order to improve the accuracy of state estimation, this involves anomaly detection, pattern identification, and data-driven modelling.
(d) **Resilience and Cybersecurity** The design of state estimation systems is emphasising cybersecurity and resilience in order to fend against cyberattacks and preserve grid stability. This covers methods for identifying and countering cyberattacks on state estimation systems.
(e) **Distributed State Estimation** To efficiently monitor and operate the grid in light of the growing deployment of distributed energy resources (DERs), researchers are developing state estimation techniques that can handle decentralized and distributed data sources.
(f) **Real-Time Operation and Control** Real-time operation and control systems are incorporating state estimates to facilitate quicker decision-making and grid management in dynamic contexts.
(g) **Improved Data Quality Assurance** To increase the precision of state estimates, methods for guaranteeing the quality of measurement data, such as data cleansing and bad data detection, are constantly being improved.
(h) **Enhanced Topology Processing** To increase state estimation resilience in real-time applications, sophisticated methods for managing network topology changes, such as line switching and breaker operations, are being developed.
(i) **Adaptive and Online State Estimation** To provide more accurate and dynamic state estimation, researchers are developing adaptive state estimation algorithms that can update and adjust the estimation model in real time.
(j) **Integration with Energy Management Systems (EMS)** Power system monitoring, control, and optimization are now possible thanks to the ongoing development of state estimation integration with EMS software (Fig. 4).

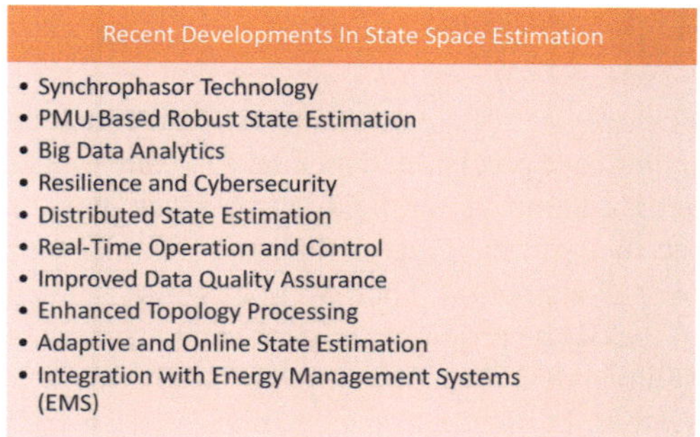

Fig. 4 Recent developments in state space estimation

7 Power System State Estimation Industrial Applications

State estimate of power systems is an essential procedure used in many different businesses and sectors to guarantee the dependable and effective operation of electrical grids and networks. Various sectors in which it finds its use are Electric Utilities, Power Generation,grid operators, renewable energy, Industrial facilities,data centers, electrical rail and transit, research and development, smart grids and energy trading and market operations.

Accurate power system state estimate is critical to preserving grid stability, maximizing power flow, and reacting quickly to disruptions and emergencies in all of these sectors and applications. It lessens energy losses, guards against power interruptions, and guarantees the electrical grid's stability.

8 Software Used by Power Companies for State Estimation of Power System

Energy Management Systems (EMS) and grid operation procedures are carried out by power corporations and utilities using a variety of software tools and systems for state estimate of power systems. The following are a few software programs that are frequently used in the power industry for state estimation General Electric PowerOn Advantage, ETAP (Electrical Transient Analyzer Program), DNV GL Synergi Electric, PowerWorld,Grid Solutions PSLF (Positive Sequence Load Flow),DIgSILENT PowerFactory, ABB Network Manager,Siemens PSS®E.

The power company's particular requirements, the power system's complexity, the available funds, and the features and capabilities that are desired all play a role in the software selection process. Beyond state estimation, these software programs provide several additional features, such as load flow analysis, contingency

They are essential to the effective and dependable operation of power systems because they provide analysis, dynamic simulation, and other features.

9 Conclusion

Power system state estimation should conclude with a focus on the continued efforts to increase precision, dependability, and efficiency as well as the dynamic and cooperative nature of the field to guarantee power systems resilience and adaptability in the face of changing challenges. Since power system state estimation had evolved over a long period, There were lots of advancements in the development of algorithms to tackle the problems of state estimation like estimation techniques, error analysis, and bad data detection however all the work is mostly confined to the Transmission system, In distribution system very less work is done hence it is in evolution stage. The importance of state estimation in transmission and distribution systems should be emphasized, as it plays a crucial role in preserving grid resilience, adapting to the changing energy landscape, and facilitating the move towards more intelligent and sustainable electrical networks. To meet the problems of contemporary power systems, state estimation approaches must be seamlessly integrated into grid operations and improved upon continuously.

References

1. Blood EA, Krogh B, Ilic M (2008) Electric power system static state estimation through Kalman filtering and load forecasting. In: Power and energy society general meeting. IEEE, pp 1–6
2. Van Cutsem, Th, Ribbens-Pavella M (1985) Bad data identification methods in power system state estimation a comparative Studybad data identification Methods in power system state estimation a comparative study. IEEE Trans Power Apparatus Syst 104(11)
3. Kalman RE (1960) A new approach of linear filtering and prediction problems. J Basic Eng 82(1):35-45
4. Meng D, Zhou N, Lu S, Lin G (2012) Estimate the electromechanical states using particle filtering and smoothing. In: IEEE power and energy society general meeting, pp 1–7
5. Roytelman I, Shahidehpour SM (1993) State estimation for electric power distribution systems in quasy real time conditions. IEEE Trans Power Del 8(4):2009–2015
6. Naka S, Genji T, Yura T, Fukuyama Y (2003) A hybrid particle swarm optimization for distribution state estimation. IEEE Trans Power Syst 18(1):60–68
7. Li H (2011) An improved state estimator in distribution systems. In: Proceedings Shanghai, China, IEEE international conference computer science automation engineering, pp 30–34
8. Primadianto A, Lu C-N (2017) A review on distribution system state estimation. IEEE Trans Power Syst 32(5)
9. D' Antona G, Perfatto L, Bad data detection and identification in power system. State estimation with network parameters uncertainty. IEEE

Artificial Intelligence and Solar Energy: Investigating Solar Power Forecasting for Sustainable Future

Neha Srivastava and Ashwani Kumar Yadav

Abstract Electricity is essential for the well-being of the human population. However, it is mostly generated via polluting, carbon-emitting traditional sources. To address these concerns, the world's attention is shifting towards renewable energy sources that provide an infinite supply, are environmental friendly, and convert directly to electrical energy. Out of these renewable energy sources, solar energy is widely available, essentially unlimited, cost-effective, and efficient. The study presented a detailed and systematic review of PV output power forecast models. For creating an effective PV power prediction model, many steps are explored in the forecasting process, such as forecasting horizon, input data selection, pre-processing and post-processing of data, and different forecasting models. A critical analysis of various forecasting models, spanning from machine learning to physical models, is also offered. Furthermore, the probable advantages of hybrid approaches for PV power prediction models are comprehensively examined.

Keywords Solar power forecasting · ANN · Hybrid forecast models

1 Introduction

The rapid increase in the world's population has increased demand for traditional energy sources. However, these conventional energy sources have several limitations, including the fact that they take years to create, are only available in certain areas, are unsustainable, and have a negative influence on the surroundings and human health. The enormous industrialisation that followed the turn of the century resulted in an overuse of these resources. Several scientists have predicted that if traditional energy sources are consumed at their current rates, they would become extinct by

N. Srivastava (✉)
Malaviya National Institute of Technology Jaipur, Jaipur, India
e-mail: 2024ree9073@mnit.ac.in

A. K. Yadav
Amity University Rajasthan, Jaipur, India

S. Kumar et al. (eds.), *Soft Computing Applications in Modern Power and Energy Systems*, Lecture Notes in Electrical Engineering 1298,
https://doi.org/10.1007/978-981-96-0058-8_8

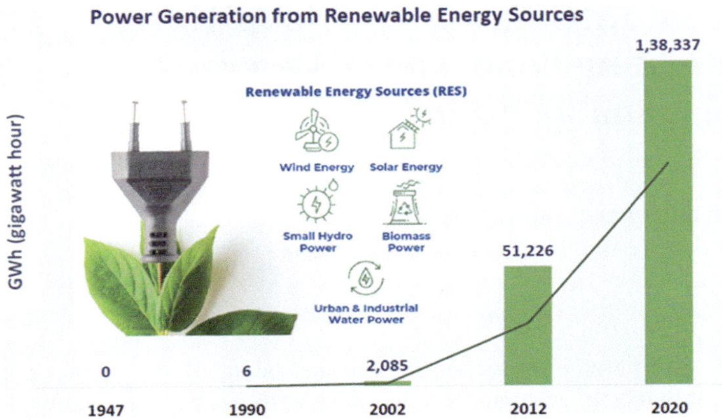

Fig. 1 Development of power generation from renewable energy sources in India [3]

2060 [1]. Furthermore, overuse of conventional energy sources has hastened other environmental challenges such as global warming, air pollution, and greenhouse gas emissions. This has already resulted in a 1 °C increase in global temperature over the pre-industrial era. A further 0.5 °C rise in temperature will result in more sea level rise, species extinction, biodiversity loss, and harsh weather events [2]. All of these difficulties can be addressed by using non-traditional energy sources instead of traditional energy sources. These non-traditional or renewable energy sources are derived from natural processes and cannot be depleted, implying that the rate of renewal exceeds the rate of use.

With non-traditional sources comprising 40% of installed capacity, India has become the world's third largest renewable energy generator (2022). Figure 1 depicts the development of electricity generation using renewable energy sources. To capitalize on the green energy revolution and expand its uses, India has established a number of renewable energy targets [4], some of which are given below:

a. By 2070, become a net zero carbon emitter.
b. By 2030, produce 500 GW from renewable energy (which includes 280 GW of solar power).
c. To meet 50% of energy demand through renewable energy.
d. By 2030, reduce the volume of emissions per unit of India's gross domestic product (GDP) by 45%.

Among all the renewable energy sources, solar energy has an advantage since it is abundant, practically unlimited, cost effective, efficient, and available in almost every corner of the world. Solar energy is defined as the sun's radiant light and heat, whereas solar power is transformation of solar energy into electricity. The total global absorption of solar energy is 1.8×10^{11} Mega Watt, of which only 0.01% is adequate to meet the whole global demand if harnessed efficiently [5]. In other words, less than one hour of incident solar energy can meet the world's annual energy requirement.

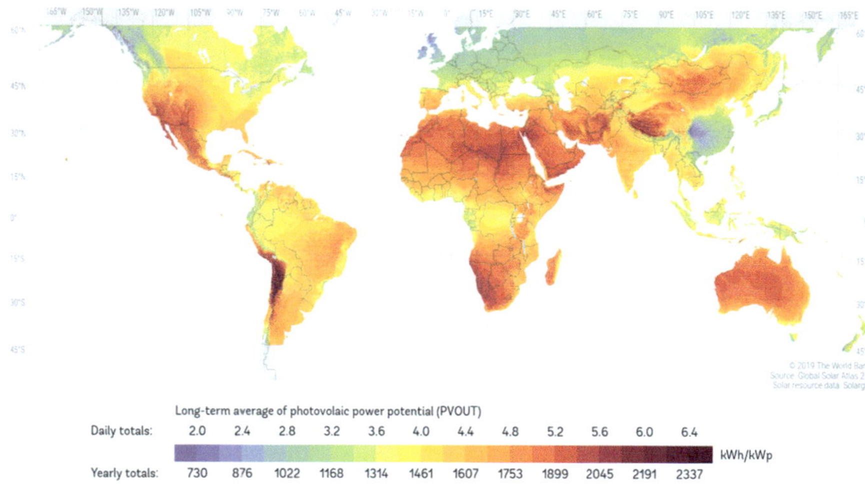

Fig. 2 The world solar resources map: Intensity of solar energy across the globe [6]

Because of these factors, solar energy has the capability to meet the world's current and future energy needs.

The countries located at latitudes above 45° N and below 45° S have numerous opportunities for solar power production. Figure 2 displays the intensity of solar energy around the world, highlighting the prospects for global solar PV capacity whereas in Fig. 3 potential of global horizontal irradiance specific to Indian origin is displayed.

The incorporation of varying renewable energy sources, such as solar photovoltaic, into the electricity grid presents grid workers with issues in ensuring grid stability. Furthermore, the output power of photovoltaic systems can vary by nearly 70% within one minute and by more than 75% between one-hour assessments. The reason for this is that big variations in atmospheric conditions result in a surge in the uncertainty of PV (photovoltaic) output power. Such changes in power production generate grid imbalance, affecting the operating frequency and local voltage [7]. As a result, an accurate PV output projection throughout the forecast horizon is necessary for independent power producers (IPPs) to maintain grid stability, limit fluctuations in power output, and minimize the influence of grid imbalances.

2 Major Factors Affecting PV Output Power Forecast

Several aspects influence PV panel power forecasting, including forecasting horizon, forecasting model input parameter selection, geographical location, forecasting methodology, and performance evaluation. Relevant data and an appropriate model should be chosen to refine the accuracy of the PV output power predicting model.

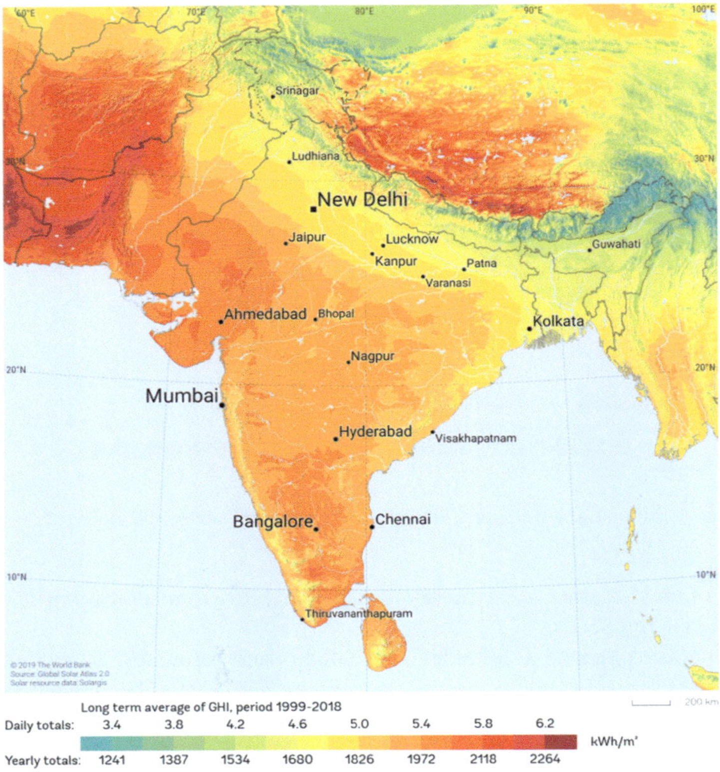

Fig. 3 GHI (Global Horizontal Irradiance) potential of India [8]

2.1 Forecasting Horizons

Forecast horizons are the expected future times for which we need to predict PV power, or, in other words, the time interval between the actual and operational time of forecast. Because there are no set prediction horizon categories, some researchers prefer very short term (1 min-several min), short-term (1h–6h), medium term (6h–24h) and long-term (greater than 24h) classification, but many researchers have developed another method of categorization: intra-hour, intra-day, and day-ahead [9]. This category frequently overlaps with forecast horizons of extremely short, short, medium, and long term.

Intra-hour. It ranges from a few seconds to an hour, and is consequently also referred to as nowcasting. Such forecasting is used by island grids and low-quality power equipment in areas with significant sun penetration [10]. It also contributes to grid quality and stability. Due to their scattered and aggregate nature, most distributed solar plants are unaffected by intra-hour forecasts.

Intra-day. It varies from 1 to 6 h and is used to control electrical load in a region as well as in electricity trading outside of a normal grid when the anticipated amount of electricity exceeds possible demand. While for most of the researches NWP is chosen for intra-day prediction since it is effective for forecast horizons exceeding 4 h, but in some cases NWP was also used for intra-hour prediction.

Almonacid [11] has utilized ANN model to accurately predict PVPF (Photovoltaic power forecasting) 1h ahead by taking global horizontal irradiance and cell temperature as input. In another paper, [12] has accurately forecasted 1h, 3h and 6h ahead using particle-swarm optimization.

Day-ahead. The duration of day-ahead predictions typically ranges from 6 to 48 h. Many of these forecasts utilize Numerical Weather Prediction (NWP) data as input. As the forecasting time frame increases, the inclusion of NWP data enhances the precision of predicting photovoltaic (PV) power output [13]. However, longer prediction windows are particularly useful for managing economic electricity transmission and scheduling plant maintenance, especially when electricity generation falls below expectations.

2.2 *Forecast Model Inputs Selection*

In the forecast model, the independent/target variable is PV output power whereas the dependent variables are mostly climatological factors such as solar irradiance, atmospheric-temperature, cloud cover, dew point temperature etc. depending upon researcher's input selection. The extraneous or confounding variable will be remaining meteorological parameters which are not selected as input parameters and affect photovoltaic power output. These parameters can be dew point, relative humidity and wind direction etc. depending upon researcher's non- input selection. Most of the photovoltaic output power forecast models using machine learning algorithms used the historical PV output data as one of the forecast model inputs [14]. Solar irradiance has the maximum correlation coefficient with photovoltaic output power so majority of the authors prefer it as one of the input parameter [15].

The correlation coefficient between climatological parameters and PV power output as presented in [9] are listed in Table 1.

Using multiple no. of inputs, the forecast accuracy of the PV output power forecast model can be improved. It will, however, increase computing complexity and expense. As a result, correct model inputs must be chosen so that optimal forecasting accuracy and model performance can be achieved while decreasing computing complexity and expense.

Table 1 Correlation between climatological parameters and PV output power [9]

S. No	Climatological parameter	Unit	Correlation co-efficient
1	Solar Irradiance	w/m^2	0.9840
2	Air temperature	°C	0.7615
3	Wind speed	m/s	0.1970
4	Relative Humidity	–	−0.4918
5	Dew point temperature	°C	0.6386
6	Precipitable water	M	0.3409
7	Wind direction	–	0.1263
8	Cloud type	–	−0.4847
9	Air pressure	Pa	0.0815

2.3 Input Data Optimization and Pre-Processing Methods

The input statistics such as PV power production and solar irradiance sometimes contain occasional static or spike components (outliers) due to weather or seasonal variations, oscillations in electricity demand, and power system failures. Several approaches are utilized to reconstruct input datasets, including wavelet transform [16], empirical mode decomposition, self-organized map (SOM), normalization, and singular spectrum analysis [17–19]. In [20], a fuzzy logic based input data pre-processing toolkit is used in conjunction with an Artificial neural network model to discover data correlations between air temperature, wind speed and direction along with solar irradiance value. This will incorporate a fuzzy logic pre-processing toolkit into an ANN model to perform real-time weather analysis and feed it into predicting output.

Following the execution of the prediction models, the anticipated values are post-processed. All anticipated values are examined and constrained to the theoretical maximum production during the post-processing step. Because potential cloud enhancement effects are imperceptible at an hourly time resolution [21], the actual turnout should not surpass the maximum production of the photovoltaic system under clear sky situations.

2.4 Photovoltaic Output Power Forecasting Techniques

There are numerous forecasting modelling methods such as physical model, benchmark model, ensemble model, statistical technique, and hybrid based prediction models as represented in Fig. 4.

Physical Model. This model is also known as the indirect forecast method since it predicts using meteorological data and atmospheric model equations rather than

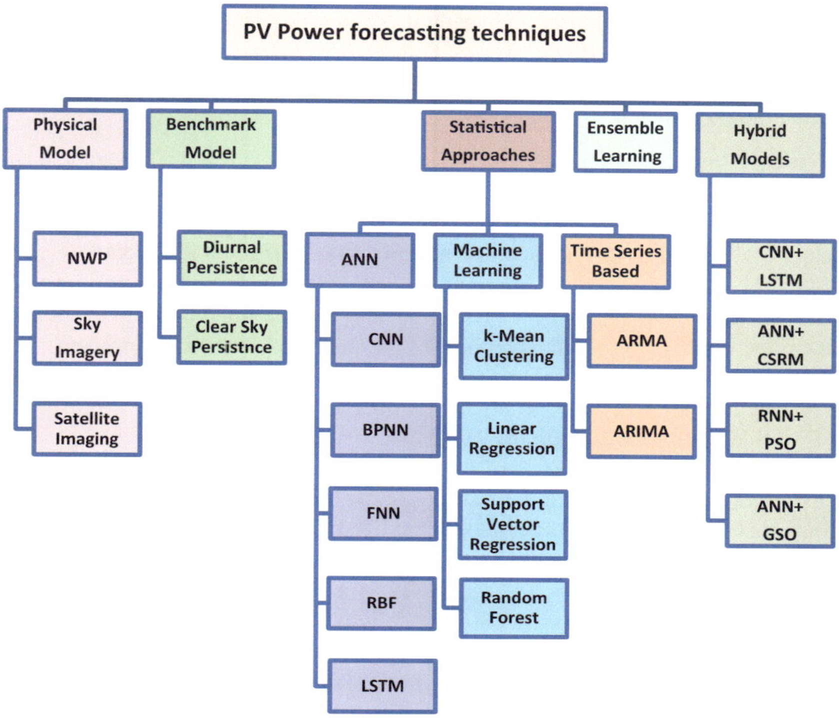

Fig. 4 Different techniques to forecast PV Power

previous training data. Air pressure, surface roughness, air and dew point temperature, GHI, orography, and other lower atmospheric meteorological data are used to forecast [22]. Long-term forecasting is often more trustworthy with this technique [23].

Benchmark Model. This model is valuable for analyzing the performance of other prediction models when used as reference, particularly for extremely short-term and short-term forecasting. It has a lower computational cost, a shorter time lag, but a bigger forecast inaccuracy than other models [9]. This model uses the notion of today equals tomorrow, which means that the climate conditions a day ahead are expected to be comparable to the day before [9].

Ensemble Learning. These techniques were created to address the shortcomings of individual approaches and improve their strength and accuracy. This can be accomplished in two ways. The forecasting tasks are separated into a few sub-tasks in the first technique, known as co-operative ensemble forecasting, and each sub-task is solved independently. The second method, known as competitive ensemble forecasting, employs numerous predictors with varying parameters to construct individual forecast models, which are then combined to generate an overall ensemble forecast model [24].

Statistical Model. This forecast model learns using time series and historical data. It tries to reduce errors by estimating the difference between the forecasted photovoltaic output power and the actual calculated data. Therefore, forecasting accuracy of this model highly depends on the quantity and quality of historical input data.

Machine Learning Models. These models are based on a computer program that employs intelligence to learn from previous historical data and train itself to anticipate future data. Figure 5 depicts an overview of the statistical forecasting process, where raw data from weather prediction models, PV panel system and production details, and solar information are used as input variables [27]. They are pre-processed before being divided into test and training data.

To develop a model, 70–80% of the data should be trained. The remaining 20–30% of data is subjected to model testing [14]. If the model's prediction accuracy is within an acceptable range, the model is finalized; otherwise, the model parameters are adjusted based on feedback (Fig. 6).

Artificial Neural Network (ANN). ANN is a prominent and commonly used AI technique for various prediction applications with a high degree of accuracy (ML method for forecasting). They are robust and adaptive in nature because they can respond to quick changes dynamically using interconnected neurons and learning processes. This attribute distinguishes ANN as an adaptable forecasting approach that may be replicated for changing initial conditions at various geographical-climatological settings [9].

The fundamental ANN structure is distributed into three layers: the input layer, the hidden layer, and the output layer, all of which are made up of artificial neurons and connections [26]. Each artificial neuron in an artificial neural network, like biological neurons, serves as an activation node for all data analysis and decision-making. A specific activation node accepts input from a preceding node and generates the weighted sum using machine learning parameters. This processed data is then sent into an activation function, which computes the combined forecast. This generated forecast is analysed incrementally till it achieves the preferred outcome [20].

Hybrid Models. Hybrid models are created by combining two or more strategies that have better qualities. It has been found that combining two or more procedures yields superior outcomes than the individual forecasting technique. One of the primary objectives of the hybrid models is to investigate the prospects using several algorithm groupings in order to improve forecasting accuracy. They can give more accurate forecasts by using each approach to its full potential [26]. There were several hybrid models utilized in a number of forecasting applications to attain the greater forecast accuracy. Another advantage of hybrid model is reduction of computational complexity and time. As a result, forecast output power obtained through hybrid models can be used in maintaining real-time grid stability and minimize the influence of grid imbalances [9].

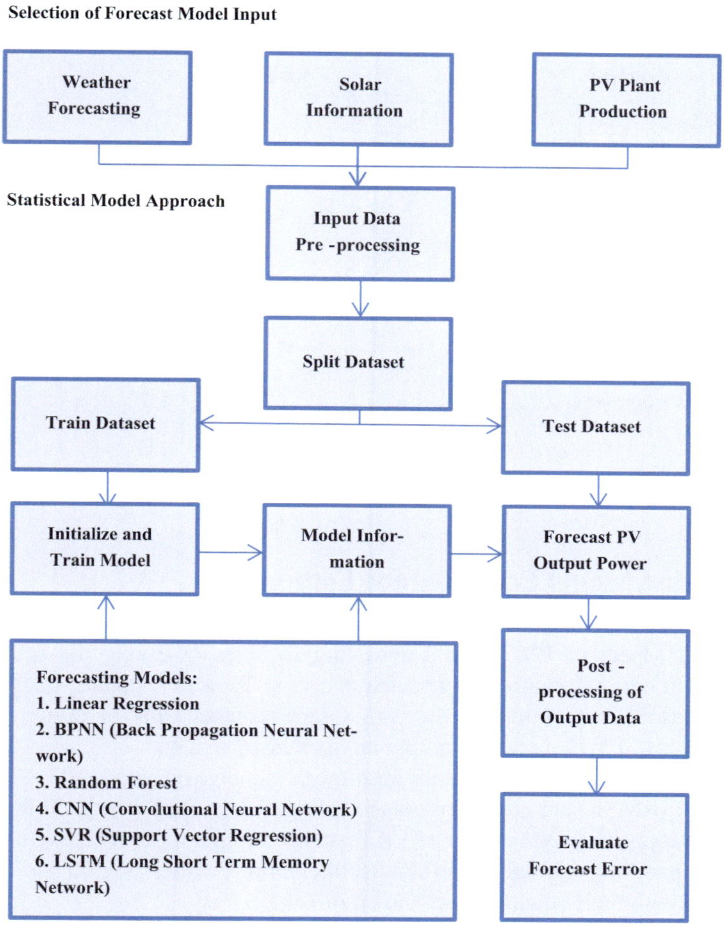

Fig. 5 Overview of process involved in statistical forecasting model [25]

Tables 3 and 4 give a systematic and complete evaluation of the literature on solar PV output power using the ANN and hybrid methods. The ANN and hybrid based literature assessment of PV power forecasting is separated by forecasting horizon. Table 3 shows the strategies for forecasting twenty-four hours or one day ahead and Tables 4 describes forecasting approaches for 1 min to less than one day in advance. Both tables provide author information, year of publication, prediction horizon, forecast error, forecast method, and analysis. The presented literature review was created after researching several databases such as Elsevier, Springer, Taylor & Francis, IEEE and others.

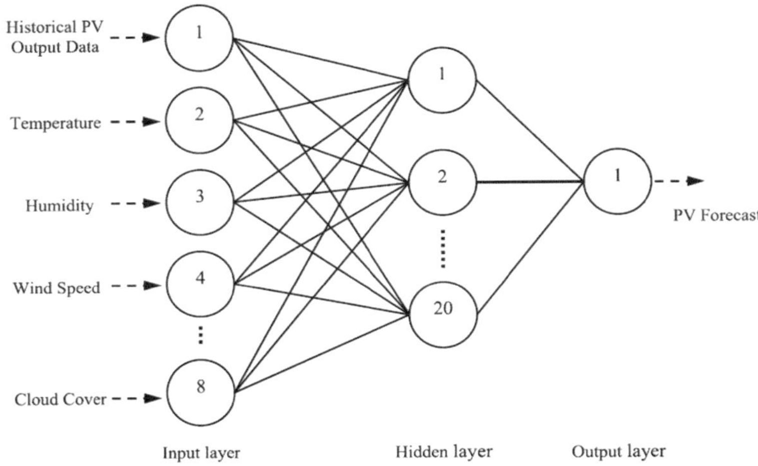

Fig. 6 Multilayer perceptron model of ANN [26]

3 Forecast Model Performance Estimation

An integral aspect of PV power forecasting involves assessing the accuracy of forecasting models through performance analysis. Typically, the accuracy error of solar predictions is quantified in terms of solar irradiance (unit in W/m^2), whereas discrepancies in PV plant power output are expressed in kW.

Mean absolute error (MAE), Mean absolute percentage error (MAPE), Root mean square error (RMSE) are common tools to calculate model accuracy. Of the various assessment tools, Root Mean Square Error stands out as a more robust and effective method for handling anomalies and sudden fluctuations within the data [9]. The other tools along with their equations are shown in Table 2.

4 Conclusion and Future Work

In order to satisfy current global energy demand, numerous forecasting methodologies and their forecast accuracy have been examined in this article. Because of the huge potential of solar energy in various parts of the world, there has been a greater emphasis on PV power during the previous decade. Large variations in weather conditions, on the other hand, can cause a variety of problems for grid power systems, either directly or indirectly. As a result, an accurate PV output estimate is required to ensure grid stability, restrict power output fluctuations, and reduce the impact of grid imbalances.

Several efforts have been made in recent years to correctly forecast output power using various methodologies. A few strategies were investigated, such as physical

Table 2 Performance evaluation indices

S. No	Performance metric	Formula	Details
1	Pearson's correlation co-efficient	$\rho(\mathrm{P}_{\mathrm{meas}},\ \mathrm{P}_{\mathrm{pred}}) = \frac{\mathrm{cov}(\mathrm{P}_{\mathrm{meas}},\ \mathrm{P}_{\mathrm{pred}})}{\sigma \mathrm{P}_{\mathrm{meas}}}.\sigma \mathrm{P}_{\mathrm{pred}}$	It measures the correlation between two parameters. Here, $\mathrm{P}_{\mathrm{meas}}$ and $\mathrm{P}_{\mathrm{pred}}$ are measured and forecasted PV output power respectively
2	Mean Absolute error	$MAE = \frac{1}{N}\sum_{i=1}^{N}\lvert \mathrm{P}_{\mathrm{pred}} - \mathrm{P}_{\mathrm{meas}}\rvert$	It denotes average difference between the measured data and predicted data. This metric is more suitable to measure consistent forecast error
3	Mean absolute percentage error	$MAPE = \frac{100}{N}\sum_{i=1}^{N}\left\lvert \frac{\mathrm{P}_{\mathrm{pred}} - \mathrm{P}_{\mathrm{meas}}}{P}\right\rvert$	It is the mean of all absolute percentage errors of forecasts. Here, P is the capacity of analysed PV plant
4	Root mean square error	$RMSE = \sqrt{\frac{1}{N}\sum_{i=1}^{N}(\mathrm{P}_{\mathrm{pred}} - \mathrm{P}_{\mathrm{meas}})^2}$	It measures overall error during entire forecasting period

Table 3 A brief review of one-day ahead forecasting technique for solar PV power using ANN and hybrid models

Author	Year	Forecast horizon	Forecast model	Forecast error	Comments
Ding et al. [28]	2011	24 h ahead	ANN	MAPE 10.06%	Backpropagation learning algorithm based ANN is used
Zhang et al. [29]	2013	N/A	RNN + PSO + EA	RMSE 0.028	It is based on a recurrent neural network (RNN) that employs a hybrid learning approach which incorporates particle swarm optimization and an evolutionary algorithm
Quan et al. [30]	2013	24 h ahead	ANN + GSO + EBP	NMAE 14%	The forecast model developed is a hybrid of ANN, genetic swarm optimisation, and evidence-based particle. When compared to traditional EBP alone, the hybrid model's convergence capability is faster
Sansa et al. [31]	2014	24 h ahead	ANN	Error 0.027 pu	Nonlinear Auto Regressive model is employed in conjunction with artificial neural networks. However, the forecast model's performance differs over forecast horizons
Dolara et al. [32]	2014	24 h ahead	ANN + CSRM	Error reduced by 50% compared to traditional method on sunny days	It combines ANN with clear sky solar radiation model which is known as PHANN (Physical hybridized ANN). The proposed model was able to show better accuracy than ANN alone on most of the days
Leva et al. [33]	2015	24 h ahead	ANN	NAME 11%	An ANN-based model is utilised to assess performance under various climatic situations, including bright days, partly cloudy days, and cloudy days
Sivaneasan et al. [20]	2017	Monthly	ANN + Fuzzy	MAPE 29.6%	The suggested model prevails over ANN only, coupled ANN, and the ANN-Fuzzy model with an error correction algorithm

(continued)

Table 3 (continued)

Author	Year	Forecast horizon	Forecast model	Forecast error	Comments
Lee et al. [34]	2018	24 h ahead	DCNN + LSTM	RMSE 9.87%	It is developed using a deep convolution neural network (DCNN) and a long short term memory (LSTM). The Twin DNN model was used to generate day-ahead solar power projections using time series data from PV inverters and national weather center reports. Two DCNNs with different filter widths were used to recover short-time local patterns from the data, while an LSTM was used to retrieve long-time features. This method had outperformed other traditional models such as linear regression, RFR, SVR
Raza et al.35]	2018	24 h ahead	NNE	MAPE 9.75%	This hybrid model is comprised of Feedforward neural network trained via particle swam optimization. It demonstrated higher model performance when compared with five conventional forecasting techniques
Liu and Chang et al. [36]	2018	24 h ahead	RBFNN	MAPE 3.71%, RMSE 4.65%	A hybrid radial basis function neural network was developed. The final model outperformed ARIMA (autoregressive integrated moving average) and BPNN (back propagation neural net-work) models
Lim et al. [37]	2022	N/A	LSTM + CNN	MAE 4.58 (sunny days), MAE 7.06 (cloudy days)	The suggested model offers exact power generation predictions based on instantaneous changes in power generation patterns

(continued)

Table 3 (continued)

Author	Year	Forecast horizon	Forecast model	Forecast error	Comments
Shalini et al. [38]	2023	N/A	ANN + SVPWM	THD 2.2%	This research utilizes ANN controller with the SVPWM (Space vector Pulse width Modulation). The result shows that this model performed better compared to traditional methods (PI controller, fuzzy controller, and ANN controller) and introduced less harmonic current (Total harmonic distortion = 2.2%) into the grid
Diaa Salman et al. [39]	2024	24 h ahead	CNN + LSTM + Transformer model	MAE of 0.551%	All the possible combinations of convolutional neural network (CNN), long short-term memory (LSTM), and transformer (TF) models are experimented. These hybrid models were also compared with the single CNN, LSTM and TF models with respect to different kinds of optimizers

Table 4 A brief review of one minute to less than 24 h forecasting technique for solar PV power using ANN and hybrid models

Author	Year	Forecast Horizon	Forecast model	Forecast error	Comments
Sharma et al. [40]	2016	hourly	WNN	nRMSE ranges from 9.42% to 15.41%	The proposed model comprises a feed-forward ANN and Wavelet Transform along with Morlet and Mexican hat activation functions. Its forecasts were found to be superior to ETS, ARIMA, persistence, and ANN for forecast horizon between 15 min to 24 h
Gutierrez-Corea et al. [41]	2016	Between 1 to 6h	ANN	nRMSE below 20%	To forecast worldwide sun irradiance, the model used multiple architectures and parameters of ANN based on MLPNN with BP algorithm
Wang et al. [42]	2017	-15 min ahead -30 min ahead -1 h ahead -2 h ahead	WT + DCNN + QR	RMSE averagely improved by 52.25%, 48.79%, and 3.78%	This hybrid model is formed by incorporating deep convolution neural network with wavelet transform and spine quantile regression. To improve prediction capability over BPNN, SVM, and SVM integrated WT approaches, a hybrid strategy combining WT, DCNN, and QR was applied

(continued)

Table 4 (continued)

Author	Year	Forecast Horizon	Forecast model	Forccast error	Comments
Jamali et al. [43]	2019	N/A	ANN + PSO-GA	RMSE 24%	The proposed study utilized an ANN forecast model along with Genetic algorithm (GA) and Particle swarm optimization (PSO) to anticipate performance of a SSHS (Solar space heating system). Benchmarking was done against High Exploration Particle Swarm Optimization (HEPSO) and Team Game Algorithm (TGA) stipulating that their hybrid model using PSO-GA-ANN was more accurate and dependable
Pitchai et al. [44]	2021	-hourly -daily -weekly	ANN	RMSE 0.25 (ELMAN), RMSE 0.30 (FFN), RMSE 0.426 (GRN)	The hourly, daily, and weekly PV output power forecasts were found to have a stronger correlation coefficient R
Amreen et al. [45]	2023	15 min	CNN + LSTM	Improved efficiency by 0.45%, 21.31% compared to LSTM and CNN alone	To anticipate solar energy, this study presents a hybrid model based on deep learning (DL) and convolution neural networks (CNN). This enables the LSTM to learn features from the input data that CNN has learned

method, ensemble models, statistical methods, artificial intelligence techniques and hybrid models. The artificial intelligence methods can be used to predict PV power output in a dynamic environment if appropriate historical data patterns are available to train the network. Among them, CNN or its hybrid variants, commonly used in ensemble arrangements, holds the most potential for forecast accuracy. ANN-based models provide superior non-linear approximation conduct and the capacity to control uncertainty by utilising superior training data and algorithms. Combining two or more models, on the other hand, produces better outcomes compared to individual forecasting strategies because each method is used to its maximum potential. As a result, ANN combined with other forecasting techniques produces better results than standalone models.

To increase the prediction accuracy of the photovoltaic output power forecast model, appropriate input parameters should be selected and sudden variations and instabilities in climatological factors and historical photovoltaic output power should be eliminated.

There is currently a need to create PV forecasting tools appropriate for rooftop systems. In order to improve accuracy, different days such as clear days, cloudy days, and rainy days should be separated for testing the performance of PV power systems. It is also important to learn more about ensemble forecast models with different combinations of predictors in order to attain improved prediction accuracy.

References

1. Zazoum B (2022) Solar photovoltaic power prediction using different machine learning methods. Energy Rep 8:19–25
2. Kannan N, Vakeesan D (2016) Solar energy for future world:-A review. Renew Sustain Energy Rev 62:1092–1105
3. PIB, Renewable Energy in India, https://pib.gov.in/FeaturesDeatils.aspx?NoteId=151141&ModuleId=2, last accessed 2023/08/23
4. Moallemi, Enayat A, Lu Aye, John M. Webb, Fjalar J de Haan, Biju A George (2017) India's on-grid solar power development: Historical transitions, present status and future driving forces. Renew Sustain Energy Rev 69, 239–247
5. Mitra, Indradip, Detlev Heinemann, Aravindakshan Ramanan, Mandeep Kaur, Sunil Kumar Sharma, Sujit Kumar Tripathy, Arindam Roy (2022) Short-term PV power forecasting in India: recent developments and policy analysis. Int J Energy Environ Eng, 13(2), 515–540
6. World Bank group, https://solargis.com/maps-and-gis-data/download/world, last accessed 2023/08/23.
7. Ghannam, Rami, Paulo Valente Klaine, Muhammad Imran (2019) Artificial intelligence for photovoltaic systems. Sol Photovolt Power Plants: Adv Control Optim Tech, 121–142
8. World Bank group, https://globalsolaratlas.info/download/india, last accessed 2023/08/23
9. Ahmed R, Sreeram V, Mishra Y, Arif MD (2020) A review and evaluation of the state-of-the-art in PV solar power forecasting: Techniques and optimization. Renew Sustain Energy Rev 124(February):109792
10. Zhang J, Verschae R, Nobuhara S, Lalonde JF (2018) Deep photovoltaic nowcasting. Sol Energy 176(June):267–276
11. Almonacid F, Pérez-Higueras PJ, Fernández EF, Hontoria L (2014) A methodology based on dynamic artificial neural network for short-term forecasting of the power output of a PV generator. Energy Convers Manag 85:389–398

12. Al Hakeem, Donna, Paras Mandal, Ashraf Ul Haque, Atsushi Yona, Tomonobu Senjyu, Tzu-Liang Tseng (2015) A new strategy to quantify uncertainties of wavelet-GRNN-PSO based solar PV power forecasts using bootstrap confidence intervals. In: IEEE Power & Energy Soc Gen Meet 2015, pp 1–5. IEEE
13. Snegirev DA, Eroshenko SA, Khalyasmaa AI, Dubailova VV, Stepanova AI (2019) Day-ahead solar power plant forecasting accuracy improvement on the hourly basis, In: Proc. IEEE Conf Russ Young Res Electr Electron Eng ElConRus 2019, pp 1088–1091
14. Voyant C, Notton G, Kalogirou S, Nivet M-L, Paoli C, Motte F, Fouilloy A (2017) Machine learning methods for solar radiation forecasting: A review. Renew Energy 105:569–582
15. Alcañiz A, Grzebyk D, Ziar H, Isabella O (2023) Trends and gaps in photovoltaic power forecasting with machine learning. Energy Rep 9:447–471
16. Cao, JCSH Cao, Cao SH (2006) Study of forecasting solar irradiance using neural networks with preprocessing sample data by wavelet analysis. Energy 31(15), 3435–3445
17. Kumar, Dhilip R, Prakash K, Abirama Sundari P, Sathya S (2023) A hybrid machine learning model for solar power forecasting. In: E3S web of conferences, 387, pp 04003. EDP Sciences
18. El-aal, Shereen A, Maha abdulkarim alqabli, Amany A Naim (2023) Forecasting solar photovoltaic energy production using linear regression-based techniques. J Theor Appl Inf Technol 101(9)
19. Anuradha K, Deekshitha Erlapally, Karuna G, Srilakshmi V, Adilakshmi K (2021) Analysis of solar power generation forecasting using machine learning techniques. In: E3S web of conferences, 309, pp 01163. EDP Sciences
20. Sivaneasan B, Yu CY, Goh KP (2017) Solar forecasting using ANN with fuzzy logic pre-processing. Energy procedia 143:727–732
21. Impacts on global horizontal irradiance (2017) Gueymard, Christian A. Cloud and albedo enhancement impacts on solar irradiance using high-frequency measurements from thermopile and photodiode radiometers. Part 1. Sol Energy 153:755–765
22. Zhao J, Guo Z-H, Su, Zhong-Yue, Zhao Z-Y, Xiao X, Liu F (2016) An improved multi-step forecasting model based on WRF ensembles and creative fuzzy systems for wind speed. Appl Energy 162:808–826
23. Yang Z, Wang J (2018) A hybrid forecasting approach applied in wind speed forecasting based on a data processing strategy and an optimized artificial intelligence algorithm. Energy 160:87–100
24. Sobri, Sobrina, Sam Koohi-Kamali, Nasrudin Abd Rahim (2018) Solar photovoltaic generation forecasting methods: A review. Energy Convers Manag 156, 459–497
25. Visser L, AlSkaif T, van Sark W (2022) Operational day-ahead solar power forecasting for aggregated PV systems with a varying spatial distribution. Renew Energy 183:267–282
26. Raza, Muhammad Qamar, Mithulananthan Nadarajah, Chandima Ekanayake (2016) On recent advances in PV output power forecast. Solar Energy 136, 125–144
27. Scott C, Ahsan M, Albarbar A (2023) Machine learning for forecasting a photovoltaic (PV) generation system. Energy 278:127807
28. Ding M, Wang L, Bi R (2011) An ANN-based approach for forecasting the power output of photovoltaic system. Procedia Environ Sci 11:1308–1315
29. Zhang, Nian, Pradeep K Behera, Charles Williams (2013) Solar radiation prediction based on particle swarm optimization and evolutionary algorithm using recurrent neural networks. In: IEEE International Systems Conference (SysCon) 2013, pp 280–286. IEEE
30. Quan, Duong Minh, Emanuele Ogliari, Francesco Grimaccia, Sonia Leva, Marco Mussetta (2013) Hybrid model for hourly forecast of photovoltaic and wind power. In: IEEE international conference on fuzzy systems (FUZZ-IEEE) 2013, pp 1–6. IEEE
31. Sansa, Ines, Sihem Missaoui, Zina Boussada, Najiba Mrabet Bellaaj, Emad M Ahmed, Mouhamed Orabi (2014) PV power forecasting using different artificial neural networks strategies. In: First international conference on green energy ICGE 2014, pp 54–59. IEEE
32. Dolara A, Grimaccia F, Leva S, Mussetta M, Ogliari E (2015) A physical hybrid artificial neural network for short term forecasting of PV plant power output. Energies 8(2):1138–1153

33. Leva S, Dolara A, Grimaccia F, Mussetta M, Ogliari E (2017) Analysis and validation of 24 hours ahead neural network forecasting of photovoltaic output power. Math Comput Simul 131:88–100
34. Lee W, Kim K, Park J, Kim J, Kim Y (2018) Forecasting solar power using long-short term memory and convolutional neural networks. IEEE access 6:73068–73080
35. Raza, Muhammad Qamar, Nadarajah Mithulananthan, Jiaming Li, Kwang Y Lee, Hoay Beng Gooi (2018) An ensemble framework for day-ahead forecast of PV output power in smart grids. IEEE Trans Ind Inform 15(8), 4624–4634
36. Liu J, Fang W, Zhang X, Yang C (2015) An improved photovoltaic power forecasting model with the assistance of aerosol index data. IEEE Trans Sustain Energy 6(2):434–442
37. Lim S-C, Huh J-H, Hong S-H, Park C-Y, Kim J-C (2022) Solar Power Forecasting Using CNN-LSTM Hybrid Model. Energies 15(21):8233
38. Anu Shalini T, Sri Revathi B (2023) Hybrid power generation forecasting using CNN based BILSTM method for renewable energy systems. Automatika 64 (1), 127–144
39. Salman, Diaa, Cem Direkoglu, Mehmet Kusaf, Murat Fahrioglu (2024) Hybrid deep learning models for time series forecasting of solar power. Neural Comput Appl 5
40. Sharma V, Yang D, Walsh W, Reindl T (2016) Short term solar irradiance forecasting using a mixed wavelet neural network. Renew Energy 90:481–492
41. Gutierrez-Corea F-V, Manso-Callejo M-A, Moreno-Regidor M-P, Manrique-Sancho M-T (2016) Forecasting short-term solar irradiance based on artificial neural networks and data from neighboring meteorological stations. Sol Energy 134:119–131
42. Wang H, Yi H, Peng J, Wang G, Liu Y, Jiang H, Liu W (2017) Deterministic and probabilistic forecasting of photovoltaic power based on deep convolutional neural network. Energy Convers Manag 153:409–422
43. Jamali B, Rasekh M, Jamadi F, Gandomkar R, Makiabadi F (2019) Using PSO-GA algorithm for training artificial neural network to forecast solar space heating system parameters. Appl Therm Eng 147:647–660
44. Kumar PM, Saravanakumar R, Karthick A et al (2022) Artificial neural network-based output power prediction of grid-connected semitransparent photovoltaic system. Environ Sci Pollut Res 29:10173–10182
45. Amreen, Sana T, Radharani Panigrahi, Patne NR (2023) Solar power forecasting using hybrid model. In 5th international conference on energy, power and environment: towards flexible green energy technologies (ICEPE) 2023, pp 1–6. IEEE

Butterfly Optimization Algorithm Tuned Fractional Order Controller for the Load Frequency Control of Multi Area Multi Fuel Power System

CH. Naga Sai Kalyan, Priyanka Joshi, and Mohit Bajaj

Abstract This paper introduces a new control approach based on a butterfly optimization algorithm (BOA)-based tilt-integral-derivative-filter (TIDF) controller for the load frequency control (LFC) of a multi area multi fuel (MAMF) interconnected power system (IPS). The optimization of the TIDF using BOA is undergone with respect to the integral square error (ISE) minimization. However, the performance supremacy of the BOA-tuned TIDF controller is demonstrated by several control approaches reported in the literature recently. Moreover, the investigation on the MAMFIPS is performed by considering the integration of the plug-in electric vehicles in Area 1 for the step load perturbation (SLP) of 10%. Further, the effect of integrating the plug-in-electric vehicles (PEVs) with the IPS in terms of frequency variations is assessed and showcased. Furthermore, MAMFIPS is targeted at the wide variations in the load disturbance to test the robustness of the adopted regulator. The simulation analysis revealed the robustness of the presented control approach, and the deviations in the system behavior for wide variations are hardly noticed.

Keywords Butterfly optimization algorithm · ISE index · TIDF controller · PEVs integration · 10%SLP

CH. Naga Sai Kalyan (✉)
Department of Electrical and Electronics Engineering, Vasireddy Venkatadri International Technological University (VVITU), Guntur, India
e-mail: kalyanchallapalli@vvit.net

P. Joshi
Department of CSE, Graphic Era Hill University, Bhimtal, Uttarakhand, India

M. Bajaj
Department of Electrical Engineering, Graphic Era (Deemed to be University), Dehradun, India

Hourani Center for Applied Scientific Research, Al-Ahliyya Amman University, Amman, Jordan

Graphic Era Hill University, Dehradun 248002, India

S. Kumar et al. (eds.), *Soft Computing Applications in Modern Power and Energy Systems*, Lecture Notes in Electrical Engineering 1298,
https://doi.org/10.1007/978-981-96-0058-8_9

1 Introduction

Electricity consumption has been rising around the world for the past two decades, and in the coming years it will rise predominantly. However, with an increase in concern over carbon monoxide emissions, power system operators are focusing more on clean energy generation sources, and their integration into the IPS is an adequate solution. While renewable energy sources (RES) are clean and eco-friendly, grid stability issues are arising in major concern with their intermittent nature and uncertainty behavior. This kind of impact would be more on lower inertia and small-size micro-grid stability. In addition to that, the usage of electric vehicles is rapidly increasing and can continue to rise exponentially in the coming years. The battery unit in the PEVs is regarded as the energy storage unit, and hence a more stable and robust control technique is necessary for securing the IPS stability and delivering high-quality of power [1].

The available literature in the LFC study remarked the various control studies that are administered for the reduction of the real power imbalance (RPI). The difference in real power generation and the available load demand leads to the development of the RPI on the electrical network. Moreover, the RPI is the considered as the direct parameter in controlling the system frequency and as a result to control the frequency and stability of the IPS the RPI is to be regulated. The traditional proportional-integral (PI), and PID control structures are remarked in the literature extensively based on several optimization algorithms. In [2], hybridized algorithm of electric filed (HAEFA) is adopted to make the performance of the PID regulator more robust for the stability of dual-area IPS. Application of the Ions motion optimizer [3] and back tracking search [4] technique tuned PID controllers are implemented for two-area IPS networks. The modified PID with COOT optimization is suggested in [5], for the stability of renewable energy penetrated IPS. A hierarchy bi-level PI controller using moth flame algorithm is presented for multi-area LFC investigation in [6]. For the IPS LFC the authors in [7] presented the PID with marine predator's algorithm and further extended to the analysis of energy storage systems. Further, the algorithms like differential search technique, leaping frog algorithm, artificial fish schooling technique, spotted hyena algorithm, cryogenic algorithm, mayfly approach, slime mold algorithm, atom search optimizer, reptile search technique, Falcon algorithm, naked mole rat approach, seagull optimizer, stochastic fractal search, gravitational search, and imperialist competitive algorithm etc. are also implemented. However, the PID is inferior in eliminating the disturbances when it is set to operate in the complex models and to overcome this; the researchers are proposed the cascade controllers. But, the applicability of the cascade controllers is also restricted to only certain extent.

Later, the major focus had been made on the implementation of intelligent based control theory such as artificial neural networks, model predictive control, artificial intelligence [8], and fuzzy logic theory etc. Fuzzy PID as the frequency regulator using improved grey wolf optimization (IGWO) is presented in [9] and the improvement in the performance is validated with traditional control techniques. Hybridized sine–cosine and GWO is applied for optimization of type-II fuzzy controller in [10]. Though the intelligent controllers have the precise regulation over the complex IPS network, they involve large data for training and the approximations in the design at various stages might affect the optimal control. This motivates the research community to look for alternative control strategies.

The degree-of-freedom (DOF) aided with integer-order controllers like 2DOFPID/3DOFPID dominated the LFC investigations over the past decade. But, these controllers also need optimization approaches for effective operation. Teaching learning based optimizer (TLBO) [11] and seagull optimization [12] are used for the 2DOFPID, 3DOFPID controllers respectively for the IPS stability control. Various algorithms like African vulture optimization [13], volley-ball premier algorithm, crow search algorithm, harmony search algorithm, pattern search, and COOT algorithm [14], etc. are found their application in tuning the DOF controllers. However, the complexity design of the DOF made their implementation difficult especially to the complex IPS networks.

Later the fractional order (FO) control (FOC) techniques are extensively reported as robust controllers for the complex IPS. The FOPI/FOPID and the cascade versions of the FO and FLC are reported based on the optimizations of sailfish algorithm [15], selfish herd optimizer, deterministic optimizer, squirrel search algorithm, equilibrium optimization, differential search algorithm, quasi-oppositional GWO, Coronavirus immunity optimizer [16] and artificial humming bird algorithm. Moreover, the TIDF is also form the FOCs but its performance study is yet to be conducted rigorously in the LFC domain. In [17], TIDF performance is tested on two-area system using the fruit fly algorithm (FFA) and extensive investigations on TIDF are not reported. However, the dominance of TIDF over other FOCs is discussed in [18] and hence this work considered the TIDF using the newest algorithm of BOA.

The contributions of this work are.

(a) The novel BOA-tuned TIDF controller is presented for MAMFIPS stability.
(b) The superiority of the BOA-tuned TIDF is demonstrated with various control techniques reported in the literature.
(c) The impact of PEV integration on MAMFIPS performance is exhibited.
(d) Conducted the sensitivity test, and the BOA-tuned TIDF controller robustness is validated.

2 Power System Model

The MAMFIPS deliberated for the LFC analysis shown in Fig. 1 has two-areas and incorporates the generation units of gas, hydro, and thermal in both areas, and for 10% SLP the investigation is initiated. Moreover, area-1 is considered with the PEV integration, whose aggregated model is shown in Fig. 2. The PEV integrated MAMFIPS is developed in version (R2019a) MATLAB/SIMULINK, and the parameters necessary are made available in [19]. The total number of PEVs (n_h) is taken as 100 to assess their impact on the stability of MAMFIPS under the suggested controller.

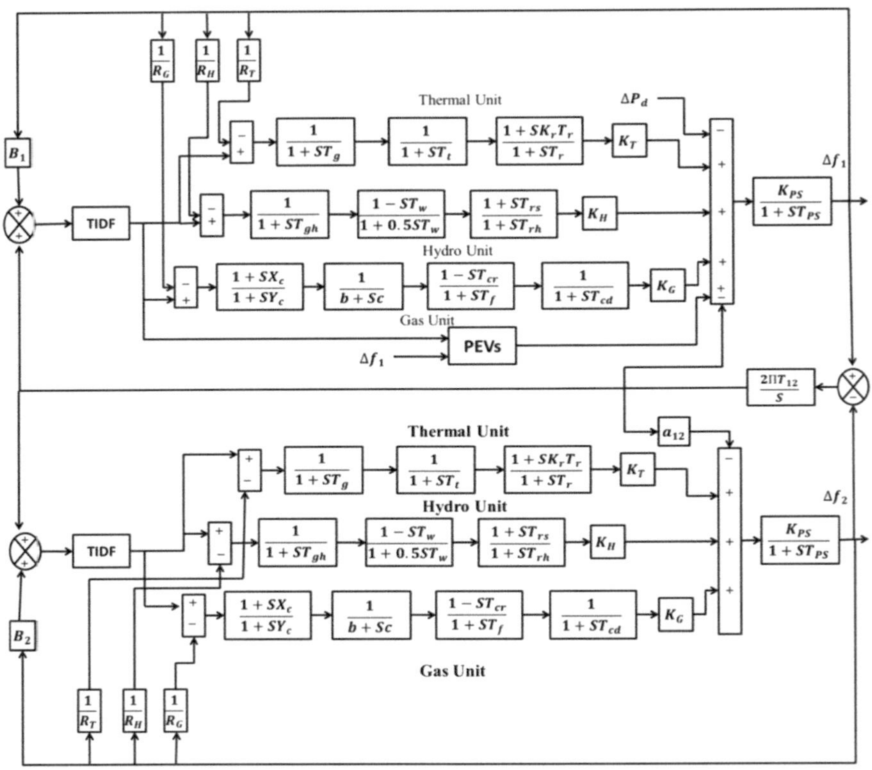

Fig. 1 Transfer function model of considered MAMFIPS for investigation

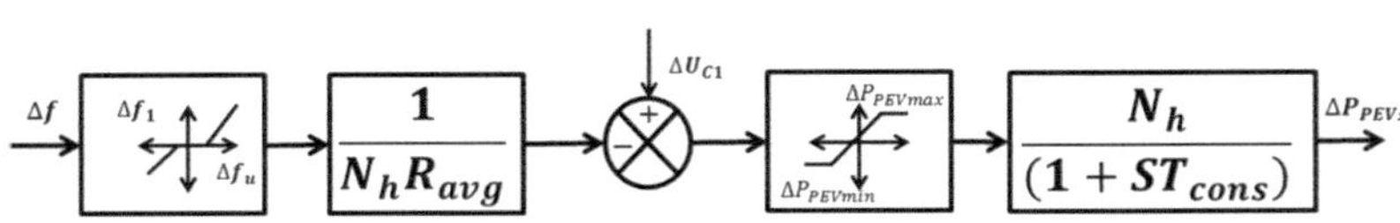

Fig. 2 Aggregated model of PEVs

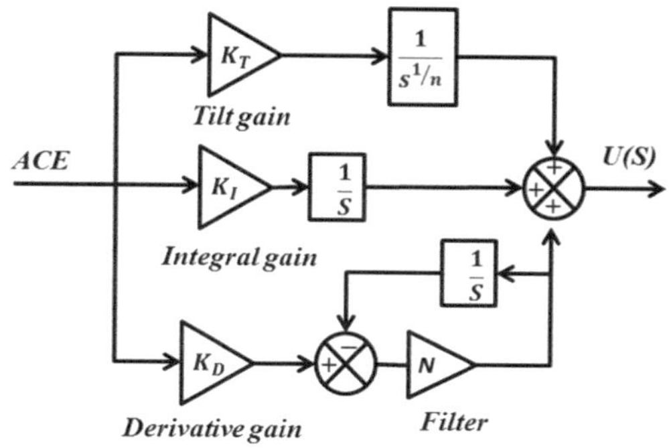

Fig. 3 TIDF controller

3 Controller and Objective Function

The integer order controllers are well established in the literature as well as in the industrial sector, and they have been operating redundantly for the past two decades. Moreover, these require some soft computing algorithms for finding the parameters in order to make sure their operation is suitable for the plant model. However, their existence is inevitable for non-linearized plant models. Thus, the alternative control mechanisms have gained interest among the research community and are being put forward for possible implementation in a real-time environment. One such alternative is FOC, which emerged from the theory of fractional calculus. There are many controllers that are aided by the fractional order in which the TIDF has the minimum tunable parameters and has the potential to exhibit efficacious performance. The structure of the TIDF is shown in Fig. 3, and its parameters are optimized using the BOA subjected to ISE [20, 21] in Eq. (1).

$$\mathrm{J_{ISE}} = \int_{0}^{\mathrm{T_{Sim}}} \left(\Delta f_1^2 + \Delta P_{\mathrm{tie12}}^2 + \Delta f_2^2\right) \mathrm{dt} \tag{1}$$

4 Butterfly Optimization Algorithm

The BOA is a recent methodology that has been proposed by [22] and finds application to numerous problems. This methodology mimics the social behavior of the butterflies that are from the family Lepidoptera in the kingdom of animals. There are currently more than 19,000 butterfly species in existence worldwide. The sensory organs in the butterflies ensure their survival for longer durations, and they can use these sense organs for smell and touch to find their mating partner. Moreover, they have sensory receptors that are extended on their body and legs in order to effectively detect the nectar, which they can feed on. However, a few assumptions are made in the modeling of the BOA, including that the butterflies are emitting a fragrance through which they can attract others, and each fly can move towards the fly with a greater fragrance. The modeling of the fragrance (f) is performed by considering the constraints of sensor modality (c), and the intensity of the stimulus (I) is given in Eq. (2).

$$f = cI^{\alpha} \tag{2}$$

The parameter 'α' depends on the modality that indicates fragrance absorption. Its value can be taken from [0–1]. In BOA, there are two crucial steps: global and local searches. The flies can travel in the direction of the best fly in the global search, which is represented by Eq. (3).

$$x_i^{t+1} = x_i^t + (r^2() \times g^* - x_i^t) \times f_i \tag{3}$$

where 'x' indicates the position, 't' is time, and the sub-script 'i' represents the butterfly. The 'g^*' is the best solution at present, and 'r ()' is assigned a value from [0, 1]. The local random walk is modeled as in Eq. (4).

$$x_i^{t+1} = x_i^t + (r^2() \times x_j^t - x_k^t) \times f_i \tag{4}$$

However, a parameter 'P' is employed to determine the implementation of the global and local searches. A detailed view of the BOA can be grabbed from the flowchart shown in Fig. 4.

5 Simulation Results

5.1 Case-I: Analysis of MAMFIPS with PEVs Integration Under Various Controllers

The BOA-tuned TIDF controller operation is tested on the dynamic behavior of the MAMFIPS, for 10% SLP on area 1 the analysis is began. Additionally, the PEVs are included for analysis in Area 1 of the MAMFIPS. The responses of the MAMFIPS with PEV integration under the BOA tuned TIDF controller are shown in Fig. 5, along with the responses that are retrieved under the regulation of FFA tuned TIDF [17], TLBO optimized 2DOFPID [11], IGWO-based fuzzy PID [9], TLBO

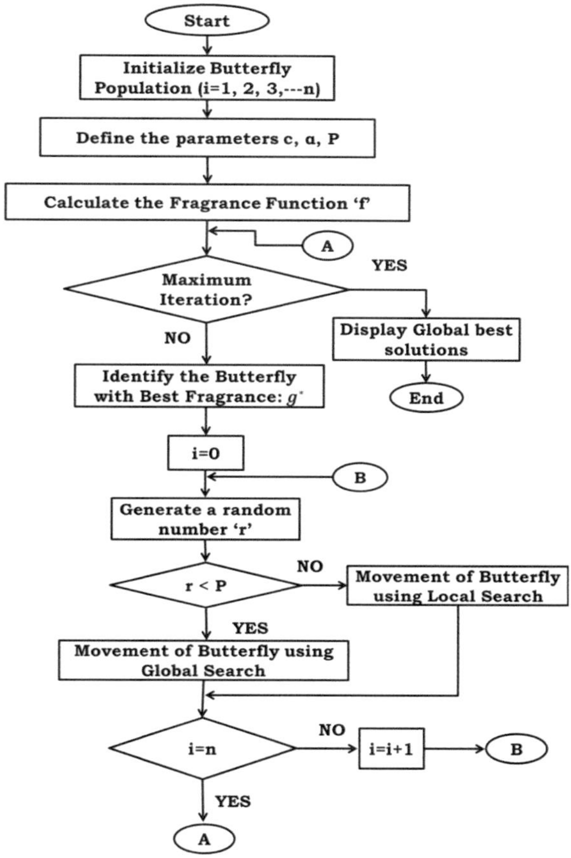

Fig. 4 BOA flowchart

[11], and HAEFA [2] optimized PID controllers. The above controllers are operated in both areas of MAMFIPS individually, and Fig. 5 shows the related responses. Additionally, the responses are examined in light of settling time in Table 1, and moreover, it is figured as the bar chart shown in Fig. 6 to clearly demonstrate the controller's efficacy. It is clearly identified from observing the Fig. 5 and noticing Table 1 that the TIDF optimized with BOA exhibits dominance in trimming the peak undershoots, eliminating the oscillations, and conceding less time in settling down the fluctuations. Moreover, the ISE value was effectively reduced by the BOA tuned TIDF controller and is enhanced by 37.65% with FFA tuned TIDF, 46.36% with TLBO tuned 2DOFPID, 57.67% with IGWO tuned fuzzy PID, 79.10% with TLBO tuned PID, and 83.19% with HAEFA tuned PID in the literature. The optimal parameters of various controllers are noted in Table 2.

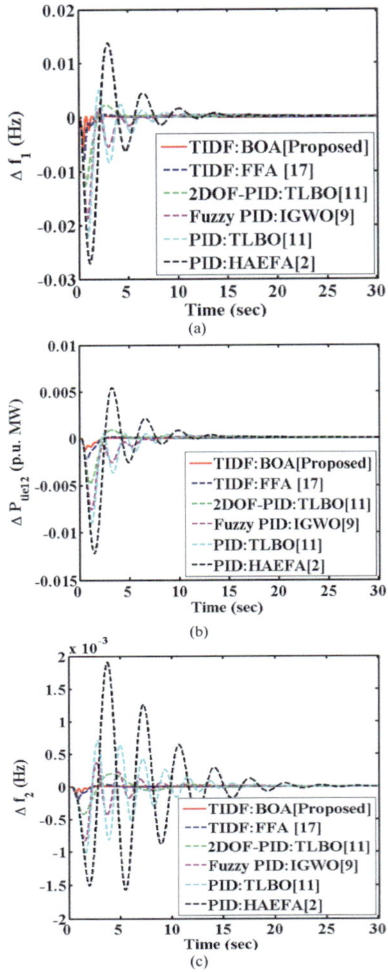

Fig. 5 Case-I responses. **a** Δf_1 **b** ΔP_{tie12} **c** Δf_2

Table 1 Responses settling time in seconds

Settling time (Seconds)	BOA:TIDF [Proposed]	FFA:TIDF [17]	TLBO: 2DOFPID [11]	IGWO: Fuzzy PID [9]	TLBO: PID [11]	HAEFA: PID [2]
Δf_1	4.13	5.85	9.22	13.17	16.54	22.02
ΔP_{tie12}	5.16	8.32	12.26	14.85	17.91	22.15
Δf_2	3.47	5.64	12.30	15.76	20.69	23.63
ISE* 10^{-3}	60.44	96.95	112.69	142.80	289.30	359.74

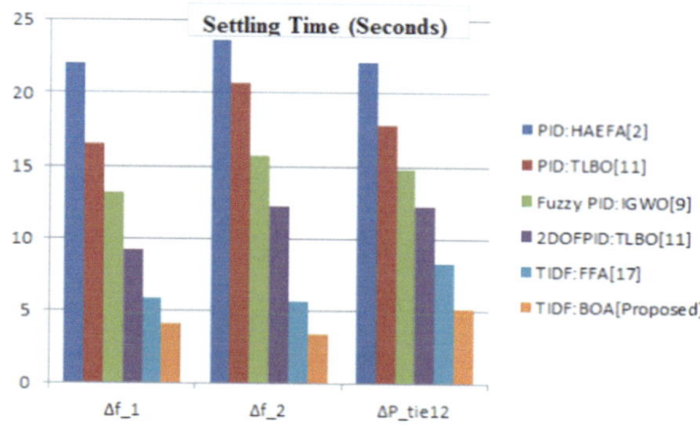

Fig. 6 Comparison of responses settling time under different control strategies

Table 2 Controller parameters

Parameters	BOA:TIDF	FFA:TIDF	TLBO: 2DOFPID	IGWO: Fuzzy PID	TLBO: PID	HAEFA: PID
K_P	1.208	1.577	0.549	1.969	1.625	1.116
K_I	0.764	0.969	0.124	0.042	1.045	0.899
K_D	1.153	0.725	0.163	0.378	0.328	0.649
N	0.037	0.163	–	–	–	–
N	233.55	240.27	230.10	–	–	–
K_1	–	–	0.044	–	–	–
K_2	–	–	0.061	–	–	–

5.2 *Case-II: Effect of PEVs on the Stability of MAMFIPS*

To exhibit the prominence of PEV integration in IPS stability, the analysis of MAMFIPS is conducted with and without integrating the PEVs under the BOA-tuned TIDF controller. The responses for this subsection are shown in Fig. 7, and after careful analysis, it is identified that the responses of the MAMFIPS are far better when considering the integration of the PEVs. The battery storage units in the PEVs have a quick response nature and can deliver the energy to the grid to regulate the RPI, and thereby, grid stability can be ensured. The storage units can charge up at off-peak demands and are ready to discharge whenever necessary. Thus, the battery storage unit in PEVs plays a pivotal role in grid stability.

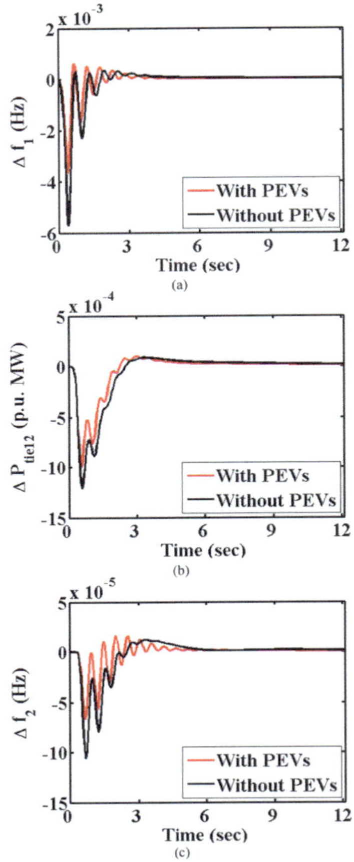

Fig. 7 Case-II responses. **a** Δf_1 **b** ΔP_{tie12} **c** Δf_2

5.3 Case-III: Performance of MAMFIPS Under Wide Range of Load Variations

To check the BOA-tuned TIDF performance robustness, the PEVs integrated MAMFIPS are targeted with deviations in the load of $\pm$ 25% from the nominal load. The responses for the variations in the load are shown in Fig. 8, and it is clear that the major changes in the system's dynamic behavior are not observed even for the wide variations in the load. Hence, it is declared that the suggested BOA-tuned TIDF is robust and effective enough to handle the stability of MAMFIPS even under wide variations in load demands.

Fig. 8 Case-III responses. **a** Δf_1 **b** ΔP_{tie12} **c** Δf_2

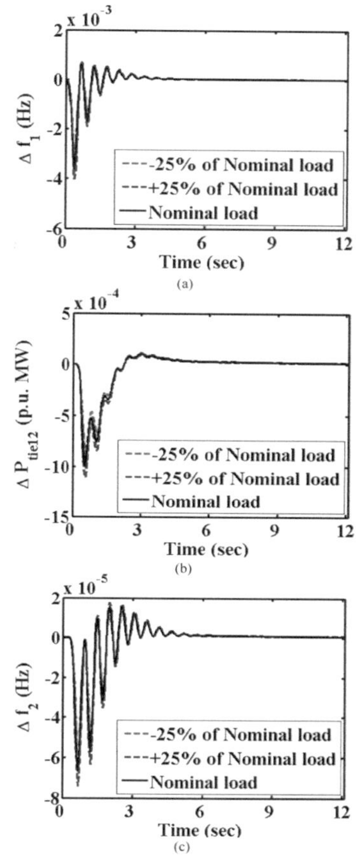

6 Conclusion

This work proposes a novel control approach for a BOA-based TIDF controller subjected to the ISE index for MAMFIPS stability. Initially, the MAMFIPS is targeted with 10% SLP, and the dynamical responses under load disturbances is regulated with several control approaches, like PID optimized with HAEFA and TLBO, fuzzy PID based on IGWO, TLBO tuned with 2DOFPID, FFA-based TIDF, and BOA-tuned TIDF individually. The simulation analysis revealed the dominance of the BOA-based TIDF over the other reported controllers. With the presented controller, the dynamic performance of MAMFIPS is regulated much better, and the deviations are quickly attained in the zero state positions, as well as the oscillations and peak undershoots, which are effectively damped. Further, the impact of PEVs on the MAMFIPS performance is showcased, and the improvement in system stability is observed. The battery storage units in the PEVs have a quick response nature, which aids in fast actions in regulating the frequency. Further, the sensitivity analysis is performed on the MAMFIPS with PEV integration under the control of a BOA based TIDF for the wide range in load of ± 25% from the nominal range. The investigative results for the sensitivity test demonstrated that the deviations in the responses for a wide range of load disturbances from the nominal values are not much fluctuated, and thus, the robustness of the BOA based TIDF performance was determined.

References

1. Ahmed EM, Elmelegi A, Shawky A, Aly M, Alhosaini W, Mohamed E (2021) Frequency regulation of electric-vehicle penetrated power system using MPA-Tuned new combined fractional order controllers. IEEE Access 9:107548–107565
2. Kalyan CNS, Rao GS (2020) Coordinated SMES and TCSC damping controller for load frequency control of multi area power system with diverse sources. Int J Electr Eng & Inform 12(04):747–769
3. Mohapatra G, Debnath M, Mohapatra K (2020) IMO-based novel adaptive dual-mode controller design for AGC investigation in different types of systems. Cogent Eng 7(01):1–19
4. Guha D, Roy PK, Banerjee S (2018) Application of backtracking search algorithm in load frequency control of multi-area interconnected power system. Ain Shams Eng J 9:257–276
5. Shukla H, Raju M (2023) Application of COOT algorithm optimised PID plus D2 controller for combined control of frequency and voltage considering renewable energy sources. E-Prime-Adv Electr Eng, Electron Energy 6, 100269
6. Kunya AB (2024) Hierarchical bi-level load frequency control for multi-area interconnected power systems. Electr Power Energy Syst 155:109600
7. Sobhy M, Abdelaziz A, Hasanien H, Ezzat M (2021) Marine predators algorithm for load frequency control of modern interconnected power systems including renewable energy sources and energy storage units. Ain Shams Eng J 12:3843–3857
8. Garia P, Mittal A, Singh A, Kumar N, Oli S (2023) A study and performance review of an On-Grid PV solar plant using artificial intelligence. International conference on power energy, environment & intelligent control (PEEIC), 484–489, Greater Noida, 19–23
9. Sahoo BP, Panda S (2018) Improved grey wolf optimisation technique for fuzzy aided PID controller design for power system frequency control. Sustain Energy, Grids Netw 16:278–299
10. Yousef M, Mosa M, Ali A, Ghany A, Ghany M (2023) Load frequency control for power system considering parameter variation using parallel distributed compensator based on Takagi-Sugino fuzzy. Electr Power Syst Res 220:109352
11. Sahu RK, Panda S, Rout UK, Sahoo DK (2016) Teaching learning based optimisation algorithm for automatic generation control of power system using 2DOF-PID controller. Electr Power Energy Syst 77:287–301
12. Kalyan CNS, Suresh CV (2022) Higher order degree of freedom controller for load frequency control of multi area interconnected power system with time delays. Global transition proceedings 3(1):332–337
13. Swain N, Sinha N, Behera M (2023) Stabilized frequency response of a microgrid using a two-degree-of-freedom controller with African vultures optimization algorithm. ISA Trans 140:412–425
14. Moschos I, Parisses C (2023) Combined frequency and voltage control of two-area multi-source interconnected microgrids via the 2DOF-TID controller. E-Prime-Adv Electr Eng, Electron Energy 5, 100268
15. Rai A, Das DK (2022) The development of a fuzzy tilt integral derivative controller based on the sailfish optimizer to solve load frequency control in a microgrid, incorporating energy storage systems. J Energy Storage 48:103887

16. Biswas J, Bera P, Chakrabarty K (2023) Determination of control area and design of fuzzy rule-tuned PID controller for LFC of multimachine power system. Electr Power Syst Res 221:109411
17. Jalendra M, Reddy KS, Lavanya M, Jaideep NV, Kalyan CNS, Suresh CV (2022) Fruit fly optimization algorithm based TIDF controller for LFC of traditional hydro thermal system with SSSC and ultra capacitors. International conference on knowledge engineering and communication systems 1–6, Chickballapur, 28–29
18. Wadi M, Shobole A, Elmasry W, Kucuk I (2024) Load frequency control in smart grids: A review of recent developments. Renew Sustain Energy Rev 189:14013
19. Tripathi S, Singh VP, Kishor N, Pandey AS (2023) Load frequency control of power system considering electric vehicles aggregator with communication delay. Electr Power Energy Syst 145:108697
20. Kalyan CNS, Rao GS (2020) Frequency and voltage stabilisation in combined load frequency control and automatic voltage regulation of multiarea system with hybrid generation utilities by AC/DC links. Int J Sustain Energ 39(10):1009–1029
21. Kalyan CNS, Rao GS, Bajaj M, Khan B, Zaidan MR (2024) Frequency regulation of geothermal power plant-integrated realistic power system with 3DOFPID controller. Cogent Engineering 11(01):2322820
22. Arora S, Singh S (2019) Butterfly optimization algorithm: a novel approach for global optimization. Soft Comput 23:715–734

Energy Management Systems for Smart Buildings: A Review

Sushrut Patankar, Raju Kumar Swami, and Sameer Nanivadekar

Abstract The conservation of energy is a critical concern due to the rapid depletion of natural resources and the need for sustainable development. This manuscript explores energy conservation technologies, particularly in the building sector, which accounts for 33% of total energy usage. The paper highlights the evolution and implementation of cutting-edge methodologies, such as machine learning algorithms and the Internet of Things (IoT), which optimize energy usage, enhance operational efficiency, and lower the carbon footprint of building operations. However, challenges such as technological barriers, high initial deployment costs, specialized expertise, and stakeholder sustainability are highlighted. The paper also discusses potential obstacles in integrating advanced technologies with existing building infrastructures and data management and privacy issues related to IoT implementations. The paper advocates for a multidisciplinary approach that incorporates technological innovation, policy formulation, stakeholder engagement, and public awareness campaigns to adopt and refine these technologies, paving the way for a more sustainable and energy-efficient future.

Keywords Artificial Intelligence · Energy management · Energy Storage · IoT · Smart Buildings

1 Introduction

Management of energy refers to the systematic planning and distribution of energy while also keeping up with the needs of the public [1]. These plans require a thorough study of the current usage of energy and then comparing it with an ideal energy-aware usage. This comparison allows to understand the amount of energy that is going to

S. Patankar (✉) · R. K. Swami
Pacific Academy of Higher Education and Research University, Udaipur, Rajasthan, India
e-mail: smpatankar19@gmail.com

S. Nanivadekar
A.P. Shah Institute of Technology, Thane, Maharashtra, India

S. Kumar et al. (eds.), *Soft Computing Applications in Modern Power and Energy Systems*, Lecture Notes in Electrical Engineering 1298,
https://doi.org/10.1007/978-981-96-0058-8_10

waste. Excessive use of energy doesn't just lead to an increase in energy bills but due to the harmful traditional methods used to produce the energy. Environmental resources are also depleting [2]. Thus, an energy management plan also includes an effective way to produce energy.

Over the years, numerous plans have been designed to combat the excessive use of energy resources. Initially, these plans started out as public awareness campaigns. This was done by educating people about the complications caused due to overuse of energy. But this method, although effective to a certain extent, did not receive the desired response. This particular method proved to work better for residential buildings as compared to commercial buildings. This could be due to the fact that humans tend to be more aware of their energy spending at home rather than at their workplaces where electricity bills are not their concern. To tackle this situation, various technologies have been developed. Technologies such as solar energy, smart inverters, thermal energy from the ground, and other control systems have been used. Apart from these, smart buildings have been designed. These buildings incorporate various technologies such as Artificial Intelligence (AI) and the Internet of Things (IoT) to control the consumption of energy.

This paper gives a comprehensive review of the different studies done in the field of energy management systems for buildings. The work provides the readers with an overview of the different technologies used to build smart buildings. Figure 1 illustrates the hierarchical tree diagram illustrating the various technologies used to construct smart buildings. The various technologies explored in the scope of this article include the Internet of Things (IoT), intelligent grids, optimization models, machine learning, and energy storage. Based on their power capacity, intelligent grids are further classified as smart, micro, and nano grids. Optimization models are further classified as stochastic and resilient. Machine learning, deep learning, and deep reinforcement learning are also used in energy management using Artificial Intelligence (AI). Finally, thermal storage integrated into structures is used in energy storage systems.

2 Energy Management Technologies

2.1 Smart Buildings

Smart buildings are buildings that are architecturally designed to be more environmentally convenient and energy- efficient. A reasonable amount of investment is involved in designing smart buildings. While designing these buildings the relationship between the structure, services and occupants is of great importance. The reason these buildings are called smart is because they have an interconnection of intelligent sensors spread out over the structure of the building that monitors different aspects of it at all times. Intelligent control systems are installed in the building to give optimized outputs about aspects such as energy, air-conditioning, lighting,

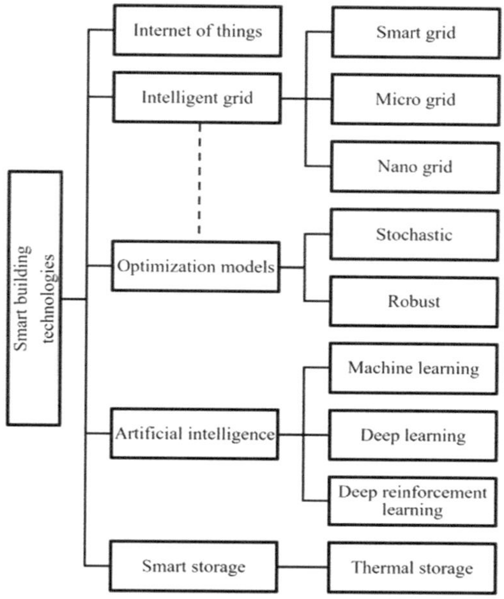

Fig. 1 Hierarchical tree diagram of the different technologies used to build smart buildings. Internet of Things (IoT), intelligent grids, optimization models, machine learning, and storage of energy are the different technologies covered in the scope of this paper. Intelligent grids are further divided into smart, micro, and nano grids based on their power capacity. Optimization models are also divided into stochastic and robust models. Energy management using Artificial Intelligence (AI) is also done by using machine learning, deep learning, and deep reinforcement learning. Finally, energy storage systems consist of thermal storage integrated into the buildings

security, network, etc. as shown in the Fig. 2. These smart embedded systems were the backbone for the development of smart buildings [3].

2.2 Optimization Model

Li et al. [3] use the flexible resources in the buildings to optimize the energy consumption of the Active Distribution Network (ADN) in office buildings. It constructed an optimization model for the Heating, Ventilation and Air Conditioning (HVAC) and the Electric Vehicle (EV) chargers in the building based on the traveling habits of the occupants. It also given an ADN optimization model using a Mixed-Integer Second-Order Cone Programming (MISOCP) model and developed a method to learn the consumption of energy in the ADN. The model were tested in the winter and the results showed a decrease in the ADN losses and energy consumption.

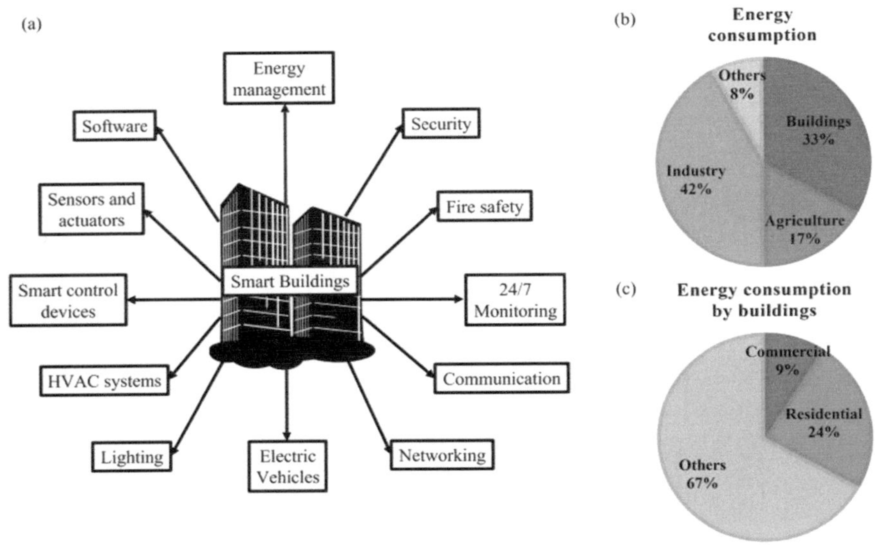

Fig. 2 **a** Smart buildings are buildings that combine multiple smart technologies under one roof to essentially automate a building and to make it more economically and environmentally friendly. These buildings integrate sensors and actuators with smart control devices. HVAC systems, lighting systems, and EVs to manage the energy in the building. The buildings have their networking and communications systems. Along with this continuous monitoring of the building thus providing safety to the occupants is also done. All these things combined make a building smart. **b** The segregation of energy consumed by the agricultural, building, industrial and other sectors. Buildings account for 33% [3]. **c** The segregation of energy consumed by the commercial and residential buildings. Residential buildings consume 24% and commercial buildings consume 9% [4]

Xu et al. [4] have developed a strategy for a renewable power supply system in a shopping mall while considering the behaviours of vehicle use and the temperature ranges. It considered two Electric Vehicles (EV) and studied their stochastic behaviour to build models. It tested the use of Retired Electric Vehicle Batteries (REVB) to achieve optimization by using a stochastic optimization method to manage the energy. The results showed that using Photovoltaic and REVB in commercial buildings reduces the consumption of energy by almost 82.2% and 83.3% respectively.

Brucke et al. [5] have developed an unsupervised algorithm that extracts data from a user's electricity meter to forecast changes in the load state (from ON to OFF). It used the Non-Intrusive Load Decomposition (NILM) technology to see the real-time consumption of electricity by 52 device profiles. This approach is useful for developing a smart grid-based strategy for efficient energy consumption. It also provide an approach for predicting the short-term use of power using the Particle Swarm Optimizer (PSO) which achieved a percentage energy error of around 1%.

Foroozandeh et al. [6] have considered a smart residential building with Photo Voltaic (PV) generation panels, EVs, and a BESS. It took a multiobjective optimization problem and represented it using two objective functions (OF). The first OF was targeted to reduce the cost of energy consumption and the second was to reduce the consumption of the peak load. The design of system is proposed by keeping the charging and discharging times of the EV and BESS in mind and then used the Pascoletti-Serafiniscalarization to get the Pareto front points of the objective functions. With this approach, the consumption cost and peak load decreased by 35.56% and 45.52% respectively.

Oskouei et al. [7] have used the robust-stochastic optimization approach to deal with the fluctuations in the electricity price, energy demands, and the effect of the power of PV systems on the predicted cost. It used a mixed-integer linear programming model to reduce the cost of operation of the Multi-Carrier Micro grids (MCMG) using the Transactive Energy Management (TEM) mechanism. In this approach, each MCMG was able to share their energies mutually rather than extracting large loads of energy from the main grid.

Rossit et al. [8] have created a stochastic optimization model that optimizes the electricity bill and user comfort in smart homes. The stochastic resolution model as user preferences can vary drastically. To achieve this goal, developed two algorithms namely simulation–optimization and greedy heuristic. For the testing of the algorithms and to understand the tradeoffs between the cost and comfort five weight vectors were used. It is observed that as the use of appliances was higher during the day, the greedy solution had the smallest time slot with high energy consumption. But for different sample sizes the simulation algorithm vector was closer to the optimal value as compared to the greedy algorithm. Thus the simulation–optimization showed better results than the greedy heuristic.

Lu et al. [9] state that dealing with uncertainties in parameters is something that an energy efficiency model must be capable of. It compared a stochastic model with a deterministic one with 13 changing input parameters. The aim is to test the robustness of the models. It used two formations of office buildings for its system. The study states that the optimization results for both stochastic and deterministic had some flaws. They also state that this technique is best suited to be used when extremely necessary as it takes a lot of computational capacity. However, they saw that when the uncertainty of the parameters was reduced by 44% then the models started giving better and more accurate results.

Bahramara et al. [10] have developed a robust optimization model to deal with the uncertainties in the electricity market and optimize the energy generated and stored by the photovoltaic cells on the roof of a smart home. They also wanted to reduce the amount of energy pulled by the cells from the main grid. Their results showed when the amount of energy extracted from the main grid was restricted between 5 kW and 1.5 kW there was a small increase in the operational cost of the system.

2.3 Intelligent Grids

Intelligent grids are technologies that monitor and control the changes in electricity usage. These grids can also be repaired in case of any faults. Smart grids aim to conserve energy while also keeping up with the energy requirements. Smart grids use computer algorithms to control the electricity network which is made up of various components. Smart grids often have an unlimited capacity and they are followed by micro grids and nano grids that vary based on their power capacity and the places in which they are installed. Kermani et al. [11] have designed an energy management architecture for a micro grid in a school building using the Supervisory Control and Data Acquisition (SCADA) system. They used a small-scale smart building that was connected to the building's electrical network.

The micro grid was composed of a photovoltaic cell, a smart switchboard, and a Battery Energy Storage System (BESS). They also had different loads connected in their grid that was divided into parts such as critical, essential, and normal. They used two international electricity standards named Modbus and Konnex. The aim was to extract less power from the main grid. Figure 3a shows the micro grid as proposed by them. When both the photovoltaic and BESS were installed in the system they were able to achieve an average monthly energy cost savings of up to 88% and in June the cost went down to 98%.

2.4 Smart Storage

The photovoltaic (PV) systems make up the largest proportion of the recently installed capacities of the RESs in the world from which the share of rooftop PVs is increasing especially regarding the development of smart homes (SHs). Using a stored form of energy that actively charges and discharges based on the temperature required in a building can reduce the amount of energy extracted by the building grid from the main grid and also reduce the peak load. It can also allow the users in the building to use the ventilation systems at their convenience. These systems are programmed to store and release energy according to the temperature of the building.

Iffa et al. [12] have tested a thermal energy storage system with active insulation that has been integrated into a wall. The stored thermal energy charges and discharges to store and use cooling and heating energy. They tested the charging, discharging, and insulation separately as well as an integrated system. Their results showed that a heat flux of almost 81.92 W/m2 was transferred to the wall from the indoor air during discharging. Further, the wall surface temperature was reduced to 10 °C from the room temperature.

Mayyas et al. [13] have designed an energy management system that powers a building using a photovoltaic cell. They used Li-ion Batteries (LIB) or Reversible Fuel Cells (RFC) to power all the electrical resources in a building by using an energy storage system. They compared LIBs and RFCs and found that unless the

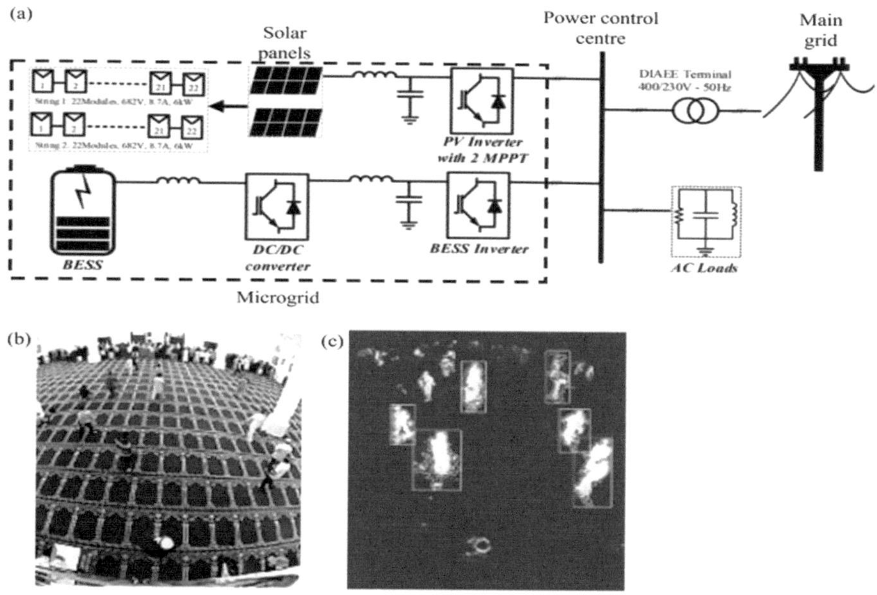

Fig. 3 **a** shows the micro grid [11]. Solar panels connected to a PV inverter are combined with a BESS. Their system reduces the total load extracted by the micro grid from the main grid. **b** The indoor of a mosque [12]. **c** The bounding boxes are created around the moving objects to count the occupancy in the mosque [13]

storage capacity was increased, LIBs did not have sufficient capabilities to satisfy the building energy demands. With the storage system, they were able to reduce the energy costs to 25.5 cents/kWh with a LIB energy storage system. Their system is shown in Fig. 4b.

2.5 *Artificial Intelligence*

Hagras et al. [7] have developed an energy management system that uses an agent-based system to intelligently control the energy in commercial buildings. They have used different techniques such as fuzzy logic, neural networks, and genetic algorithms to study the different temperature changes in a building. The thermal changes were measured concerning the outside temperature, occupant needs, and the responses of the building's plants. The data was all stored in a database and was then processed on a core CPU. The CPU decides the accuracy of the proposed algorithm against the BMS and then provides actions based on that. Even the old responses of the occupants to the energy commands were given to the algorithm. Their system was combined with the building's existing Building Management System (BMS). Their

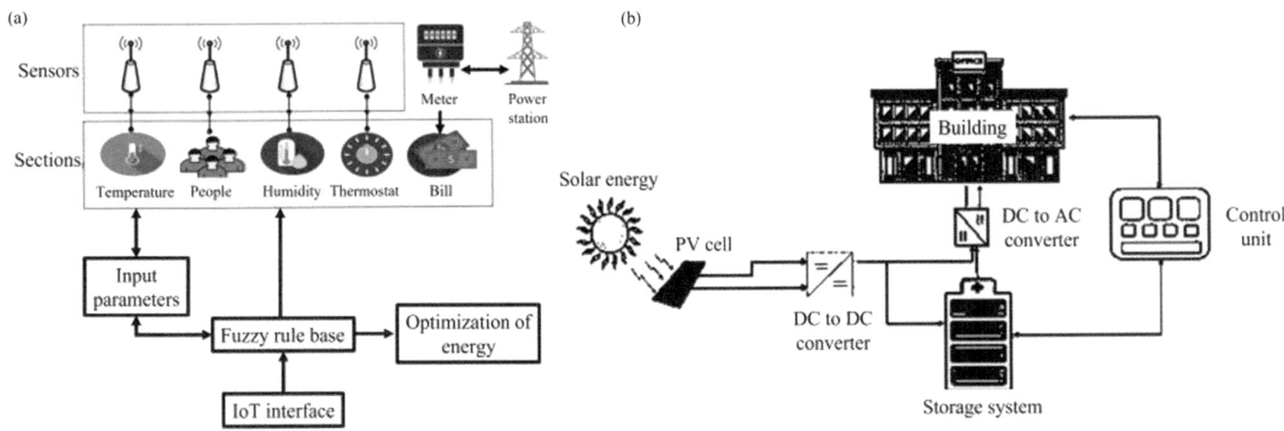

Fig. 4 **a** IoT-based fuzzy inference system [19]. A fuzzy rule base that gives the optimization decision gets its input parameters based on the different parts of a residential building. Each part namely, temperature, number of people, humidity, the thermostat is connected to a sensor. The bill is generated from the meter readings. Finally, the entire system is connected to an IoT interface. **b** Smart storage system [13]. A photovoltaic cell stores DC and then it is converted to AC as per the building requirements

algorithm, while also keeping the occupant needs in check, obtained a 7.7% savings for the gas and electricity bills in the commercial building.

Aftab et al. [14] have designed an energy management system and implemented it in an indoor mosque environment. They used a fisheye lens and thermal simulator to control the HVAC system. The fisheye lens was used to count the number of people in the room. This was done by first finding the number of moving objects in the frames and then creating bounding boxes around them. The width of the bounding boxes helped them see the number of people in one bounding box. Figure 3b shows the real-time video footage and Fig. 3c shows the bounding boxes created around the moving objects. Then, they also used a frosted camera lens to maintain privacy but achieved 7% less accuracy than a normal lens. They applied machine learning to the footage to obtain a prediction model about the occupancy patterns. A reference line was created to tell if the person was entering or leaving. Finally, they used an EnergyPlus simulator and OpenEnergyMonitor framework to check the energy consumption of the HVAC system. They achieved an accuracy of 90% and 85% for their occupancy recognition and prediction models respectively. The HVAC control system also saved around 30% energy.

Pinto et al. [15] have designed a deep reinforcement learning (DRL) approach to improve the energy fluctuations occurring in a cluster of buildings. They have considered the coordination factor that two buildings in a cluster may have with each other. The DRL optimizes the storage of thermal energy from a cluster of four buildings that have distinct energy systems. They aimed to reduce the overall load of the cluster while optimizing the energy use of individual buildings. Their approach showed an operational cost reduction of 4% and a 12% decrease in the energy demand. Syed et al. [16] have proposed a hybrid deep learning model to predict the amount of energy consumption in a building. Initially, the raw data collected from the sensors in the buildings are cleaned and preprocessed to add a few additional features. Next, a model which has layers of dense layers with, unidirectional Long Short-Term Memory (LSTMs), and bi-directional LSTMs is trained using the processed data. They compared their method to different models and found that the proposed model had the least a mean absolute percentage error of 2%.

2.6 Internet of Things

Rafsanjani et al. [17] have developed a smartphone application that keeps a track of the energy usage and behaviour of ten occupants in a building over 12 weeks. The method was divided into seven parts. In the first part, they used the advanced internet-connected meters that are installed in the electric panels of most commercial buildings to get the data of energy consumption. Next, they used the Wi-Fi network in the buildings to study the number and behaviour of the occupants.

In addition, the Wi-Fi network also gives information about the behaviour of the last leaving occupant. Then, they use the collected data to calculate the Energy-use Efficiency Index (EEI) for each day. This value of EEI is given to a trained deep learning model to identify whether the energy consumption for the particular day was efficient or not. Next, they also studied the behaviour of the occupants and combined it with the EEI value to give personalized feedback to the occupants. Their results showed a 34% improvement in the energy use in the commercial building. One of their observations was that the last people to leave the building often neglect the lighting which results in an excess of energy wastage. Marinakis et al. [18] have proposed and designed an architecture for an IoT-based intelligent energy management system for buildings. They utilized the data collected from different platforms and provided daily and weekly feedback to the end-users as well as personalized feedback for individual buildings using their developed prediction models. Their proposed architecture consisted of three distinct parts namely, the data integration layer, then came the prediction rules, and finally the suggestions/feedback. Various parameters were considered in the data integration layer such as occupant behaviour, weather, energy bills, amount of energy produced, and the other details of the building. They also proposed a semantic framework based on the Ztreamy system.

Ain et al. [19] have developed an energy management system based on the fuzzy inference models (FIS) and incorporated it with IoT technology. Their method uses a FIS that takes the temperature as well as the humidity in the air as input parameters and gives the output in accordance with the comfort of the user. They evaluated their FIS using the Mamdani and Sugeno systems. They implemented the system on MATLAB software. Further, they also proposed a method to automatically generate a rule base thus reducing the chance of manual errors when the volume of the rules increases. They incorporated temperature and humidity sensors to get the readings. Their approach reduces the consumption of energy by 28%. As seen in Fig. 4a the input parameters for the fuzzy rule base that makes the optimization decision are based on the various sections of a residential building. Each component, such as temperature, number of people, humidity, and the thermostat, is linked to a sensor. Meter readings are used to create the bill. Finally, the complete system is linked to an Internet of Things interface.

Karthick et al. [20] have designed an energy management system using a Smart Compact Energy Meter (SCEM) to monitor the usage of energy and the quality of the power in a commercial building. They have also used a Demand Side Management (DSM) using IoT. They have employed an ARM microcontroller that is robust enough to carry out tasks such as acquiring vast amounts of sensor data and processing it. The system can detect disturbances and classify them. It can also check the consumption and the data exchange inside the building. These tasks all require the continuous optimization of the system. Further, they tested their application in a commercial building and used the Blynk mobile application to monitor and control the appliances. The use of their system showed a significant decrease in energy consumption. Figure 5c shows the proposed system. The energy usage is measured using a voltage and current sensor (metre). This sensor is linked to the network's two relays, which control the primary and secondary appliances. The primary load is controlled by an ESP32, while the secondary load is controlled by an ESP8266. An ESP8266 is also connected to the metre, and the system includes a Raspberry Pi. This Raspberry Pi interfaces with the ESP modules, resulting in an IoT interface.

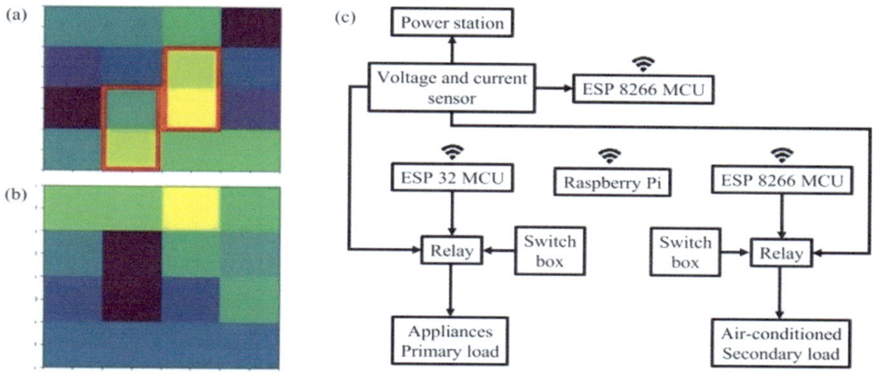

Fig. 5 **a** Output of thermal sensor when there was no occupant in the room [20]. **b** Output of thermal sensor when there were two occupants in the room [11]. **c** The IoT-based energy management system [21]. A voltage and current sensor (meter) take the readings of the energy usage. This sensor is connected to the two relays in the network which control the primary and secondary appliances. An ESP32 controls the primary load and an ESP 8266 controls the secondary load. An ESP8266 is also connected to the meter and finally, a Raspberry Pi is also installed in the system. This raspberry pi communicates with the ESP modules and creates an IoT interface

Kaushik et al. [21] have used IoT with a thermal sensor to monitor the occupants and the electrical loads in a building. They optimize the energy consumption based on occupancy. The thermal sensor as shown in Fig. 5a and b detects the presence of humans using a machine learning algorithm. Upon detection, the IoT system optimizes the electricity resources in the building. They reduced the energy consumption by 38%.

3 Discussions

A comparative analysis of the studies reviewed in this paper has been presented in Table 1. Most of the recent studies in the literature were done using a combination of two methods such as combining deep learning with IoT or by combining optimization algorithms we intelligent grids. A value of energy and money saved using the particular technique when compared to other methods was obtained. Xu et al. had the highest energy saving of83.3% when they used a stochastic optimization model in a shopping mall setting. An important point to remember here is the scale of resources that were considered when building an energy efficiency model. Some studies such as the one done by Karthick et al. aimed to save energy by targeting all the primary and secondary load appliances in a building. Whereas some studies only focused on optimizing one resource such as HVAC or lighting Aftab et al. [14]. Hence when considering which technique had the most optimal results we'll first establish which method could tackle the highest number of resources in a building. To answer this

Table 1 Comparison of energy savings and cost savings for different energy management techniques studied in literature

Method	Technology	Energysavings (%)	Costsavings (%)
Kermani et al. [11]	Microgrid	–	98
Aftab et al. [14]	ML	30	–
Foroozandeh et al. [6]	Optimization	45.52	35.56
Pinto et al. [15]	DRL	–	4
Syed et al. [16]	DL	–	–
Xu et al. [4]	Optimization	83.3	–
Brucke et al. [5]	Optimization	–	–
Hagras et al. [25]	FIS + NN	–	7.7
Rafsanjani et al. [17]	IoT	34	–
Ain et al. [19]	IoT	28	–
Kaushik et al. [24]	IoT	38	–

question, it can be seen that IoT was most frequently used to optimize a combination of multiple resources in buildings from lighting systems, HVAC, EVs, and others. In the field of IoT Rafsanjani et al. [17] saved up to 34% energy. Energy management of the hybrid wind PV system is discussed in [22].

Further, different parameters such as user demand and occupants' behaviour played a major role in most studies. Studies like the one done by Ain et al. [19] incorporated the demands of the users in their system to ensure that energy efficiency can be achieved even while maintaining the indoor temperature to suit the occupant's needs. Marinakis et al. [18] studied the behaviour of the occupants in order to provide personalized feedback to save energy while, Aftab et al. [14] studied the number of people in the room. This shows us the importance of understanding the occupants in the building. As the preferences and behaviour of occupants vary from building to building it is necessary to develop systems that can dynamically adjust to these changes. More work is required in this domain as in the existing studies the occupants were studied for short durations in specific buildings. Rafsanjani et al. [17] state that once the energy management system stops sending alerts to the occupants then they go back to behaving in the same energy-unaware way as before. This highlights the importance of building robust models to deal with a wide variety of demands and habits [23].

From the review, it can be seen that IoT is capable of combining smart sensors to observe environmental changes, occupants' behaviour, and numbers to maintain energy. It is also possible to combine micro grids, optimization models, and smart storage systems with IoT technology to take advantage of the different energy-efficiency systems under one roof. [24] This is the main advantage of IoT technology. Also, due to the developments in the sensors and microcontroller boards, it is now possible to develop large-scale building energy management systems that can be used in commercial, industrial as well as residential buildings [25, 26].

With smart buildings, there are a few challenges that need to be tackled in the future. One challenge discussed earlier was the need for robust models to predict the behaviour patterns of the occupants in all buildings. Renewable energy systems developed as the need of sustainable energy is increased [27]. Studies in the literature were limited to one or a cluster of buildings and thus the need for methods that can be deployed in any building is required. With these methods, their accuracy needs to be high. Another challenge to consider with IoT systems is the life of the sensors. Buildings are constructed to last for anywhere between 40 to 80 years but sensors have a limited life span. Hence, a method to tackle this problem needs to be introduced. Another challenge to consider would be the dependence of these systems on the atmosphere [28]. The temperature and humidity parameters of a geographical area do not have a predictable pattern. Hence sensors that can deal with a wide fluctuation of environmental changes while also maintaining accuracy need to be developed. Considering these challenges, the need for research that can tackle these is highly required. The solutions to these problems although difficult are well within reach by using the right sensors and training robust models. Universally compatible models may be difficult to achieve due to changes in the habits of the population and climate but these challenges can be tackled on a more local level such as single towns.

4 Conclusion

The manuscript provided a comprehensive review of the different technologies that have been used to manage energy usage in buildings such as residential and commercial. An overview of the need to save energy in buildings and its advantages was provided. This was followed by an overview of what smart buildings are and what are the different components that make them truly smart. Also elaborated the different types of technologies. The study covered technologies such as intelligent grids, artificial intelligence, smart storage systems, optimization models, and the Internet of Things (IoT). Also provided an insight into findings after performing this study and also introduced a few challenges that will need to be overcome to make the most efficient energy conservation system.

Study also give emphasis on machine learning algorithms, some of the most frequently used methods for energy optimization followed by IoT. In some cases, two technologies were combined into one system to provide better results. In the majority of studies, while energy saved was a constant metric, the amount of money saved on the electricity bill was also used to determine the efficiency. The use of these technologies demonstrated good conservation of energy in buildings. Most studies dealt with the energy consumption by HVAC and lighting systems while some dealt with all the resources in the building. The occupants, their behaviour, and their needs were also regularly used parameters to generate efficient systems. With the development of IoT sensors, there is capability of combining a variety of techniques under one roof. But as stated above, IoT and other techniques may bring certain challenges that need to be tackled. Occupants had a positive response to the feedback on energy-aware behaviour that was being given to them.

References

1. Huber M, Dimkova D, Hamacher T (2014) Integration of wind and solar power in Europe: Assessment of flexibility requirements. Energy 69:236–246
2. Kesavan P, Swaminathan M (2008) Strategies and models for agricultural sustainability in developingAsian countries. Philos Trans R Soc B: Biol Sci 363(1492):877–891
3. Li Z, Su S, Jin X, Chen H (2021) Distributed energy management for active distribution network considering aggregated office buildings. Renew Energy 180:1073–1087
4. Xu X, Hu W, Liu W, Du Y, Huang R, Huang Q, Chen Z (2021) Risk management strategy for a renewable power supply system in commercial buildings considering thermal comfort and stochastic electric vehicle behaviors. Energy Convers Manag 230:113831
5. Brucke K, Arens S, Telle JS, Steens T, Hanke B, von Maydell K, Agert C (2021) A non-intrusive load monitoring approach for very short-term power predictions in commercial buildings. Appl Energy 292:116860
6. Foroozandeh Z, Ramos S, Soares J, Vale Z (2021) Energy management in smart building by a multi-objective optimization model and pascoletti-serafinisecularization approach. Process 9(2):257
7. Oskouei MZ, Baboli PT, Babazadeh D (2021) Robust stochastic optimization for energy sharing between multi-carrier micro grids using transactive energy management system. In: ETG Congress 2021, pp 1–6. VDE
8. Rossit DG, Nesmachnow S, Toutouh J, Luna F (2021) Scheduling deferrable electric appliances in smart homes: a bi- objective stochastic optimization approach (2021)
9. Lu S, Wang C, Fan Y, Lin B (2021) Robustness of building energy optimization with uncertainties using deterministic and stochastic methods: Analysis of two forms. Build Environ 205:108185
10. Bahramara S (2021) Robust optimization of the flexibility-constrained energy management problem for a smart home with rooftop photovoltaic and an energy storage. J Energy Storage 36:102358
11. Kermani M, Adelmanesh B, Shirdare E, Sima CA, Carn'ı DL, Martirano L (2021) Intelligent energy management based on scada system in a real micro grid for smart building applications. Renew Energy 171, 1115–1127
12. Iffa E, Hun D, Salonvaara M, Shrestha S, Lapsa M (2022) Performance evaluation ofa dynamic wall integrated with active insulation and thermal energy storage systems. J Energy Storage 46:103815
13. Mayyas A, Chadly AA, Khaleel I, Maalouf M (2021) Techno-economic analysis of theli-ion batteries and reversible fuel cells as energy-storage systems used in green and energy-efficient buildings. Clean Energy 5(2):273–287
14. Aftab M, Chen C, Chau CK, Rahwan T (2017) Automatic HVAC control with real-time occupancy recognition and simulation-guided model predictive control in low-cost embedded system. Energy Build 154:141–156
15. Pinto G, Piscitelli MS, Va´zquez-Canteli JR, Nagy Z, Capozzoli A (2021) Coordinated energy management for a cluster of buildings through deep reinforcement learning. Energy 229, 120725
16. Syed D, Abu-Rub H, Ghrayeb A, Refaat SS (2021) Household-level energy forecasting insmart buildings using a novel hybrid deep learning model. IEEE Access 9:33498–33511
17. Rafsanjani HN, Ghahramani A, Nabizadeh AH (2020) Idea: IoT-based smartphone energy assistant for prompting energy-aware behaviors in commercial buildings. Appl Energy 266:114892

18. Marinakis V, Doukas H (2018) An advanced iot-based system for intelligent energy management in buildings. Sens 18(2):610
19. Ain QU, Iqbal S, Khan SA, Malik AW, Ahmad I, Javaid N (2018) IoToperating system based fuzzy inference system for home energy management system in smart buildings. Sens 18(9), 2802
20. Karthick T, Chandrasekaran K et al (2021) Design of IoT-based smart compact energy meter for monitoring and controlling the usage of energy and power quality issues with demand side management for a commercial building. Sustain Energy, Grids Netw 26:100454
21. Ali SS, Ruchi T (2019) The role of energy conservation building code 2017 in Indianenergy policy. Int J Recent Technol Eng (IJRTE) 9(1):1799–1806
22. Nagaraju K (2017) An overview of the energy sector in India. Int J Sci Res 6(3):1589–1594
23. Jangid R, Maherchandani JK, Kumar V, Swami RK (2022) Energy management of standalone hybrid Wind-PV aystem. In Intell Renew Energy Syst
24. Kaushik S, Srinivasan K, Sharmila B, Devasena D, Suresh M, Panchal H Ashokkumar R, Sadasivuni KK, Srimali N (2021) Continuous monitoring of power consumption in urban buildings based on internet of things. Int J Ambient Energy pp 1–7 (2021)
25. Hagras H, Packharn I, Vanderstockt Y, McNulty N, Vadher A, Doctor F (2008) An intelligent agent based approach for energy management in commercial buildings. In: 2008 IEEE international conference on fuzzy systems (IEEE world congress on computational intelligence), pp 156–162. IEEE
26. Maherchandani JK, Tirole R, Swami RK (2022) Micro grid energy management strategy for low dynamics load applications using simulink. In: Tripathi A, Soni A, Shrivastava A, Swarnkar A, Sahariya J (eds). Intell Comput Tech Smart Energy Systems Lect Notes Electr Eng, 862. Springer, Singapore
27. Swami RK, Samuel P, Gupta R (2016) Power Control in Grid-Connected Wind Energy System Using Diode-Clamped Multilevel Inverter. IETE J Res 62(4):515–524
28. Maherchandani JK, Joshi RR, Tirole R, Swami RK, Ganthia BP (2022) Performance comparison analysis of energy management strategies for hybrid electric vehicles. In: Kumar S, Singh B, Singh AK (eds). Recent Adv Power Electron Drives Lect Notes Electr Eng, 852. Springer, Singapore

Revolutionizing Microgrid Optimization: The Power of Meta-Heuristic Algorithms in Addressing Operational Challenges

G. Swetha Shekarappa

Abstract Microgrids, or MGs, use renewable energy sources to provide the increasing need for energy. On the other hand, operational issues come from intermittent renewables and low power quality. To solve these problems, heuristic optimization strategies have been created by researchers. However, addressing different operational parts of MG is challenging due to mistakes in non-linear and nonconvex optimization induced by local minima and the inability of heuristic approaches to attain a global minimum. The importance of meta-heuristic optimization algorithms (MHOAs) in enhancing MG operational performance is reviewed in this work. The principles of MG optimization are covered, together with current developments in load forecasting, MG techno-economic analysis, resilience enhancement, and energy management. Nearly 25% of research uses particle swarm optimization, while 10% and 5% use genetic and grey wolf algorithms. According to this outcome, MHOA offers a system-agnostic optimization strategy that opens up new possibilities for raising the efficacy of MGs in the future. Lastly, we draw attention to a few difficulties that arise when integrating MHOAs into MGs, which may inspire more study in this field.

Keywords Renewable energy · Optimization · Microgrid · Energy storage · Meta-heuristic optimization

1 Introduction

Large-scale energy consumption is rising, which means more electricity needs to be produced and transmitted. Conventional fossil fuel-based power plants are expensive to produce electricity on a big scale and release greenhouse gases into the atmosphere. However, renewable energy sources (RESs) including biogas, wind, water, and solar have garnered a lot of interest recently [1] because to their ability to produce power

G. Swetha Shekarappa (✉)
Alliance University, Banagalore, India
e-mail: sg.swethasg88@gmail.com

S. Kumar et al. (eds.), *Soft Computing Applications in Modern Power and Energy Systems*, Lecture Notes in Electrical Engineering 1298,
https://doi.org/10.1007/978-981-96-0058-8_11

that is safe, dependable, economical, and highly satisfying. To integrate renewable energy sources (RESs) into energy generation, the traditional architecture must be redesigned to accommodate various components such as electronic converters, loads, coupling points for power exchange with the utility grid, source-load controllers, and energy storage systems (ESSs) [2].

The improved framework that makes it easier for these components to link is known as a "microgrid." Microgrids can function in two ways: either linked to the main grid or on their own (known as "islanded mode"). The microgrid may produce, distribute, and oversee its own power supply in both modes. Microgrids provide an incredibly dependable and durable electricity source, particularly in places where grid power is expensive or unstable. In addition, microgrids provide cities, organizations, and homes with benefits including cost savings, flexibility, and energy security, which makes them a desirable choice for reaching sustainability objectives [3]. In [4], a hybrid moth-flame optimization (MFO) and particle swarm optimization (PSO) method based on several scenarios and mathematical validation on distinct benchmark functions is presented for the purpose of predicting transmission line parameters. This algorithm logically incorporates the ideas of MFO and PSO to get beyond their drawbacks and enhance their capacity for global search. The idea of transmission line parameter optimization is the focus in [5]. For this reason, several transmission line test situations are subjected to an enhanced moth flame optimization technique. Improved moth flame optimization starts with the movement of the moth in relation to the light source. Grey wolf optimization (GWO), a novel optimization technique, is used by authors in [6] for study effort to compute the overhead transmission line parameter. Analysis is also done on the impact of various bundle conductor configurations, such as two, three, and four bundle conductors, as well as the radius and spacing between the conductors on the transmission line. In order to compute capacitance and inductance per unit length for single-phase and three-phase systems with varying numbers of bundle conductors, the proposed work provided modified whale optimization techniques in [7]. Both single-phase and three-phase AC transmission lines is evident that alterations to one parameter in an AC transmission line will result in matching variations in other parameters. Using suggested whale optimization algorithm (WOA) methods for various bundle conductor layouts, the capacitance and inductance per unit length for three-phase systems have been computed in [8]. The objective in [9] is to determine the transmission-line parameters and examine the impact of various bundle conductor configurations, including two, three, and four bundle conductors. In both single- and three-phase transmission lines, it has been found that increasing the conductor area in square millimeters results in a decrease in inductance and an increase in capacitance. The average percentage differences between capacitance and inductance are 55.49 and 99.55%. The authors in [10] portray the outcome of the attained to without Distribution Generation, with Distribution Generation using the optimization techniques. In [10] the authors present a variety of methods for locating faults in subterranean cable systems, including sectionalizing, thumper, and time domain reflectometry, as well as MATLAB simulation technique-based reuse and factors that influence the fault finding. The work reflects the literature review of the various optimization techniques related to micro-grids. Further the work projects with the various sections related to the various trends in microgrid using optimization techniques, further the scope of the microgrid, various opportunities related to metaheuristic techniques and with the challenges of Microgrid in current scenario using optimization techniques.

2 Recent Trends in Microgrid Using Optimization Techniques

Microgrid (MG) systems have a number of optimization challenges, which have been addressed by a number of solutions that ensure redundancy, scalability, adaptability, and efficient component placement and arrangement [11]. Some of the features of MG functioning that these strategies strive to improve are energy management, cost effectiveness, fault detection, energy storage systems (ESSs), load forecasting, and managing MG components to maintain voltage, current, and power levels [12].

- Because of their accuracy and efficiency, optimization techniques like linear, quadratic, and mixed-integer linear programming (MILP) are frequently used in microgrid (MG) optimization. MG optimization efforts have made heavy use of these strategies. However, there are obstacles to overcome in order to optimise MG operation, including weather changes, non-linear and non-convex optimization difficulties, and market pricing volatility. These problems may be difficult for traditional optimization methods like gradient descent, quasi-Newton, and Powell's approach, which frequently become stuck in local optima and are unable to find the global optimum [13].
- One approach that shows promise for improving microgrid (MG) performance is Model Predictive Control (MPC). With consideration for system dynamics and restrictions such as energy storage and renewable energy sources, it solves optimization problems at each time step. According to research, MPC can lower greenhouse gas emissions, improve system reliability, and save operating expenses. For example, an MPC-based energy management plan with energy storage, electric vehicles, and RESs greatly lowered running costs without sacrificing dependability for an MG [14].
- For microgrid (MG) systems, meta-heuristic optimization techniques (MHOAs) provide adaptable answers to challenging optimization problems. MHOAs such as Particle Swarm Optimization (PSO) and Ant Colony Optimization (ACO), which draw inspiration from natural occurrences, are quick and easy ways to handle a variety of MG optimization problems, such as anomaly identification and control [15].

3 Framework, Requisites, and Scopes for Microgrid Optimization Technique

Microgrids, also known as mini power plants, are small power sources that can work autonomously or in tandem with the main power grid to guarantee a steady supply of electricity even in the event of unanticipated circumstances. Given the current dynamics of power generation, which are being pushed by the growing need for reliable, inexpensive, and environmentally friendly energy, microgrids (MGs) present a viable solution that prioritises environmental sustainability while being both innovative and cost-effective [16]. Through the utilisation of several benefits for power production, MGs can successfully address the issue of CO_2 emissions. Their small-scale generation capability enables for installation closer to consumers or in locations without access to the main grid, making them a viable option to large investments in building electrical distribution and transmission infrastructure to reach rural areas. Several essential MG components, including the MG central controller, source-load controller, energy storage systems (ESSs), and renewable energy sources (RESs), guarantee the effective operation of the grid system. The technical framework of a microgrid system as shown in Fig. 1 [17].

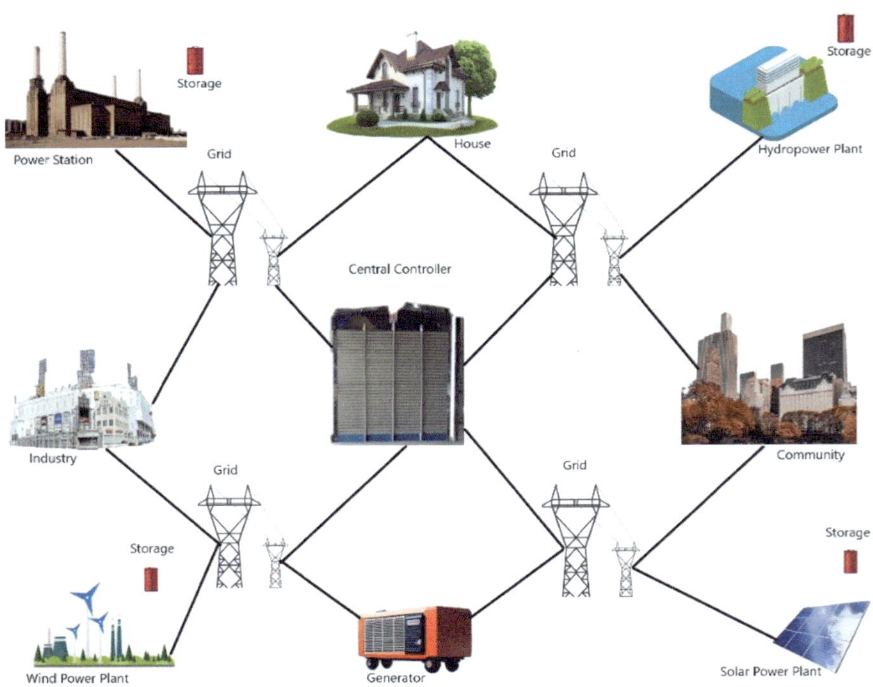

Fig. 1 Technical specifications for a microgrid system

Energy security can only be achieved by microgrid optimization, which ensures independent microgrid operation, effective resource allocation, and backup power availability. Microgrids can improve overall reliability by distributing energy resources during outages and efficiently managing peak energy demand by fine-tuning their operations [18]. Moreover, optimization guarantees long-term affordability and sustainability by reducing operational expenses through the identification of inefficiencies and the streamlining of energy usage patterns.

Furthermore, optimization is essential for tailoring microgrid solutions to particular demands and specifications, which increases their efficacy and efficiency [19]. Microgrids can minimize waste and lower overall costs while adapting to changing demands and environmental conditions through customized energy solutions. By optimizing the utilization of renewable energy sources, this customized strategy not only promotes environmental sustainability but also maximizes energy efficiency. All things considered, microgrid optimization is essential to guaranteeing dependable, economical, and ecologically friendly energy solutions [20].

4 Meta-Heuristic Techniques and Opportunities

Strong search methods called meta-heuristic algorithms are made to locate the best answers to challenging optimization issues [21]. Even with imperfect information, these algorithms are excellent at effectively navigating search areas in the real world, when resources, computing power, and time are scarce. Using a range of techniques, from simple local searches to complex learning algorithms, meta-heuristics function in an approximate and non-deterministic manner. As a result, their versatility and wide applicability in settings involving problem-solving make them valuable [22].

- In complicated and dynamic optimization issues, MHOAs effectively search solution spaces and steer clear of local optima.
- They can be applied to a variety of problems without the need for domain-specific knowledge because they don't require a mathematical model of the problem.
- The accuracy and efficiency of computation are improved through hybridization with other optimization techniques.
- Scalable solutions to large-scale optimization problems are made possible by parallelizability.
- MHOAs are robust and adaptive due to their high efficacy and efficiency in producing accurate solutions.
- Their investigation of solution spaces is improved by the diverse search strategies they use, which are inspired by natural systems.
- Because of their adaptability, MHOAs can alter their search tactics in reaction to shifting limitations or problem settings.
- They are effective at solving complicated issues with several restrictions, which makes them useful tools for solving optimization problems in the real world.

5 Trends and Challenges in Microgrids

An algorithm that uses meta-heuristics is a kind of system-independent optimization. It employs a trial-and-error technique to determine which system operations have the best fitness value. But in order to use this strategy, the algorithm must provide a population-based iteration policy. This can cause the following problems, which could jeopardize efficiency [23].

- The transitory time for optimization may grow if high-level optimization is carried out through a high number of iterations. This could be harmful to the system, particularly for those that need to be optimized in a certain amount of time. Serious alerts in the system and surrounding area may be triggered if the threshold time is surpassed. Fast processing components can be incorporated into the system to solve this problem. This will enable the system to process large amounts of data quickly, cutting down on the interim time needed for optimization [24].
- The MHOA is a global algorithm that can anticipate system behaviour and make adjustments as necessary. If precise behaviour forecasting is not accomplished, the algorithm could run into problems that result in total system failure. Optimizing output and ensuring accurate behaviour prediction can be achieved by varying the number of iterations and population proclamation to prevent system failures [25].
- Data inputs for MHOA include load, weather forecasts, and apparatus accessibility, all of which might not be easily obtainable in real time or might come at a high cost. For this reason, it is necessary to use efficient data management strategies to guarantee database persistence.
- Because of the shifting dynamics of production and popularity, MGs operate in a dynamic manner. To address dynamic optimization issues, the MHOA would need to be modified. For example, online optimization strategies or model predictive control could be used [26].

6 Conclusion

Microgrids (MGs) cannot function well without the incorporation of renewable energy sources (RESs) and the resolution of a number of issues, including cyberattacks, energy imbalances, and financial losses. While heuristic optimization techniques have long been utilized to tackle such challenges, they often struggle with pinpointing the global minimum and scaling up to tackle large-scale problems. However, meta-heuristic optimization algorithms (MHOAs) offer a promising alternative due to their capability to locate global minima.

This study delves into the realm of MHOAs, exploring their diverse applications in the microgrid (MG) domain. Through an exhaustive examination, we unveil the myriad ways MHOAs can revolutionize techno-economic modelling of MGs, facilitate anomaly investigation, streamline control and management processes, and enhance accuracy of the load forecasting. Researchers in this field find that MHOAs are effective tools because they can handle multiple objectives optimization problems with convex and nonlinear constraints. They offer a viable substitute for traditional optimization methods due to their adaptability and efficiency.

Particle swarm optimization (PSO) is a popular technique in load forecasting because of its simplicity, resilience, and fast convergence to optimal solutions. This is evident from our analysis of the implementation of optimization techniques in different MG applications. In MG optimization, hybrid techniques in particular are becoming more popular than single MHOAs and are outpacing them in terms of utilization intensity. Even though some MHOAs have drawbacks such early convergence and scalability issues, the general trend shows that using MHOAs to solve MG optimization problems is becoming more and more popular.

References

1. Mariam L, Basu M, Conlon MF (2016) Microgrid: architecture, policy and future trends. Renew Sustain Energy Rev 64:477–489
2. Aguado JA, Quintana VH (2001) Inter-utilities power-exchange coordination: a market-oriented approach. IEEE Trans Power Syst 16(3):513–519
3. Badal FR, Das P, Sarker SK, Das SK (2019) A survey on control issues in renewable energy integration and microgrid. Prot Control Mod Power Syst 4(1):1–27
4. Shaikh MS, Raj S, Babu R, Kumar S, Sagrolikar K (2023) A hybrid moth–flame algorithm with particle swarm optimization with application in power transmission and distribution. Decis Anal J 6:100182
5. Shaikh MS, Raj S, Ikram M, Khan W (2022) Parameters estimation of AC transmission line by an improved moth flame optimization method. J Electr Syst Inf Technol 9(1):25
6. Shaikh MS, Hua C, Jatoi MA, Ansari MM, Qader AA (2021) Application of grey wolf optimisation algorithm in parameter calculation of overhead transmission line system. IET Sci Meas Technol 15(2):218–231
7. Shaikh MS, Hua C, Raj S, Kumar S, Hassan M, Ansari MM, Jatoi MA (2022) Optimal parameter estimation of 1-phase and 3-phase transmission line for various bundle conductor's using modified whale optimization algorithm. Int J Electr Power Energy Syst 138:107893
8. Shaikh MS, Hua C, Hassan M, Raj S, Jatoi MA, Ansari MM (2022) Optimal parameter estimation of overhead transmission line considering different bundle conductors with the uncertainty of load modelling. Optim Control Appl Methods 43(3):652–666
9. Shaikh MS, Hua C, Jatoi MA, Ansari MM, Qader AA (2021) Parameter estimation of AC transmission line considering different bundle conductors using flux linkage technique. IEEE Can J Electr Comput Eng 44(3):313–320
10. Ansari MM, Guo C, Shaikh MS, Chopra N, Haq I, Shen L (2020) Planning for distribution system with grey wolf optimization method. J Electr Eng Technol 15:1485–1499
11. Shuai Z, Sun Y, Shen ZJ, Tian W, Tu C, Li Y, Yin X (2016) Microgrid stability: classification and a review. Renew Sustain Energy Rev 58:167–179
12. Tushar MHK, Assi C, Maier M, Uddin MF (2014) Smart microgrids: optimal joint scheduling for electric vehicles and home appliances. IEEE Trans Smart Grid 5(1):239–250
13. Kumar RS, Raghav LP, Raju DK, Singh AR (2021) Intelligent demand side management for optimal energy scheduling of grid connected microgrids. Appl Energy 285:116435
14. Saad A, Faddel S, Youssef T, Mohammed OA (2020) On the implementation of IoT-based digital twin for networked microgrids resiliency against cyber attacks. IEEE Trans Smart Grid 11(6):5138–5150

15. Wu H, Li H, Gu X (2020) Optimal energy management for microgrids considering uncertainties in renewable energy generation and load demand. Processes 8(9):1086
16. Sharma S, Bhattacharjee S, Bhattacharya A (2018) Probabilistic operation cost minimization of micro-grid. Energy 148:1116–1139
17. Zulu MLT, Carpanen RP, Tiako R (2023) A comprehensive review: study of artificial intelligence optimization technique applications in a hybrid microgrid at times of fault outbreaks. Energies 16(4):1786
18. Sedhom BE, El-Saadawi MM, Hatata AY, Elhosseini MA, Abd-Raboh EE (2020) Robust control technique in an autonomous microgrid: a multi-stage H∞ controller based on harmony search algorithm. Iran J Sci Technol Trans Electr Eng 44(1):377–402
19. Naz A, Javaid N, Rasheed MB, Haseeb A, Alhussein M, Aurangzeb K (2019) Game theoretical energy management with storage capacity optimization and photovoltaic cell generated power forecasting in micro grid. Sustainability 11(10):2763
20. Yaprakdal F, Baysal M, Anvari-Moghaddam A (2019) Optimal operational scheduling of reconfigurable microgrids in presence of renewable energy sources. Energies 12(10):1858
21. Semero YK, Zhang J, Zheng D (2020) EMD–PSO–ANFIS-based hybrid approach for short-term load forecasting in microgrids. IET Gener Transm Distrib 14(3):470–475
22. Islam QNU, Ahmed A (2020) Optimized controller design for islanded microgrid employing nondominated sorting firefly algorithm. In: Nature-inspired computation and swarm intelligence, Elsevier, pp 247–272
23. Hao Y, Dong L, Liang J, Liao X, Wang L, Shi L (2020) Power forecasting-based coordination dispatch of PV power generation and electric vehicles charging in microgrid. Renew Energy 155:1191–1210
24. Aslam S, Khalid A, Javaid N (2020) Towards efficient energy management in smart grids considering microgrids with day-ahead energy forecasting. Electr Power Syst Res 182:106232
25. Muleta N, Badar AQ (2023) Designing of an optimal standalone hybrid renewable energy micro-grid model through different algorithms. J Eng Res 100011
26. Montoya OD, Gil-Gonzalez W, Rivas-Trujillo E (2020) Optimal location-reallocation of battery energy storage systems in DC microgrids. Energies 13(9):2289

Application of HVDC Link in a RES Integrated Multi-area Power System Using MOA Optimized I-TD Controller

Naladi Ram Babu, Surampudi Satyanarayana, Areti Bala Sai, Mede V. V. S. Subrahmanyeswara Reddy, Tadi Venkata Satyanarayana Reddy, Pooja, Manjeet Kumar, and Tirumalasetty Chiranjeevi

Abstract This paper investigates the solicitation of an integral minus tilt derivative controller (I-TD) in an automatic generation control (AGC) system that integrates renewable energy sources (RES) into three areas. A Mayfly optimization algorithm (MOA) approaches were used to enhance the system and controller parameters. The effectiveness of the suggested I-TD controller is shown by the way it outperforms conventional TID and PID controllers in terms of overshoots (OS) by 53%, undershoots (US) by 50%, and settling time (ST) by 58%. The advantages of incorporating HVDC and RES lines into the dynamics of the system are also examined. It is discovered that dynamics in terms of OS, US, and ST is improved by combining RES by 23% and HVDC links by 45%.

Keywords AGC · MOA · HVDC · RES · Controller

1 Introduction

The modern electrical grid is intrinsically complex, mainly because of its ongoing growth. This intricacy makes it difficult to keep the frequency of the system within allowable limits and to guarantee an appropriate ratio of power supply to demand. These problems are addressed by AGC. The objectives of AGC are to supply a stable control mechanism, preserve a regular frequency profile throughout all control zones, and help balance the supply and demand of electricity [1, 2].

N. R. Babu · S. Satyanarayana · A. B. Sai · M. V. V. S. Subrahmanyeswara Reddy · T. V. S. Reddy
Department of Electrical and Electronics Engineering, Aditya University, Surampalem, India

Pooja · M. Kumar
Department of Electrical Engineering, Sandip University, Sijoul, Madhubani, Bihar, India

T. Chiranjeevi (✉)
School of Electrical Engineering, Shri Mata Vaishno Devi University, Katra, India
e-mail: chirupci479@gmail.com

S. Kumar et al. (eds.), *Soft Computing Applications in Modern Power and Energy Systems*, Lecture Notes in Electrical Engineering 1298,
https://doi.org/10.1007/978-981-96-0058-8_12

Research on AGC has been mostly focused on single-area thermal systems. Nonetheless, there has been a discernible move in favor of multi-area-source systems [3–5]. A number of particle nonlinearities, including the GDB and the GRC, were also the subject of earlier research. With time, this body of knowledge has expanded to include a wider range of complex factors to take into account while interpreting and applying AGC.

A number of research have looked into how to include renewable energy sources (RES) into AGC frameworks, like wind and DSTS [6–11]. Interestingly, using constant solar irradiation and two areas, researchers in [12, 13] looked into an AGC scenario that included DSTS. It is noteworthy, therefore, that the irregularity of solar radiation contributes to the erratic output of DSTS.

One of the main objectives of AGC research has been the advancement of controllers. In addition to simple conventional controllers [14–16], a variety of tilt (T) type controllers have appeared, such as TID [17, 18], TIDN [19], and I-TD [20]. Even with these developments, comparative analysis is still a crucial tool for assessing these controllers' performance in multi-area AGC systems. This emphasizes how crucial it is to conduct more study in order to completely understand and maximize the efficiency of controllers in such complicated systems.

More recently, studies on AGC have demonstrated a clear preference for using a variety of metaheuristic techniques to optimize controller parameters. In this context, methods like Anti-Lion Search [5], Crow Search [7], Bird Swarm [10], firefly [14], Coyote [17], Cuckoo Search [20], and Particle Swarm Optimization [21], Archimedes optimization [22], have become more popular. The Mayfly Optimization Algorithm (MOA), a more recent addition to this collection of optimization methods, combines swarm intelligence elements and evolutionary algorithms, taking inspiration from mayflies mating [23]. MOA has not yet been investigated for AGC optimization, despite its successful application in other domains; hence, a comprehensive analysis is required to assess its efficacy.

The performance of the system is distorted under abnormal circumstances. Furthermore, in a lengthy transmission line, the presence of transients would aggravate the dynamic. A HVDC link can effectively reduce this [6, 7]. Additionally, the HVDC quashes flowing outages and maintains control over power flow amongst connected areas [20]. Numerous academic works have integrated HVDC links to investigate the dynamic performance and assess its efficacy [11, 17, 24]. The previous AGC studies were limited to two-area systems, but there is significant potential for expanding these studies to include systems with more than two areas. Summarizing the above, the objectives are.

(a) To foster a three area AGC system with thermal-DSTS units in all areas.
(b) Application of I-TD controller and compare its dynamics with PID and TID.
(c) Employing the MOA for optimizing the controller and additional system parameters with the aim of attaining optimal performance in the AGC system.
(d) Compare the dynamic performance while considering DSTS unit in all areas.
(e) Compare the dynamic performance while considering HVDC unit in all areas.

2 System Investigated

Figure 1 depicts the transfer function model of a thermal system with three unequal areas, constrained by a 3% per minute Generation Rate Constraint (GRC). The nominal parameters for the steam system are detailed in the appendix, referenced from sources [6, 7]. Various controllers, such as TID, PID, and I-TD controllers, are considered as independent secondary controllers, and their impact on the system's dynamic responses is extensively analyzed. Furthermore, the MOA, a recently introduced metaheuristic algorithm, is introduced for optimizing controllers and other pertinent parameters. Equation (1) defines the Integral Squared Error (ISE), which is the performance index for optimization. Additionally, when evaluating dynamics of the system related to frequency deviation and tie-line power variation, a 1% Step Load Perturbation (SLP) in area-1 is taken into account.

$$\eta_{ISE} = \int_0^t \left\{ (\Delta F_m)^2 + (\Delta P_{m-n})^2 \right\} .dt \tag{1}$$

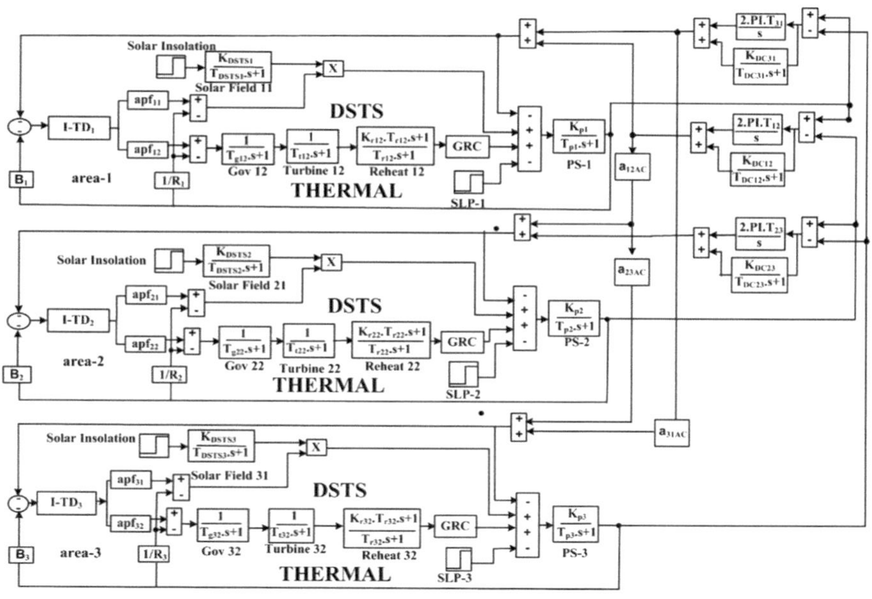

Fig. 1 Transfer function model of proposed system

3 Proposed I-TD Controller

Tilt Controller with the benefits of both integer and fractional-order controllers are combined in this controller [20]. A key difference is that the Tilt Controller preserves a structure that is very similar to the PID controller by multiplying the proportional term of the PID [22] controller by 1/s^ni. Comparably, the Tilt-Integral-Derivative (TID) controller's structure is almost exactly the same as the PID controller's, except that it adds a derivative control term to the forward path, which causes derivative kick, a phenomenon that may not be ideal for use in electronic circuits. The integral term in the TID controller is positioned in the feedback path, whilst the proportional and derivative terms are located in the forward path [14]. Equations (2) give the transfer functions for the I-TD controller, and Figs. 2, 3 and 4 show the various controller configurations used in this study.

$$F(s) = \frac{K_I}{S} - \frac{K_T}{1/s^n} + K_D S \tag{2}$$

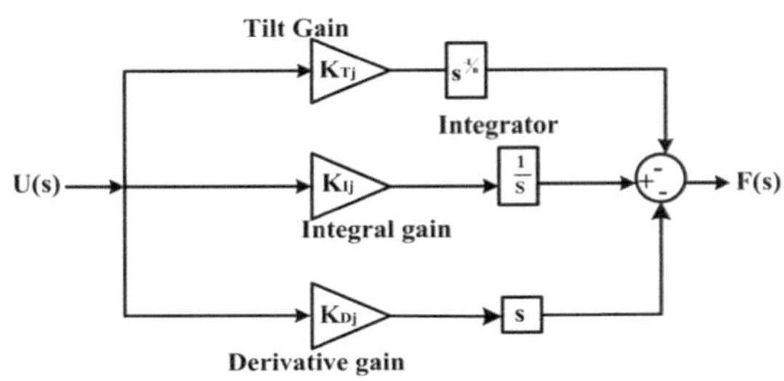

Fig. 2 I-TD controller

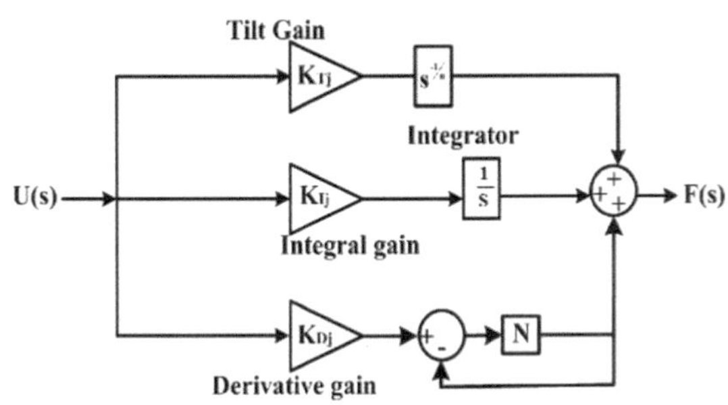

Fig. 3 TID controller [24]

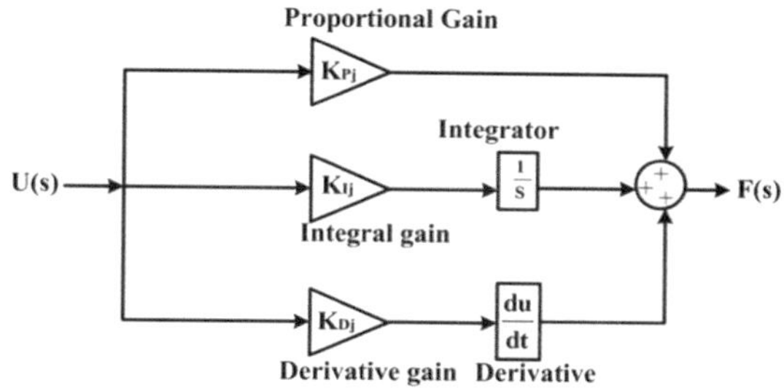

Fig. 4 PID controller [14]

4 Mayfly Optimization Algorithm

The Mayfly Optimization Algorithm (MOA), introduced by Konstantinos et al. [23], draws inspiration from the mating behavior of mayflies, known for their one-day lifespan, to efficiently tackle complex optimization problems.

MOA generates a population of candidate solutions, each representing a potential fix for the problem at hand, mirroring the iterative enhancements observed in mayfly mating behavior. The algorithm organizes the population into males and females, with males randomly positioned in the search space and females attracted to their emitted pheromones. The females' fitness levels are determined by evaluating their fitness based on improvement in their solutions. After choosing a partner, mating, and crossing over, the offspring solutions experience mutation to acquire characteristics from both parents, encouraging further exploration of the search space. Higher fitness values in offspring are preferred by selection criteria for moving on to later iterations. Until a termination condition is satisfied, usually after a set amount of iterations, iterations continue. The study's optimized parameters for MOA comprise a cap of 100 iterations, a population size of 30, and an inertia weight of 0.8. The MOA flowchart, which shows its iterative approach for the optimization, is shown in Fig. 5.

5 Simulation Results and Discussion

5.1 Comparison of Dynamics for I-TD, TID, and PID Controllers

Figure 1 illustrates a system comprising supplementary controllers, namely I-TD, TID, and PID, each evaluated independently. Through the utilization of MOA, adjustments are made to the controller gains. The controller's dynamic response are compared and visually presented in Fig. 6.

Table 1 extracts key indices parameters such as overshoots (OS), settling time (ST) and undershoots (US) and from the data presented in Fig. 6. From Fig. 6 and

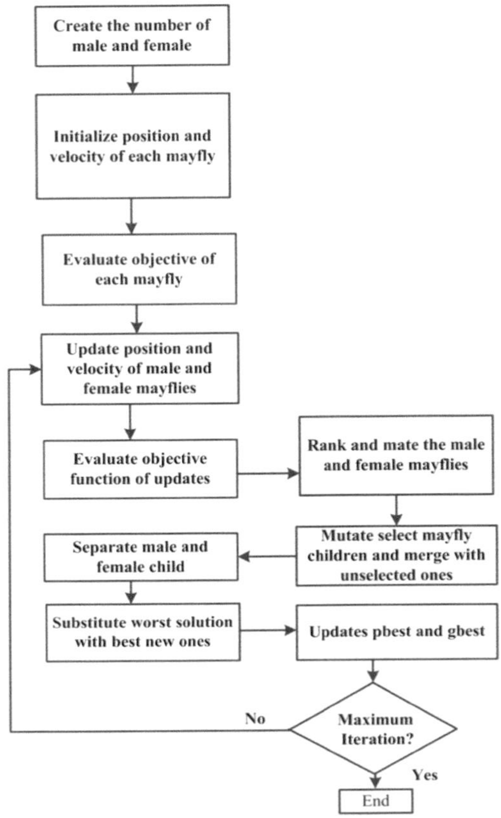

Fig. 5 Working flow chart of MOA

indices in Table 1, the suggested system with I-TD delivers improved dynamics over TID, and PID controller in term of ST, PUs and POs.

5.2 *Effect of DSTS Integration on System Dynamics*

Here, the DSTS unit is integrated in all the three area. The dynamic profiles are analyzed with the I-TD controller, and its optimal parameters are determined through MOA. The comparison of the system dynamics with and without DSTS units depicted in Fig. 7. The dynamic responses obtained indicate a noticeable enhancement in system dynamics with DSTS unit.

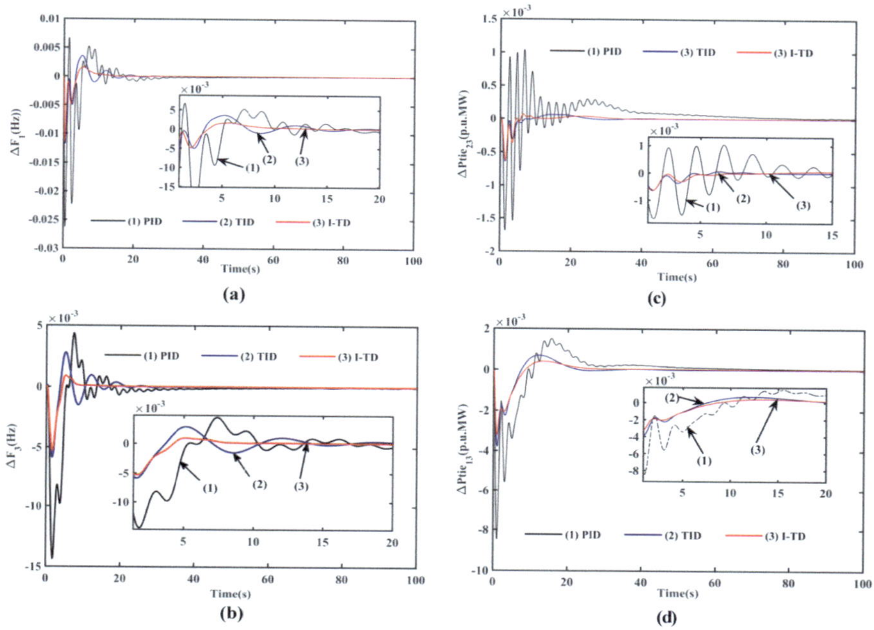

Fig. 6 System dynamic response comparison for I-TD, TID, PID, controllers **a** ΔF_1, **b** ΔF_3, **c** ΔP_{tie13}, **d** ΔP_{tie23}

Table 1 Performance indices value comparison of system dynamic responses in Fig. 7

Responses	Indices	I-TD	TID [24]	PID [24]
ΔF_1	OS	0.0036	0.0051	0.0059
	US	0.0140	0.0160	0.0240
	ST	26.540	33.040	48.130
ΔF_3	OS	0.0014	0.0038	0.0042
	US	0.0048	0.0052	0.0158
	ST	36.042	40.340	52.300
$\Delta Ptie_{13}$	OS	0.0001	0.0012	0.0015
	US	0.0036	0.0039	0.0075
	ST	42.050	44.550	52.340
$\Delta Ptie_{23}$	OS	0.0002	0.0030	0.0040
	US	0.0005	0.0006	0.0130
	ST	41.760	47.026	59.055

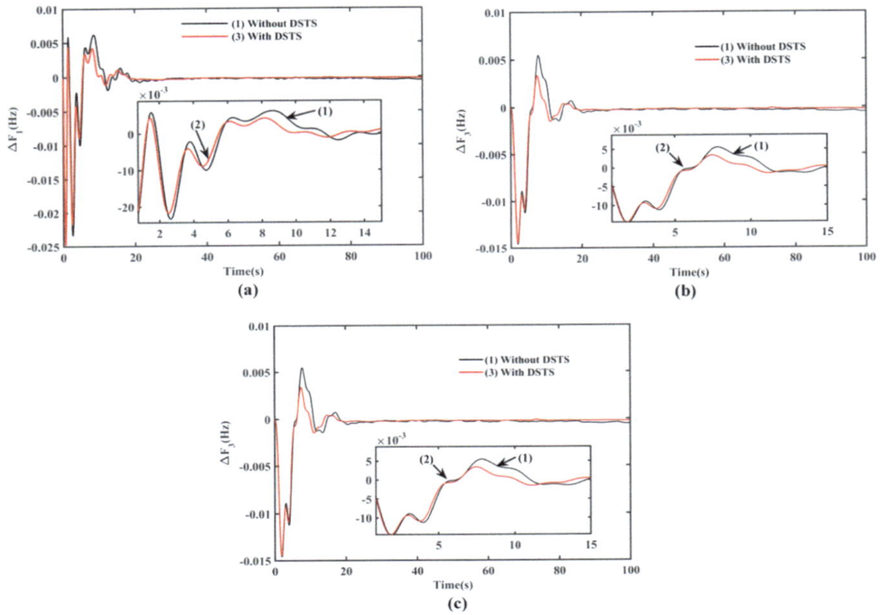

Fig. 7 System dynamic response comparison with and without DSTS unit **a** ΔF_1, **b** ΔF_3, **c** ΔP_{tie12}

5.3 Effect of HVDC Tie-Line on System Dynamics

The system under consideration in scenario-B is connected to an AC tie-line in parallel via an HVDC link. Using MOA, the suggested controller parameters are optimized. Figure 8 compares and illustrates the dynamic profiles corresponding to the scenarios with and without an HVDC link. It is clear from the acquired dynamic response that the system's dynamic greatly improves and it takes much less time to mitigate the steady state error when it is integrated with an HVDC link. Moreover, the POs and POs are relatively low in the absence of the HVDC tie-line.

6 Conclusions

In this study several secondary controllers used in the study of the three-area Automatic Generation Control (AGC) system, including TID, PID, and PI. Through the use of MOA approaches, the parameters of these controllers were adjusted, leading to successful deployment. A comparative study of the dynamics revealed that the I-TD controller outperformed the PID and TID controllers by (50–60) %. Additionally, the investigation of Distributed Static DSTS integration revealed enhancements in the dynamics. The system's overall performance were further improved by the inclusion

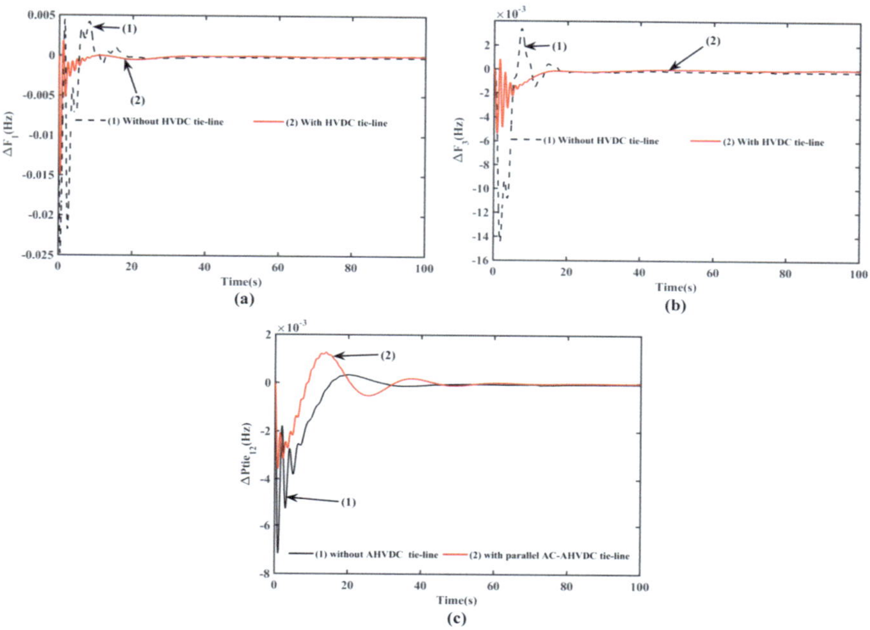

Fig. 8 System dynamic response comparison **a** ΔF_1, **b** ΔF_3 and **c** ΔP_{tie12}

of parallel AC-HVDC, which resulted in significant reductions in OS, US, and ST by 45%.

Appendix

1. Nominal system Parameters: $F = 60$ Hz, $T_{jk,AC} = 0.086$pu MW/rad, $H_j = 5$ s, $K_{pj} = 120$ Hz/MW pu, $D_j = 8.33 \times 10^{-3}$ pu MW/Hz, $B_j = 0.425$ pu MW/Hz, $R_j = 2.4$pu MW/Hz, $T_{gj} = 0.08$ s, $T_{tj} = 0.3$ s, $T_{rj} = 10$ s, $K_{rj} = 0.5$ s, $T_{pj} = 20$ s, SLP = 1%, loading = 50%, $P_{r1} = 1000$ MW, $P_{r2} = 2000$ MW, $P_{r3} = 4000$ MW, $K_{DC} = 0.5$, $T_{DC} = 0.03$ s.
2. DSTS system: $K_{DSTS} = 1$,$T_{DSTS} = 5$.

References

1. Ram Babu N, Bhagat SK, Saikia LC et al (2023) A comprehensive review of recent strategies on automatic generation control/load frequency control in power systems. Arch Computat Methods Eng 30:543–572. https://doi.org/10.1007/s11831-022-09810-y

2. Bhagat SK, Saikia LC, Babu NR (2023) Application of an optimal tilt controller in a partial loading schedule of multi-area power system considering HVDC link and virtual inertia. ISA Transactions
3. Gouran-Orimi S, Ghasemi-Marzbali A (2023) Load frequency control of multi-area multi-source system with nonlinear structures using modified grasshopper optimization algorithm. Appl Soft Comput 137:110135
4. Panda G, Panda S, Ardil C (2009) Automatic generation control of interconnected power system with generation rate constraints by hybrid neuro fuzzy approach. Int J Elect Power Energy Syst Eng 2(1):13–18
5. Shukla H, Raju M (2023) Combined frequency and voltage regulation in multi-area system using an equilibrium optimiser based non-integer controller with penetration of electric vehicles. Int J Ambient Energy 44(1):1522–1548
6. Babu NR, Chiranjeevi T, Saha A, Bhagat SK (2023) Comparative analysis of various energy storage systems in a conventional LFC system considering RDSTS, PWTS and AHVDC models. J Eng Res 11(4):425–436. https://doi.org/10.1016/j.jer.2023.100123
7. Babu NR, Bhagat SK et al (2023) Frequency control of a realistic dish stirling solar thermal system and accurate HVDC models using a cascaded FOPI-IDDN-based crow search algorithm. Int J Energy Res 9976375. https://doi.org/10.1155/2023/9976375
8. Debnath MK, Jena T, Mallick RK (2017) Optimal design of PD-Fuzzy-PID cascaded controller for automatic generation control. Cogent Eng 4(1):1416535
9. Daraz A, Malik SA, Basit A, Aslam S, Zhang G (2023) Modified FOPID controller for frequency regulation of a hybrid interconnected system of conventional and renewable energy sources. Fractal Fract 7(89)
10. Bhagat SK, Saikia LC, Babu NR (2023) Inertia emulation control strategy of an accurate model of HVDC link in a multi-area automatic generation control system integrated precise wind turbine system. Iran J Sci Technol Trans Electr Eng
11. Bhagat SK, Saikia LC (2023) Application of inertia emulation control strategy with energy storage system in multi-area hydro -thermal system using a novel metaheuristic optimized tilt controller. Electr Power Syst Res 222:109522
12. Tan R et al (2017) Modeling and mitigating impact of false data injection attacks on automatic generation control. IEEE Trans Inf Forensics Secur 12(7):1609–1624
13. Rao PK, Raju M, Gupta S, Dharawat K (2023) Frequency regulation in microgrid considering virtual inertia with firefly algorithm optimized controller. In: Murari K, Prasad Padhy N, Kamalasadan S (eds) Soft computing applications in modern power and energy systems. lecture notes in electrical engineering, vol 975. Springer, Singapore
14. Jagatheesan K, Anand B, Samanta S, Dey N, Ashour AS, Balas VE Design of a proportional-integral-derivative controller for an automatic generation control of multi-area power thermal systems using firefly algorithm. IEEE/CAA J Autom Sinica
15. Shiva V, Vedik CK et al (2023) Comparative analysis of PID and fractional order PID controllers in automatic generation control process with coordinated control of TCSC. Energy Syst 14:133–170
16. Biswas S, Kumar Roy P, Chatterjee K (2023) FACTS-based 3DOF-PID controller for LFC of renewable power system under deregulation using GOA. IETE J Res 69(3):1486–1499
17. Babu NR et al (2019) Maiden application of coyote optimizer algorithm with TIDN controller in AGC of a multi-area multi-source system. In: 2019 IEEE 16th India Council International Conference (INDICON), Rajkot, India, pp. 1–4
18. Manohar A, Yadav J (2023) A systematic study of controller and optimization techniques in automatic generation control system. In: 2023 International conference on computational intelligence, communication technology and networking (CICTN), Ghaziabad, India, pp 339–343
19. Khare P, Raju M, Gupta SC, Shukla H (2023) Liberalized automatic generation control of interconnected thermal-hydro-gas system using firefly algorithm optimized PID controller. Int J Numer Model e3136

20. Sreeramulu Mahesh G, Chandrashekar MJ, Muni Sankar A, Mohan SB, Chandra Sekhar S, and Ranjit PS (2023) An optimized filter design with cuckoo search algorithm for industrial microgrids. In: 2023 International conference on signal processing, computation, electronics, power and telecommunication (IConSCEPT), Karaikal,India, pp 1–5
21. Mahto RK, Mishra A, Bansal RC (2024) Controller design using PSO and WOA algorithm for enhanced performance of vector-controlled PMSM drive. Int J Model Simul 1–11
22. Ahmed M, Magdy G, Khamies M, Kamel S (2022) Modified TID controller for load frequency control of a two-area interconnected diverse-unit power system. Int J Electr Power Energy Syst 135(107528)
23. Zervoudakis K, Tsafarakis S (2020) A mayfly optimization algorithm. Comput Indus Eng 145:106559
24. Bhagat SK, Babu NR, Saikia LC et al (2023) A review on various secondary controllers and optimization techniques in automatic generation control. Arch Computat Methods Eng 30:3081–3111. https://doi.org/10.1007/s11831-023-09895-z

Multistep-Ahead Prediction of PM2.5 Using Nonlinear Autoregressive Machine Learning Model

Ashwani Kharola, Kapil Kumar Sharma, Amir Shaikh, Ravi Kanojia, Shristi Kharola, Deepak Juyal, Vishwjeet Choudhary, Tarun Kumar Dhiman, Kiran Sharma, and Vinayak Sharma

Abstract The article demonstrates application of nonlinear autoregressive neural network for multi-step ahead hourly prediction of concentration of Particulate Matter (PM2.5) at Anand Vihar location in New Delhi, India. The hourly PM2.5 concentration data of the area has been collected for a duration of 10 days and used for training of proposed machine learning model. The autoregressive neural network based machine learning model has been trained using three different training algorithms namely Levenberg–Marquardt (LM), Bayesian Regularization (BR) and Scaled Conjugate Gradient (SCG). The prediction output of these algorithms have been analysed for a maximum hourly prediction duration of 5 h in terms of root mean squared error (RMSE), mean absolute error (MAE) and mean absolute percentage error (MAPE). The results clearly indicate that BR-algorithm outperforms other machine learning algorithms considered in the study.

A. Kharola (✉) · K. K. Sharma · A. Shaikh · R. Kanojia · S. Kharola · D. Juyal · T. K. Dhiman · K. Sharma · V. Sharma
Graphic Era Deemed to be University, Dehradun, Uttarakhand, India
e-mail: ashwanidaa@gmail.com

R. Kanojia
e-mail: ravikanojia@geu.ac.in

D. Juyal
e-mail: deepakjuyal.mgt@geu.ac.in

T. K. Dhiman
e-mail: tkdhiman@gehu.ac.in

K. Sharma
e-mail: kiransharma@geu.ac.in

V. Choudhary
Birla Institute of Technology and Science (BITS), Pilani, India

T. K. Dhiman
Graphic Era Hill University, Dehradun, Uttarakhand, India

S. Kumar et al. (eds.), *Soft Computing Applications in Modern Power and Energy Systems*, Lecture Notes in Electrical Engineering 1298,
https://doi.org/10.1007/978-981-96-0058-8_13

Keywords AQI · Multistep-ahead predictions · Machine learning · Training algorithms · Neural networks

1 Introduction

Due to recent advancements through industrialization and urbanization, Air quality index (AQI) levels have shown a significant rise in the past few decades [1]. Some of the most common air pollutants that have a wide variety of adverse health concerns includes carbon monoxide, ground level ozone, particulate matter (PM2.5/PM10), nitrogen dioxide and sulphur dioxide [2]. The deteriorating air quality has a disastrous effect on the environment thereby affecting the life of the people. A precise prediction of AQI has therefore become important for monitoring environmental threats [3]. Machine learning models are playing an important role in data analysis, prediction and control applications [4–6]. In order to monitor, control and maintain an adequate level of AQI, several machine learning techniques have been successfully adopted by researchers in the past few years. For an instance, Janarthanan et al. [7] combined SVR and LSTM based machine learning techniques for predicting AQI in Chennai, Tamil Nadu. The data set on AQI was initially collected and pre-processed for replacing noises and missing values. The results indicate superior performance of SVR-LSTM model compared to other existing techniques.

Sarkar et al. [8] evaluated different data predicting techniques for predicting AQI in the area around New Delhi, India. The study combined two different deep learning techniques namely LSTM and Gated Recurrent Unit (GRU) for prediction. The performance of proposed LSTM-GRU model has been further compared with some other popular machine learning models. The results highlight superior performance of LSTM-GRU model giving lower error responses. Further, Yin and Liu [9] suggested a hybrid approach for hourly, 3-step-ahead forecasting of AQI. The approach constitutes three different modules incorporating feature engineering. The AQI data of four different Chinese cities were considered for analysis. The findings indicate that feature engineering approach significantly boosted the prediction performance of proposed model. Moreover, Maltare and Vahora [10] compared different machine learning models for predicting AQI for Ahmedabad city in Gujarat, India. The study considered different preprocessing techniques for managing data before supplying to the machine learning models. The outputs highlight superior performance of model designed using support vector regression (SVR) machine algorithm with RBF kernel model.

Recently, Mishra and Gupta [11] compared different deep learning methodologies for monitoring AQI levels. The performance of these models have been analysed for both hourly and daily predictions. The findings highlight that the long short-term memory (LSTM) network outperforms other machine learning models particularly with hourly data prediction. Moreover, for daily data prediction autoregressive integrated moving average (ARIMA) model yields better results compared to other machine learning models. This primary aim of this study is to present a nonlinear autoregressive neural network model designed using different machine learning algorithms for multi-step ahead hourly prediction of concentration of Particulate Matter (PM2.5) at Anand Vihar location in New Delhi, India. The study further aims at comparing the prediction performance of these machine learning algorithms in terms of different error indicators. PM2.5 is the one of the predominant air pollutant directly affecting the air quality index (AQI). The hourly PM2.5 concentration data of the area for a duration of 10 days has been collected from Central Pollution Control Board (CPCB) portal of Government of India. The autoregressive neural network based machine learning model has been trained using three different training algorithms namely Levenberg–Marquardt (LM), Bayesian Regularization (BR) and Scaled Conjugate Gradient (SCG). The performance of these algorithms have been monitored for a maximum hourly prediction duration of 5 h in terms of root mean squared error (RMSE), mean absolute error (MAE) and mean absolute percentage error (MAPE).

2 Nonlinear Autoregressive Neural Network Design for Multistep-Ahead Prediction of PM2.5 Concentration

Studies have revealed that machine learning models have been gained significant attention particularly in the field of data analysis and forecasting [12]. Numerous machine learning techniques have been considered by researchers for optimal prediction of air quality indicators [13–15]. The nonlinear autoregressive neural network are popular machine learning models primarily used in multistep-ahead predictions [16]. These models are capable of learning from a particular time-series to predict future values of the same series [17]. Mathematically, a nonlinear autoregressive neural network model can predict particular values of data series 'x' using 'k' previous values of same series as shown with the help of Eq. (1).

$$x(t) = g(x(t-1), x(t-2), \ldots x(t-k)) + e(t) \tag{1}$$

In the above equation g (.) and $e(t)$ represents the unknown function and error term respectively. The model is initially developed in an open-loop and then converted into a close-loop as shown in Figs. 1 and 2 respectively. The model has been designed using 2 feedback delays and 20 neurons in the hidden layer.

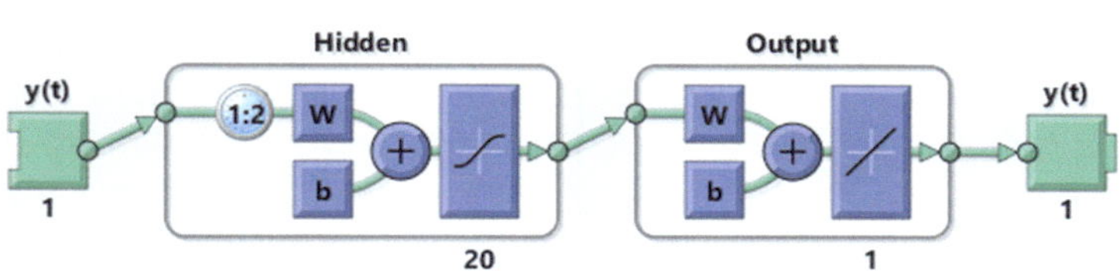

Fig. 1 Open-loop design of nonlinear autoregressive neural network

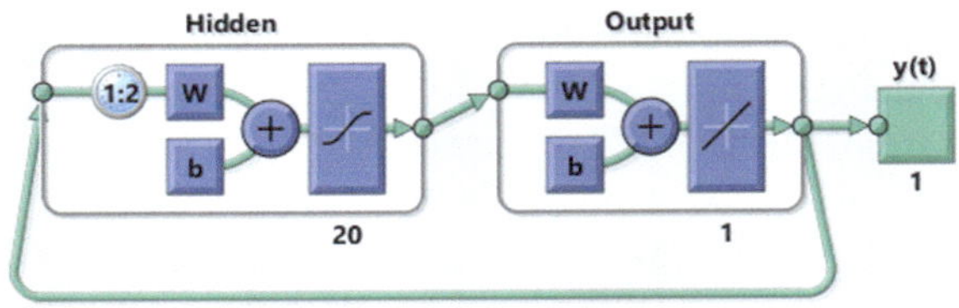

Fig. 2 Close-loop design of nonlinear autoregressive neural network

The nonlinear autoregressive neural network model has been designed using three different machine learning algorithms as already mentioned in the previous section of the article. The autocorrelation plot shows the autocorrelation of errors across varying lags. The autocorrelation plot obtained using different training algorithms are shown with the help of Fig. 3.

It can be clearly observed from the above results that using BR-algorithm all the correlation values except for the one at zero lag lie well within the confidence limit. Further, for LM-algorithm all the correlation values except for the ones at zero lag, lag (−10, 10) and lag (−16, 16) were well within the confidence limit. Lastly using SCG-algorithm all the correlation values except for the ones at zero lag, lag (−4, 4) and lag (−1, 1) were well within the confidence limit. Thus it may be concluded that BR-algorithms better autocorrelation responses compared to other training algorithms. Further, the regression results obtained for training and test samples using different machine learning algorithms are shown with the help of Fig. 4.

It can be observed from the above results that the best regression results were obtained using BR-algorithm, followed by LM-algorithm and SCG-algorithm respectively. The regression results obtained for training and test samples using BR-algorithm were 0.96876 and 0.95186 respectively. Similarly, the regression results obtained for training and test samples using LM-algorithm were 0.94983 and 0.9317 respectively. Lastly, the regression results obtained for training and test samples using SCG-algorithm were 0.8617 and 0.85137 respectively. Thus, it can be easily concluded that BR-algorithm gives the least error responses between the outputs and target values.

3 Results and Discussion

The performance of all the three algorithms have been monitored in terms of different error indicators for mapping the errors between the output and target values. The responses obtained for RMSE between outputs and targets are shown in Fig. 5.

It can be concluded from above results that the RMSE values increases with increase in duration of predicted hours. Moreover, the BR-algorithm provided the least RMSE responses, followed by LM-algorithm and SCG-algorithm respectively. Lastly, the RMSE values obtained for a maximum prediction duration of 5 h using BR, LM and SCG algorithms were 6.59 $\mu g/m^3$, 16.52 $\mu g/m^3$ and 38.29 $\mu g/m^3$

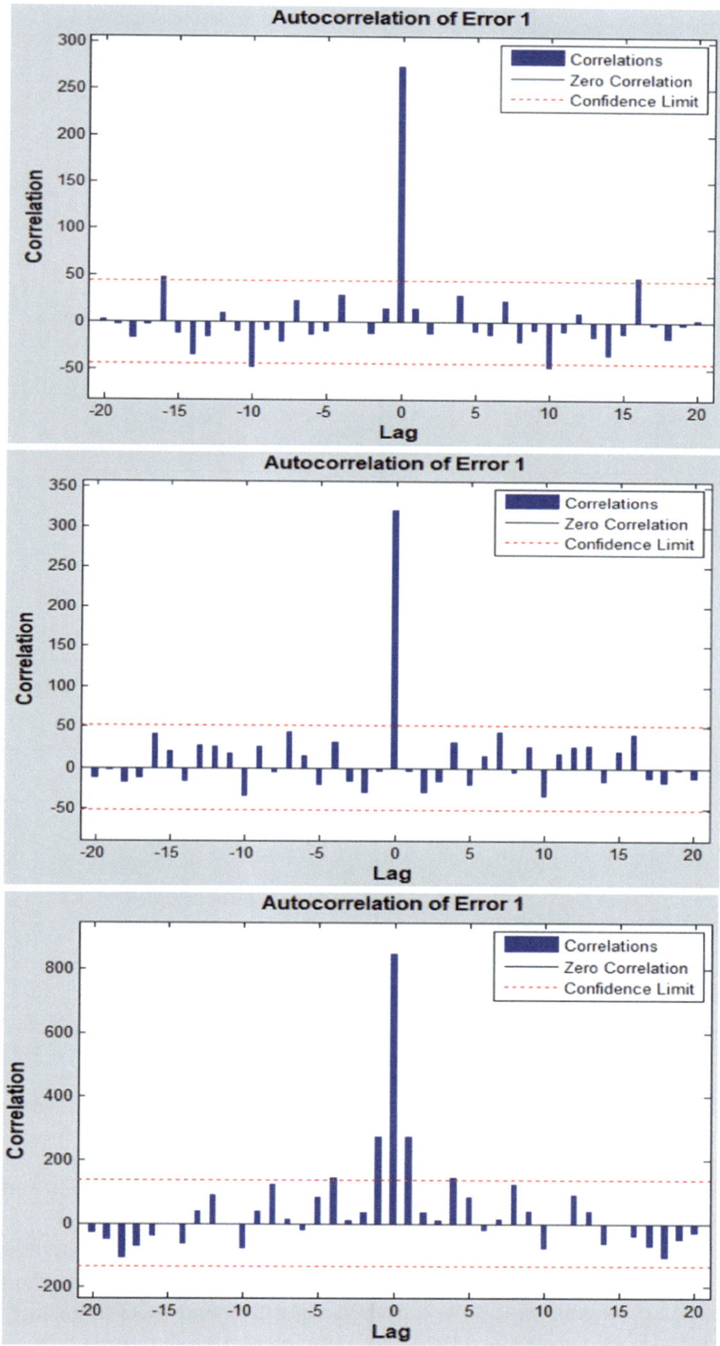

Fig. 3 Plot for autocorrelation of error obtained using **a** LM-algorithm, **b** BR-algorithm, c SCG-algorithm

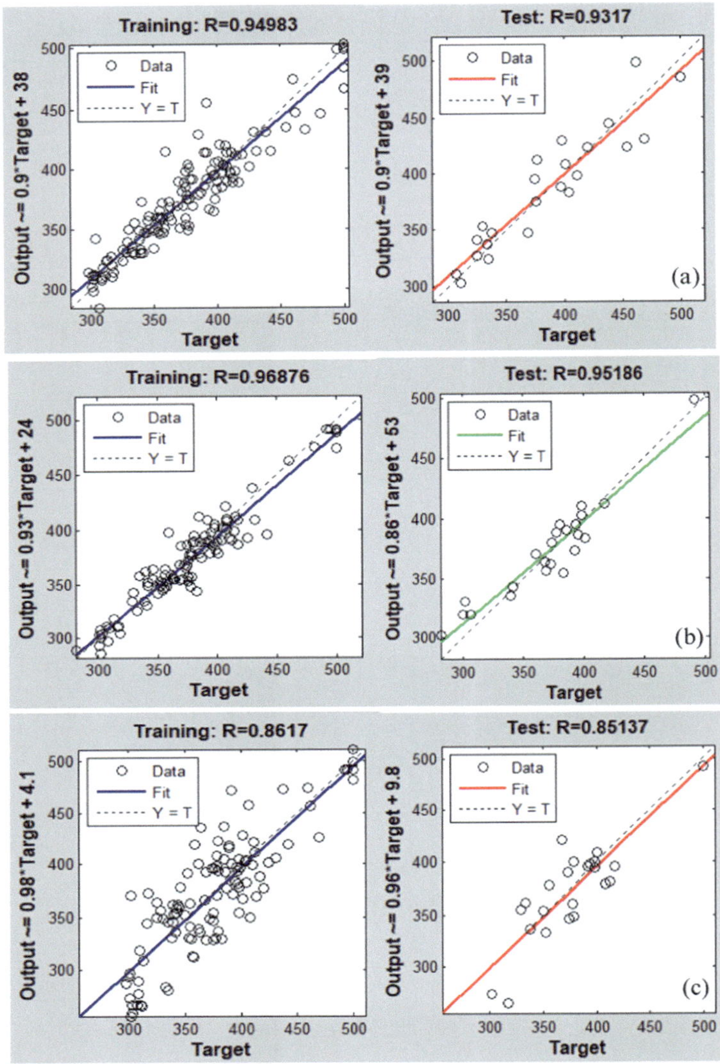

Fig. 4 Regression results obtained using **a** LM-algorithm, **b** BR-algorithm, **c** SCG-algorithm

respectively. Further, the responses obtained for MAE between outputs and targets are shown in Fig. 6.

It can be concluded from above results that the MAE values also increases with increase in duration of predicted hours. Moreover, the BR-algorithm provided the minimum MAE responses, followed by LM-algorithm and SCG-algorithm respectively. The MAE values obtained for a maximum prediction duration of 5 h using BR,

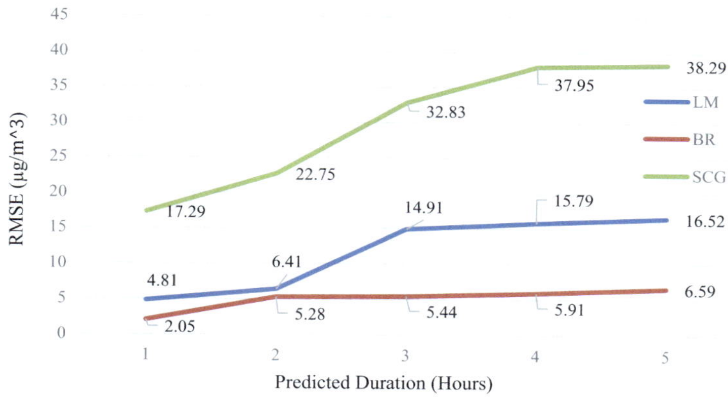

Fig. 5 RMSE responses obtained using different algorithms

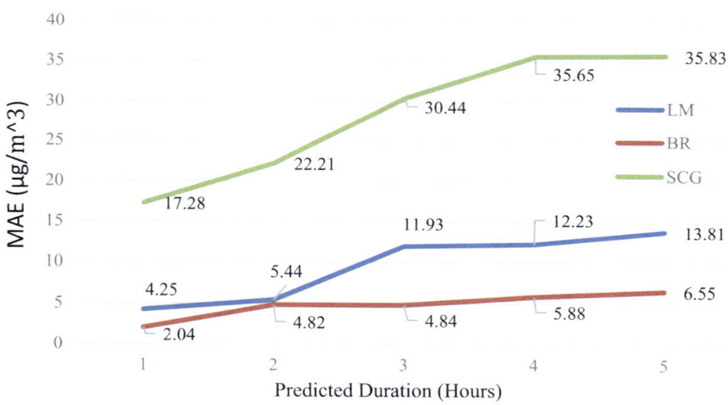

Fig. 6 MAE responses obtained using different algorithms

LM and SCG algorithms were 6.55 $\mu g/m^3$, 13.81 $\mu g/m^3$ and 35.83 $\mu g/m^3$ respectively. Lastly, the responses obtained for MAPE between outputs and targets are shown in Fig. 7.

It can be concluded from above results that similar to RMSE and MAE responses, the MAPE values also increases with increase in duration of predicted hours. Moreover, the BR-algorithm provided the minimum MAPE results, followed by LM-algorithm and SCG-algorithm respectively. The MAPE values obtained for a maximum prediction duration of 5 h using BR, LM and SCG algorithms were 1.81%, 3.87% and 10.1% respectively.

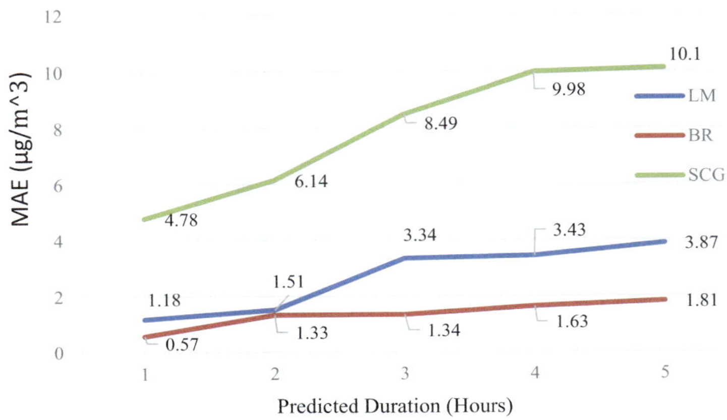

Fig. 7 MAPE responses obtained using different algorithms

4 Conclusion

The study examines application of nonlinear autoregressive neural network for hourly prediction of PM2.5 concentrations at Anand Vihar location in New Delhi, India. The machine learning model has been trained using hourly PM2.5 concentration data of past 10 days. The model has been designed using three different machine learning algorithms. The performance of these models have been analysed for a maximum duration of 5 h of multistep-ahead predictions. The findings clearly indicate superior performance of BR-algorithm, followed by LM-algorithm and SCG-algorithm respectively. The results also highlight that the values of all the error indicators increases with increase in time duration of prediction. The RMSE obtained for a maximum prediction duration of 5 h using BR, LM and SCG algorithms were 6.59 $\mu g/m^3$, 16.52 $\mu g/m^3$ and 38.29 $\mu g/m^3$ respectively. Similarly, the MAE obtained for a maximum duration of 5 h ahead prediction using BR, LM and SCG algorithms were 6.55 $\mu g/m^3$, 13.81 $\mu g/m^3$ and 35.83 $\mu g/m^3$ respectively. Lastly, the MAPE obtained for a maximum duration of 5 h ahead prediction using BR, LM and SCG algorithms were 1.81%, 3.87% and 10.1% respectively. Moreover, the results of this study were further compared with the results of Deep Ensemble Neural Network model developed by Guo et al. [18] for PM2.5 predictions. A comparison of results clearly highlight that the MAPE obtained using Deep Ensemble Neural Network model was 16.20% which were comparatively greater compared to BR (1.81%), LM (3.87%) and SCG (10.1%) algorithms considered in this study. As an extension for future work, the nonlinear autoregressive neural network with exogenous input can also be examined for multistep-ahead prediction of AQI or other significant environmental pollutant concentrations.

References

1. Aram SA, Nketiah EA, Saalidong BM, Wang H, Afitiri AR, Akoto AB, Lartey PO (2024) Machine learning-based prediction of air quality index and air quality grade: a comparative analysis. Int J Environ Sci Technol 21:1345–1360
2. Ahmad S, Ahmad T (2023) AQI prediction using layer recurrent neural network model: a new approach. Environ Monit Assess 195:1180
3. Wang W, Tang Q (2023) Combined model of air quality index forecasting based on the combination of complementary empirical mode decomposition and sequence reconstruction. Environ Pollut 316(2):120628
4. Kansal V, Jain U, Pant B, Kotiyal A (2023) Comparative analysis of convolutional neural network in object detection. Lect Notes Netw Syst 520:87–95
5. Dhiman G, Kumar AV, Nirmalan R, Sujitha S, Srihari K, Yuvaraj N, Arulprakash P, Raja AR (2023) Multi-modal active learning with deep reinforcement learning for target feature extraction in multi-media image processing applications. Multimed Tools Appl 82(4):5343–5367
6. Kharola A (2022) Novel adaptive control for avoiding fuzzy rule explosion in nonlinear systems. Int J Autom Control 17(4):377–396
7. Janarthanan R, Partheeban P, Somasundaram K, Elamparithi PN (2021) A deep learning approach for prediction of air quality index in a metropolitan city. Sustain Cities Soc 67:102720
8. Sarkar N, Gupta R, Keserwani PK, Govil MC (2022) Air quality index prediction using an effective hybrid deep learning model. Environ Pollut 315:120404
9. Yin Y, Liu H (2023) Air quality index prediction based on three-stage feature engineering, model matching, and optimized ensemble. Air Qual Atmos Health 16:1871–1890
10. Maltare NN, Vahora S (2023) Air Quality Index prediction using machine learning for Ahmedabad city. Digital Chem Eng 7:100093
11. Mishra A, Gupta Y (2024) Comparative analysis of air quality index prediction using deep learning algorithms. Spat Inf Res 32:63–72
12. Li G, Tang Y, Yang H (2022) A new hybrid prediction model of air quality index based on secondary decomposition and improved kernel extreme learning machine. Chemosphere 305:135348
13. Bekkar A, Hssina B, Douzi S, Douzi K (2021) Air-pollution prediction in smart city, deep learning approach. J Big Data 8:161
14. Zhang X, Jiang X, Li Y (2023) Prediction of air quality index based on the SSA-BiLSTM-LightGBM model. Sci Rep 13:5550
15. Wu H, Yang T, Li H, Zhou Z (2023) Air quality prediction model based on mRMR–RF feature selection and ISSA–LSTM. Sci Rep 13:12825
16. Baccoli R, Sollai F, Medda A, Piccolo A, Fadda P (2022) An adaptive nonlinear autoregressive ANN model for high time resolution traffic noise predictions. Experimental results for a port city waterfront. Build. Environ. 207(B):108551
17. Blasques F, Koopman SJ, Lucas A (2020) Nonlinear autoregressive models with optimality properties. Economet Rev 39(6):559–578
18. Guo C, Liu G, Chen CH (2020) Air pollution concentration forecast method based on the deep ensemble neural network. Wirel Commun Mob Comput 2020:8854649

Ensuring Grid Stability: Comprehensive Analysis of Mechanical Protection Systems for Power Transformers and Reactors

Pankaj Kumar Jha, Brijendra Singh, and Sumit Ray

Abstract Mechanical protections such as Buchholz, elevated winding temperature, high oil temperature, pressure relief valves, and oil surge relays play a crucial role in protecting transformers and reactors. The contacts of these protective devices are connected to the binary inputs of numerical relays or auxiliary relays, facilitating the tripping of transformers and reactors when necessary. Typically installed outdoors on the transformer and reactor bodies, these mechanical protection devices are susceptible to potential mis-operations caused by various external factors, including cable damage, malfunctions in mechanical relay contacts, issues with transducers, intrusion of external objects, subpar workmanship, and moisture ingress. Transformers and reactors are pivotal components of power transmission networks, and any inadvertent tripping resulting from mechanical protection malfunctions can lead to widespread disruptions in power system networks. This paper delves into a comprehensive analysis of noteworthy field events involving the tripping of transformers and reactors due to mechanical protection mis-operations. Through these experiences, valuable lessons have been gleaned, and this paper aims to share practical insights. Furthermore, innovative techniques to prevent mis-operations in the mechanical protection of transformers and reactors are explored. The implementation of these innovative methods, leveraging various digital and numerical relays, are discussed in detail.

Keywords Buchholz · High oil temperature trip (OTI trip) · High winding temperature trip (WTI trip) · Oil surge relay (OSR) · Pressure relief valve (PRV) · Sud-den pressure rise (SPR) · Transformers · Shunt reactor · DPS · SPS

P. K. Jha (✉)
Power Grid Corporation of India Ltd, Vadodara, Gujrat, India
e-mail: pankaj.jha@powergrid.in

B. Singh
Power Grid Corporation of India Ltd, Gurugram, India

S. Ray
Power Grid Corporation of India Ltd, Jaipur, Rajasthan, India

S. Kumar et al. (eds.), *Soft Computing Applications in Modern Power and Energy Systems*, Lecture Notes in Electrical Engineering 1298,
https://doi.org/10.1007/978-981-96-0058-8_14

1 Introduction

Transformers and reactors represent the pivotal components of power transmission networks. Unintentional tripping of these elements due to mis-operations in mechanical protection can potentially lead to widespread disturbances across power system networks.

Mechanical protection mechanisms, including Buchholz, OLTC Buchholz, Pressure Relief devices, etc., fall within the category of mechanical protection. Their design specifically addresses low-magnitude internal faults such as turn-to-turn faults, which are not covered by other protection schemes. Additionally, these mechanisms offer limited protection against insulation failure faults within the transformer tank. However, they are incapable of detecting faults outside the transformer or within the transformer bushings. All mechanical protection schemes operate based on pressure differential or pressure rise of the oil or gas inside the transformer. Detection of internal tank faults relies on this pressure, enabling swift tripping before a major fault occurs [1].

Exceeding the nameplate rating by overloading power transformers can result in elevated temperatures in both transformer oil and windings. If the temperature rise in the windings surpasses prescribed limits, insulation deterioration may occur, leading to premature failure. Prolonged thermal heating weakens the insulation over time, accelerating the loss-of-life of the transformer [2]. Consequently, monitoring transformer oil and winding temperatures is crucial, and remedial measures must be taken if these settings exceed specified limits.

According to POWERGRID's technical specifications, auxiliary protection contacts (Buchholz, PRD, SPRD, Oil Temperature, Winding Temperature, OLTC Buchholz, etc.) can be appropriately wired into numerical protection IEDs. Alternatively, separate flag relays or auxiliary relays are provided as required by the scheme for the tripping of transformers and reactors [3].

The tripping of transformers and reactors remains widespread due to installation and wiring errors, as well as malfunctions in the contacts of mechanical protection devices and auxiliary trip relays caused by battery ground faults. This paper aims to enhance understanding regarding the fundamental concepts and requirements associated with the mechanical protection of transformers and reactors. Detailed discussions on multiple case studies involving the tripping of transformers and reactors due to mechanical protection issues will be presented.

Section 2 of this paper provides a brief overview of the commonly used protection schemes for transformers and reactors in POWERGRID substations. In Sect. 3, the basic concepts of various mechanical protections for transformers and reactors will be discussed. Section 4 will specifically focus on the concept of the Auxiliary DC system in POWERGRID substations, addressing the concepts of battery ground fault relay and issues related to auxiliary trip relays in protection schemes.

Further exploration in Sect. 5 will involve in-depth case studies of transformers and reactors tripping due to mechanical protection mis-operations. Section 6 will concentrate on the techniques employed by POWERGRID to mitigate the tripping of transformers and reactors resulting from mechanical protection mis-operations. The paper will conclude by discussing the benefits and limitations of the various techniques adopted by POWERGRID to prevent inadvertent tripping of transformers and reactors due to mechanical protection mis-operations.

2 Protection Schemes of Transformers and Reactors

To ensure redundancy and eliminate the risk of common-mode protection failure, the protection scheme is divided into two parts, namely Group-A (fed from Station DC-A) and Group-B (fed from Station DC-B). Additionally, main and backup protection functions are housed in separate relay hardware. For enhanced reliability, alarm and trip signals for specific mechanical protections, such as OTI, WTI, Bucholtz, etc., are wired independently to Group-A and Group-B protection schemes. Figure 1 illustrates a single-line diagram of an autotransformer protected by two numerical multifunction relays.

Transformer protections at POWERGRID are organized so that the following are on DC source-A: differential trip, Buchholz-I trip, oil temperature alarm, and winding temperature trip. On the other hand, DC source-B accommodates REF trip,

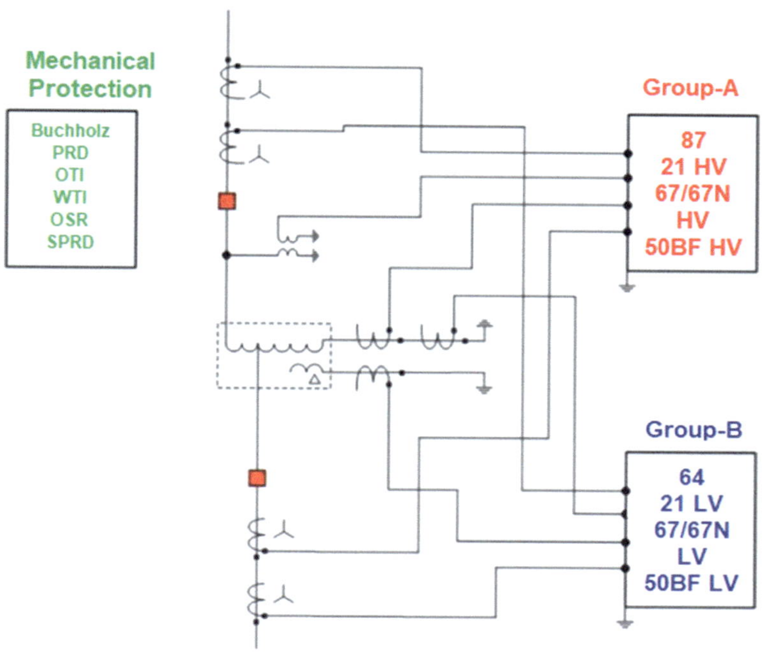

Fig. 1 Protection schemes of auto transformer

Buchholz alarm, PRD-I trip, winding temperature alarm, and oil temperature trip. This arrangement allows for the division of mechanical protection between Group-A and Group-B, routing trip commands through the numerical relays. This setup enables the automatic utilization of numerical relay features, such as disturbance recording, remote access, and breaker failure protection initiation, among others [4].

The protection schemes for shunt reactors at POWERGRID are depicted in Fig. 2. Auxiliary protection contacts for reactors (Buchholz, PRV, oil temperature, winding temperature, etc.) can be appropriately wired into numerical protection relays. Alternatively, separate flag relays or auxiliary relays are provided based on scheme requirements.

Reactor protections are further categorized, placing differential protection, Buchholz-I trip, and Winding temperature trip on DC-A, while REF protection, backup impedance protection, PRD-I, and Oil temperature trip are on DC-B. The tripping schemes for line reactors vary depending on whether they are switchable or non-switchable types [5]. Detailed discussions on tripping schemes for shunt reactors are beyond the scope of this paper.

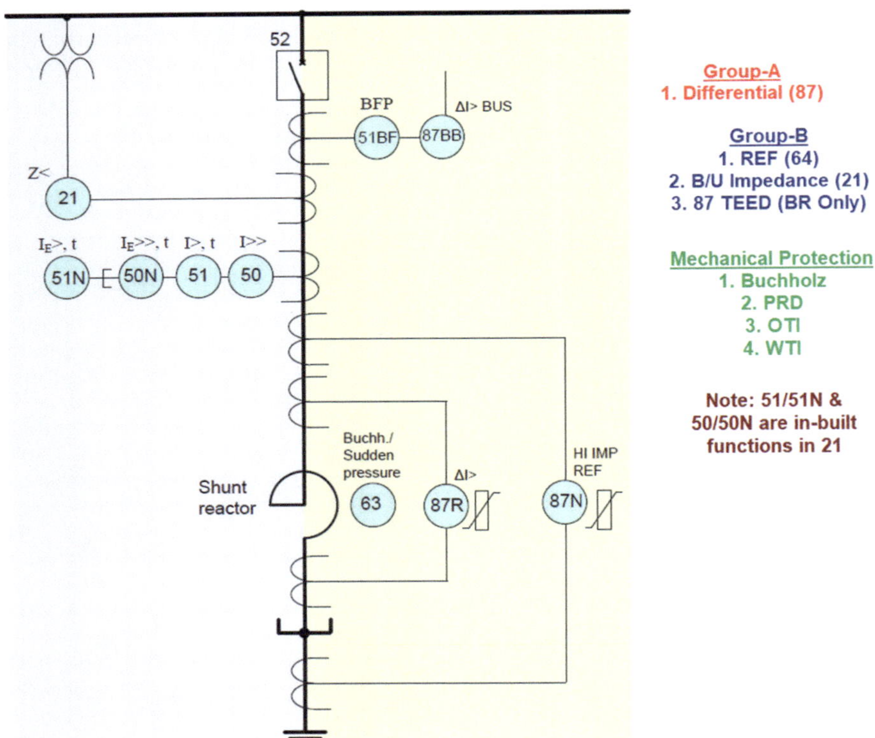

Fig. 2 Protection schemes of shunt reactors

3 Basics of Mechanical Protection of Transformers and Reactors

The presence of gas accumulation or alterations in pressure within the transformer tank serves as reliable indicators of internal issues. Mechanical protection is recommended as excellent supplementary protection wherever applicable. These mechanisms often exhibit higher sensitivity, allowing them to operate on subtle internal faults that might go unnoticed by differential or other relays. It's crucial to note that their functionality is confined to issues inside the transformer tank and does not extend to faults in the transformer bushings or connections to external CTs. Consequently, their protective scope is limited to the tank itself [6].

In this section, we will provide a brief overview of the fundamental concepts of mechanical protection devices for transformers and reactors, such as Buchholz, PRD, SPRD, OTI, WTI, etc. It's acknowledged that some faults in transformers and reactors may escape detection when relying solely on the electrical protections described in Sect. 2. For instance, a turn-to-turn fault can induce significant current in the shorted turn, while the current in the remaining winding remains relatively unchanged. Since there is minimal or no alteration in the current monitored by the CTs, there is no differential current for the relays to activate. Over time, the turn-to-turn fault transforms into a ground fault, providing the necessary change in current for the protective relays to operate.

There are two methods for detecting transformer faults beyond electric measurements [7]:

- Accumulation of gases due to slow decomposition of the transformer insulation or insulating liquid. These relays can detect heating due to high-resistance joints or due to high eddy currents between laminations.
- Sudden or rapid increases in tank pressure caused by internal transformer faults.

Sudden-pressure and gas-accumulation relays provide partial redundancy for transformer differential protection, because they respond only to faults inside the transformer tank [8].

3.1 Gas Accumulation Protection (Buchholz Relay)

The gas accumulator relay, commonly referred to as the Buchholz relay, is specifically suitable for transformers featuring conservator tanks. Positioned in the pipe connecting the main tank to the conservator tank (refer Fig. 3), this relay is engineered to capture any gas that might ascend through the insulating liquid. It functions by accumulating gas gradually for minor faults or responds to significant faults that propel the insulating liquid through the pipe at a high velocity. This device is capable of detecting a minimal volume of gas generated by low-energy arcs, overheating, and insulation decomposition.

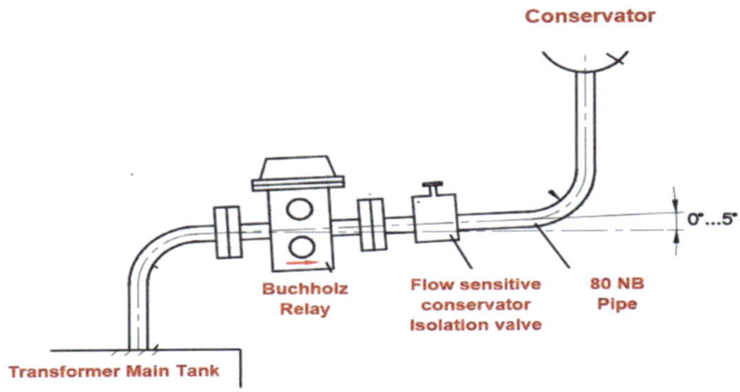

Fig. 3 Buchholz relay arrangement

Each transformer is equipped with a double float/reed type Buchholz relay positioned between the transformer and the conservator tank, adhering to the POWERGRID technical specifications. This relay serves as a collector for any gas generated within the transformer. The device incorporates two electrically independent ungrounded contacts—one for issuing an alarm in the event of gas accumulation and the other for triggering a response to a sudden rise in pressure or oil surge. Various switches exhibit diverse reaction times to the steepness of pressure waves during impulse conditions, such as waves lasting 2–3 ms, and also demonstrate sensitivity to vibrations. The characteristics of different types of switches are outlined in Table 1 [9].

In view of the superior characteristics of reed-type switches, they are currently preferred over mercury contact switches. During normal operation, they are completely filled with insulating liquid. Due to buoyancy, the floats of the double-float relay are positioned at their topmost position. Figure 4 illustrates the various types of floats provided by different manufacturers.

Table 1 Characteristics of different types of switches

	Reed type	Mercury type
Sensitivity to vibration	>2 g/10–15 Hz	0.25 g/10 Hz
Peak value of flow speed	120–200 cm/sec	40–70 cm/sec
Peak value of oil pressure	6.9 Bar	0.9 Bar
Tripping time calculated from first front of pressure wave	6 ms	100 ms

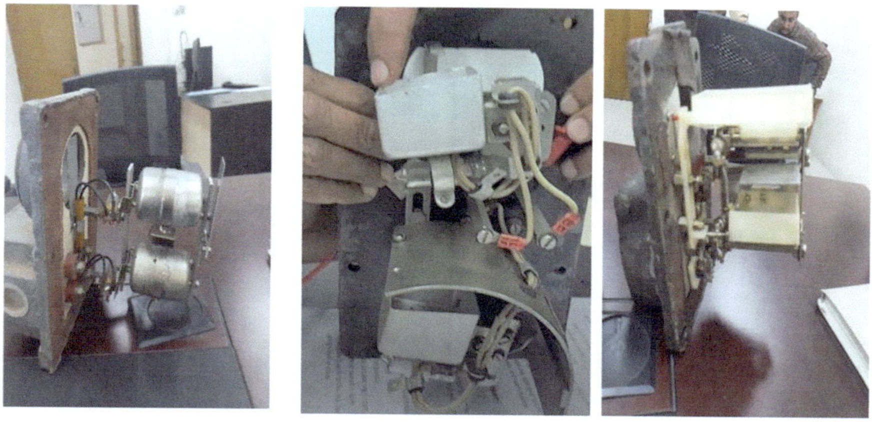

Ball/ Cylindrical type Float

Bucket Type float

Fig. 4 Different types of floats in Buchholz Relay

3.2 *Sudden Pressure Protection*

Sudden-pressure relays utilize a diaphragm to monitor the pressure within the transformer tank. A small orifice in the diaphragm allows pressure to equalize as it changes with thermal cycling inside the tank. In the event of a fault occurring inside the transformer tank, resulting in a sudden pressure change, the diaphragm moves, causing a microswitch attached to it to close a contact. Due to the transient nature of the monitored parameter, the relay incorporates either a seal-in auxiliary relay circuit or a trip-and-seal-in auxiliary relay within the tripping circuit.

3.3 *Thermal Protection System*

The loading of a transformer is typically overseen by relays that simulate heat generation in the transformer and its dissipation through the cooling system. A straightforward implementation of thermal protection involves a temperature relay set to trip if the hot spot temperature of the transformer winding exceeds a predetermined value established by well-known standards or specified limits set by the utility.

Oil-immersed reactors are occasionally constructed with forced cooling to diminish size and cost. In such reactors, cooling is typically crucial and should be operational whenever the reactor is energized. The loss of cooling can be identified by monitoring oil flow using flow indicators, checking the AC supply voltage to the cooling fans and oil pumps, and monitoring temperature with temperature relays. Generally, the temperature relays are linked to trip and remove the reactor

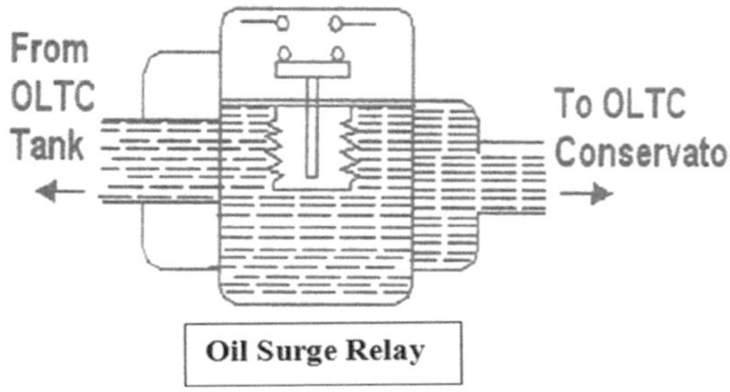

Fig. 5 Mounting arrangements of OLTC Surge Relay

from service. To ensure adequate protection for the reactor, a combination of all the aforementioned indicators is usually recommended [10].

3.4 *Oil Surge Relay (OSR)*

The Oil Surge Relay (OSR) is an auxiliary component of an oil-immersed type transformer, refer Fig. 5. Its primary purpose is to mitigate damage to the on-load tap changer in the event of a failure. The OSR is positioned between the On Load Tap Changer (OLTC) tank and the OLTC conservator. It reacts solely to the rate of rising pressure caused by internal arcing. When the rate of rising pressure in the oil exceeds a specific value, the alarm and trip contacts are activated.

3.5 *Pressure Relief Device (PRD)*

Sudden and violent short circuits within the metal tanks of oil-cooled transformers instantly generate a significant amount of gas, leading to a substantial increase in internal pressure. If this pressure cannot be externally released, there is a risk that the transformer could explode, posing potential harm and causing damage. In the event of a sudden and uncontrolled surge in pressure within the transformer, the pressure relief device facilitates the discharge of insulating fluid in milliseconds.

4 Auxiliary Dc System, Battery Earth Faults and Problems of Auxiliary Trip Relays

At POWERGRID, the DC system for all substations rated at 220 kV and above consists of two float cum-boost chargers and two battery sets for each of the 220 V and 48 V systems, respectively [11]. This configuration includes two independent

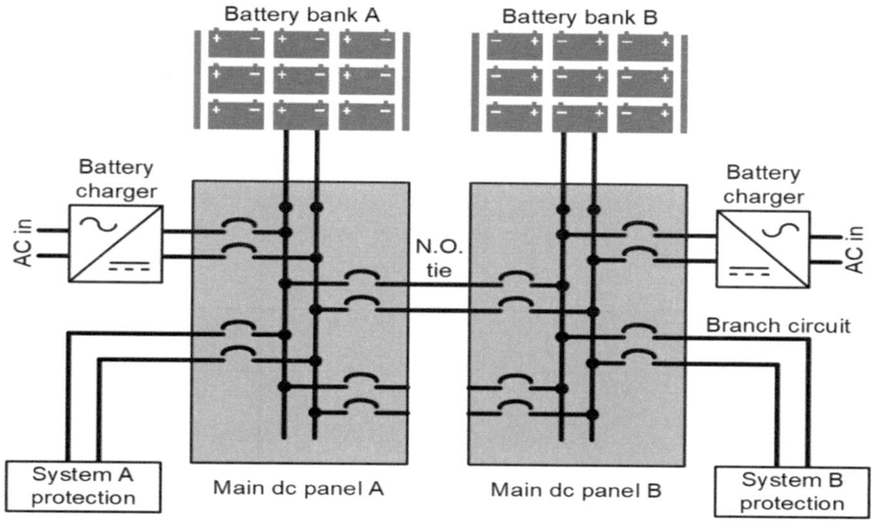

Fig. 6 Typical arrangement of auxiliary DC system

battery banks, each equipped with its own battery charger. System A and System B protection are powered by their respective battery bank, battery charger, and DC distribution panel. This setup ensures control power availability to detect and trip for system events, even if one DC system fails [12]. The illustration of this auxiliary DC system arrangement is depicted in Fig. 6.

Battery systems are typically grounded at a single point through resistors to establish a known reference voltage between positive, negative, and ground. Battery chargers may include this circuitry internally, or external resistors or lamps may be used, as shown in Fig. 7. Whether balanced resistors are employed to establish equal voltage above and below ground potential, or unbalanced resistors are used to offset the ground reference, monitoring the DC rail-to-ground voltage is a common method to detect extraneous battery grounds. Resistance values are generally selected to limit the current drain to several milliamps or less [13].

In Fig. 7, capacitors C1 and C2 represent the stray capacitance of the DC circuitry, surge capacitance, and power supplies. Under normal conditions, the voltage across each capacitance is one-half of the battery voltage. The first extraneous ground does not impair the battery system. However, it is crucial to detect and set off an alarm for the first extraneous ground because the second ground could potentially short the entire battery system.

Detecting DC battery system ground is relatively straightforward by monitoring the voltages between ground and the positive and negative rails. Any shift in these voltages indicates the presence of an extraneous battery system ground.

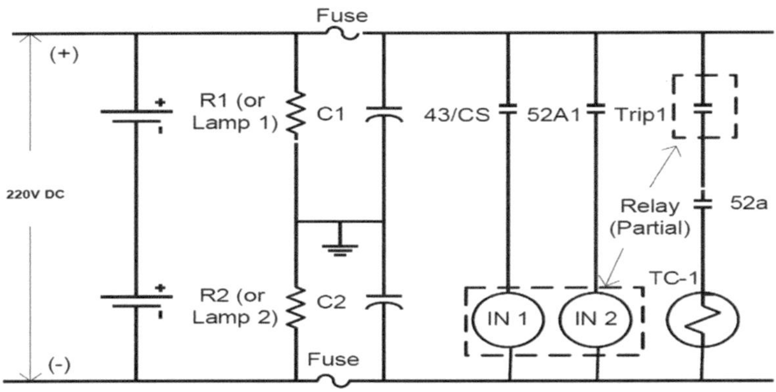

Fig. 7 Portion of a typical dc battery system and relay connections

4.1 Effects of Battery Earth Faults

Similar to any electrical circuit, substation battery wiring may encounter faults from time to time. While short circuits are typically cleared by the fuses, an earth fault may persist without being cleared due to the circuit earth impedance. This leaves the circuit in a faulted state, which may result in the following:

a. An impulse discharge through the relay coil/opto due to wiring capacitance
b. A steady-state voltage across the relay coil/opto
c. A steady-state 'trickle' current through the relay coil/opto.

The battery alarm can effectively be utilized to 'high impedance' earth the DC supplies. This provides greater security, as the integrity of the DC auxiliary supply will not be compromised by a single earth fault.

The earthing is facilitated by the battery monitoring device. This device restricts the earth fault current to a relatively low magnitude and also limits the voltage across the device to typically less than 50% ($R1 = R2 = 2.2\ K\Omega$) of the nominal battery voltage (VDC), refer Fig. 8. When calculating the maximum current and voltage that the device is subjected to, it is prudent to assume that the battery is on boost charge at 125% of nominal (i.e., 1.25 VDC).

4.2 Impact of the Auxiliary Trip Relays Burden

The burden of a protective device determines the power required for its operation. Devices with high input impedance draw lower current and consequently require less power for operation, classifying them as low burden. Conversely, high burden devices have significantly lower input impedance, resulting in a higher operating current requirement.

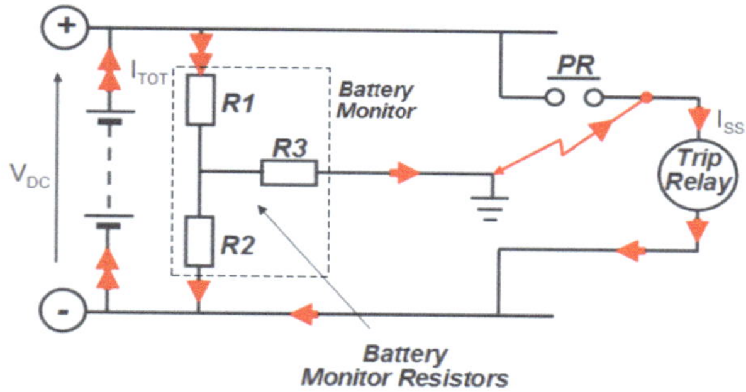

$$R_{TOTAL} = \frac{R2\,(R3 + R_{RELAY})}{R2 + R3 + R_{RELAY}} + R1$$

$$I_{TOT} = \frac{1.25\ V_{DC}}{R_{TOTAL}}$$

$$I_{SS} = \frac{1.25V_{DC} - (I_{TOT} \times R1)}{R3 + R_{RELAY}}$$

Fig. 8 Centre Tapped Battery Earthing Arrangements at POWERGRID

The burden of the device directly influences the relay's stability during battery earth faults and induced AC signals. It can be demonstrated that high burden devices are inherently more stable than their low burden counterparts, albeit at the expense of increased battery drain. Striking a balance between immunity and limiting battery drain is crucial, especially for continuously operated devices.

In general, low burden relays are suitable for applications where the trip wiring remains in the same room (i.e., short distances). On the other hand, high burden relays are designed for applications where the trip relay is remote from the protective relay or where the trip wiring extends beyond the confines of the relay room.

4.3 Auxiliary Trip Relays and Its Coordination with Battery Earth Fault Relay

After experiencing multiple undesired tripping incidents at various substations of POWERGRID, an analysis was conducted on the operation of the auxiliary trip relay. The tripping times and pickup voltage for the two most commonly used trip relays are tabulated in Table 2.

A DC earth fault was simulated to analyze its effect on the performance of auxiliary trip relays, assuming that the DC voltage is balanced prior to the earth fault

Table 2 Tripping time and pickup voltage of the auxiliary relay

Parameters	Model-A	Model-B
Operating voltage	85 V	110–112 V
Operating time	<5 ms	20–25 ms
Coil resistance	7.3 kΩ	40 kΩ
Pickup current at U = Ur	30 mA	5.5 mA
Power consumption at U = Ur	6.6 W	1.2 W

(+118 V & −118 V). A low-impedance earth fault was created in the floating positive wires carrying tripping signals from outdoor auxiliary protection devices to auxiliary tripping relays. The voltage across the operating coil of these auxiliary trip relays just after the DC earth fault (DCEF) and the current through the DCEF relay are tabulated in Table 3.

In the above table, it is evident that in the case of an earth fault in the floating positive wires of auxiliary tripping relays, the voltage across the operating coil of the tripping relay is 118 V (>85/112 V), leading to the operation of tripping relays. At this time, the leakage current through the battery earth fault relay is <3 mA. According to prevailing POWERGRID practices, the battery earth fault relay is set to operate at a 3 mA current, and therefore, during this event, the battery earth fault relay does not operate. This situation is undesirable because even though the auxiliary trip relay operates due to DCEF, the low value of leakage current may not reach the threshold of the DCEF relay pickup. It is evident that the battery earth fault relay needs to be coordinated with these fast-operating tripping relays in a way that ensures the battery earth fault relay operates when the auxiliary tripping relays operate. However, the battery earth fault relay cannot be made more sensitive (setting low threshold pickup <3 mA) as it may lead to frequent Earth Fault alarms. These fast-operating auxiliary tripping relays are utilized to extend the tripping in the case of mechanical protection relays for transformers and reactors. This relay operates at much less than 50% of the voltage of the battery system.

These auxiliary relays are susceptible to incorrect operation in the case of minor leakages or minor earth faults in the DC system. Furthermore, the fast operation and low operating voltage of these auxiliary trip relays result in no recording of events in BCU/SCADA in some cases. This, in turn, creates problems during event analysis. In the subsequent section, we will discuss several cases where transformers and reactors tripped inadvertently due to these auxiliary trip relays. Additionally, we will delve into the mitigation techniques adopted by POWERGRID to avoid this type of tripping, along with simulation results.

Table 3 Voltage across the operating coil of these auxiliary trip relays just after the DC earth fault (DCEF)

Conditions	Voltage across auxiliary trip relays	Current through battery EF relay	Remarks
Healthy system	0	–	
DCEF in floating positive wire	118 V	118/40000 = < 3 mA	Auxiliary trip relays operated but no alarm of DCEF in the system were recorded

5 Field Cases of Transformers and Reactors Tripping

In this section we will discuss some of the actual field events of transformers & reactors due to mechanical protection mis-operations.

5.1 400/220 kV ICT (Inter Connecting Transformer) Tripped Due to Operation of OSR

At one of the substations of POWERGRID, 400/220 kV ICT tripped approximately after six (06) minutes of Single line to ground fault in one of the 400 kV transmission line connected to this substation. DR is shown in Figs. 9 and 10.

The tripping of transformer on Operation of OSR was not desirable. During analysis, it was observed that the contacts of OSR has mis-operated and leads to undesirable tripping of ICT, refer Fig. 11.

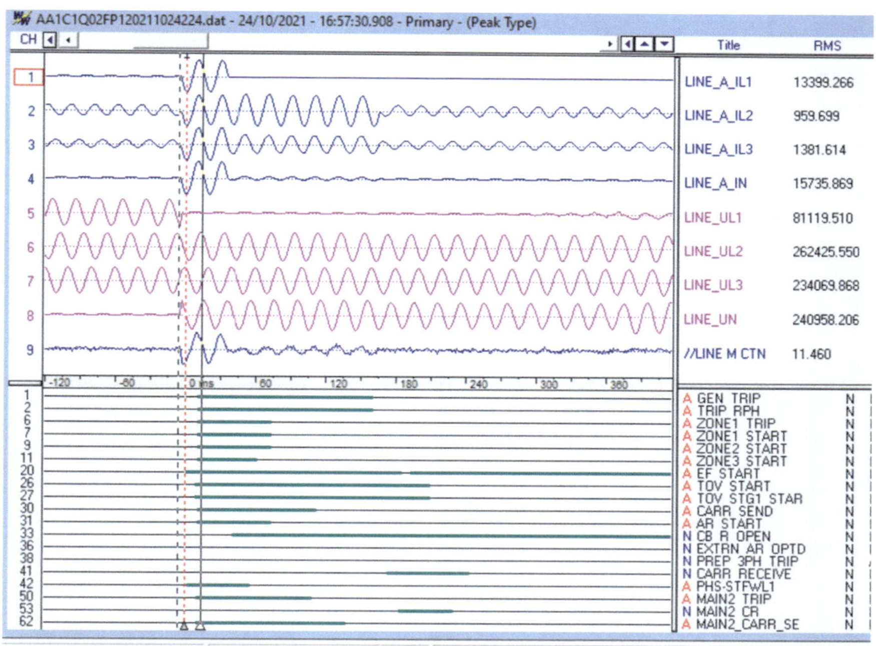

Fig. 9 Disturbance report for SLG fault on transmission line

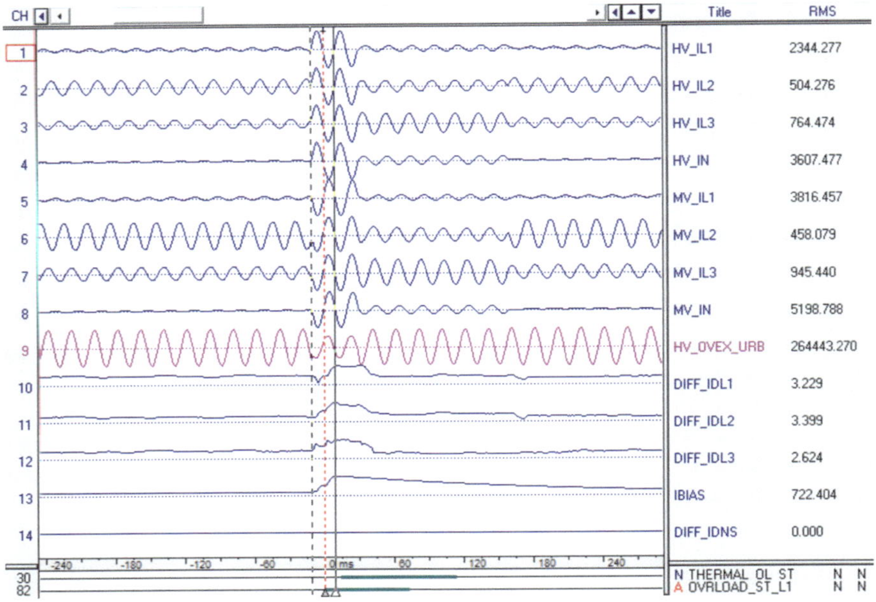

Fig. 10 Disturbance report transformer trip

Fig. 11 Malfunctions of auxiliary contacts in OSR

5.2 *Tripping of 765 kV Non-switchable Line Reactor on Buchholz Trip*

At one of the substations of POWERGRID, 765 kV non-Switchable line reactor tripped due to problem in the auxiliary trip relays of Buchholz Trip. During analysis of the event, it was observed that there was no operation of Buchholz relay. No gasses or air-trapping were observed in the Buchholz relay. The Buchholz trip signal, in the SCADA system was activated only for 110 ms duration. There was no DCEF recorded in the system at the time of event. All these things were suggesting that the auxiliary trip relays for Buchholz trip must have mis-operated leading to tripping of the Reactor as well as 765 kV line. The schematic diagram of this particular case is as per Fig. 12.

During analysis it was observed that the auxiliary trip relay (30A-SP) was operating whenever there is any DCEF in wire no. K921. However, during this condition DCEF was not being recorded in the system. The reason for this type of behavior was as discussed in Sect. 4.3 of this paper.

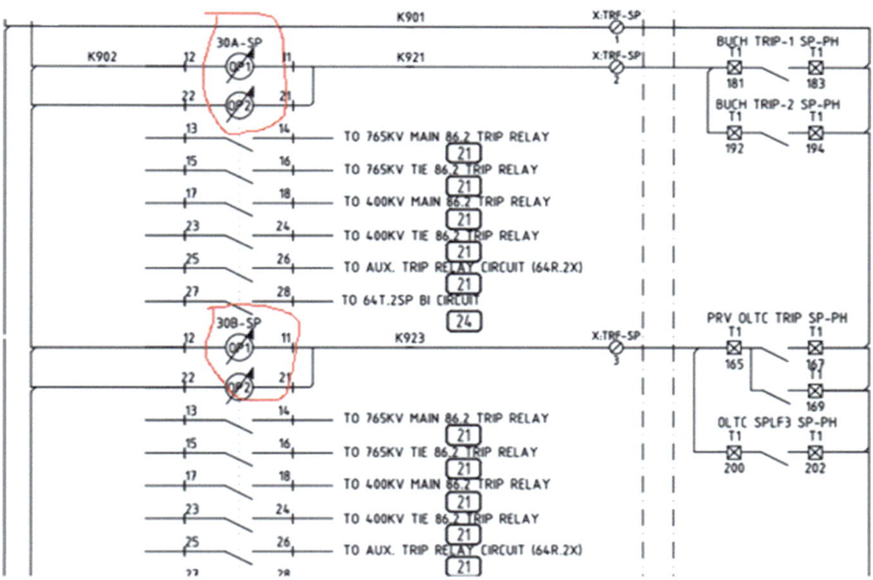

Fig. 12 Scheme drawing for reactor buchholz trip

5.3 Tripping of Transformers Due to Buchholz Relay Mis-Operation During Pump Operation

There have been numerous incidents where the Buchholz trip initiated during the switch-on or switch-off operation of the pump, indicating an oil surge exceeding the rating of the Buchholz relay. The flow rate range of different makes of Buchholz relays varies among manufacturers and can be adjusted according to requirements. However, site adjustment is not advisable as it cannot be tested. In modern times, all Buchholz relays have a rating of more than 65 cm/sec, as illustrated in Table 4.

Any Buchholz operation during pump operation may be attributed to either the oil velocity exceeding the specified limit in the specification or the lower flow rate rating of the Buchholz relay. In many cases, transformers equipped with a 4500 LPM discharge pump experience tripping due to the Buchholz relay. The use of a higher-rated pump, such as a 4500 LPM pump, results in a reverse flow oil surge exerted more on the lower contacts of the Buchholz relay and less on the upper contact. Nowadays, all transformers/reactors are provided with two Buchholz Relays, and the position of relays has little impact on the tripping operation of the individual relay. Figure 13 shows the option for adjustment of flow rate in Buchholz Relay.

Table 4 Alarm and trip contact settings of different Buchholz relay for 80NB pipe dia

Make	Alarm switch gas volume	Trip switch gas volume
Sukrut	200 to 300 CC	90 to 160 Cm/sec
VIATS instrument	200 ± 100 CC	100 ± 15 cm/sec (Standard)
CEDASPE	200 ± 100 CC	100 ± 15 cm/sec (Standard)

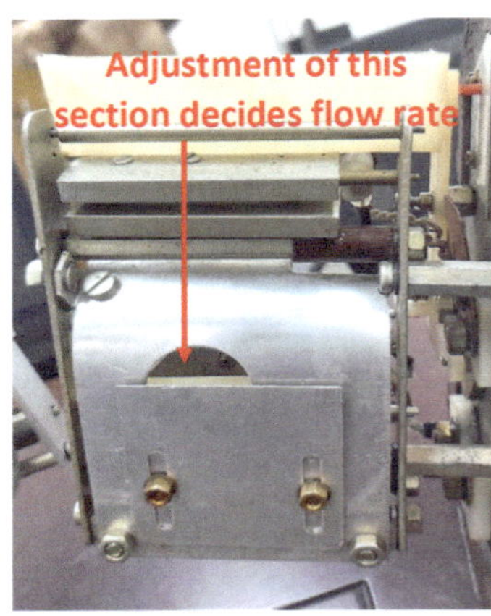

Fig. 13 Manual flow rate adjustment in Buchholz relay

5.4 Tripping of Transformers and Reactors Due to Low Flow Sensitivity of TCIV (Transformer Conservator Isolation Valve) and Operation of Buchholz Due to TCIV Operation

All Transformers/Reactors supplied to POWERGRID, India after 2010–11 are equipped with flow control conservator isolation valves. The function of the flow-sensitive conservator isolation valve is to restrict the supply of oil in the event of a fire in the transformer by isolating the conservator oil from the main tank. The valve is designed to be flow-sensitive and will shut off when the flow in the pipe exceeds the expected flow under permissible normal operating conditions. The flow rate of the TCIV varies from manufacturer to manufacturer; for Manufacturer-A, it is 60–100 LPM, and for Manufacturer-B, it is 40–80 LPM. Once the flow rate in the reverse direction surpasses the pre-set limit, the TCIV operates, blocking the oil to the main tank, and manual resetting is required.

In many transformer cases at POWERGRID, it was observed that TCIV operated during pump operation, leading to the tripping of the transformer. In the case of a Reactor at one of the substations, TCIV malfunctioned, resulting in a low oil level in Buchholz and, consequently, the tripping of the Reactor. Malfunctions of TCIV are also evident under normal conditions. At the same time, the effectiveness of TCIV during a fire incident in Transformer/Reactor is negligible. Therefore, the provision of TCIV in Transformers/Reactors needs to be reviewed.

5.5 Tripping of Transformer and Reactor Due to Buchholz Relay—Effect of Wafer Swing Check Valve/RFBD on Buchholz Relay Tripping

Transformers are equipped with pumps for the forced cooling of oil. Alongside the pumps, a wafer Swing Check valve system/RFBD (reverse flow blocking device) is also provided. The primary function of the wafer Swing Check valve system is to restrict the reverse flow of oil (Fig. 14). In some transformers, the wafer swing check valve is always positioned in the closed condition to restrict the reverse flow, thereby not reducing the reverse oil flow surge. Another type of wafer swing check valve or RFBD may be effective for lower-rated pumps (i.e., 1800 LPM), as experimented by one manufacturer of transformers in India that had Buchholz trip issues with higher-rated pumps. However, the efficacy of RFBD/wafer swing check valve for blocking the reverse flow of oil is doubtful for higher discharge rates.

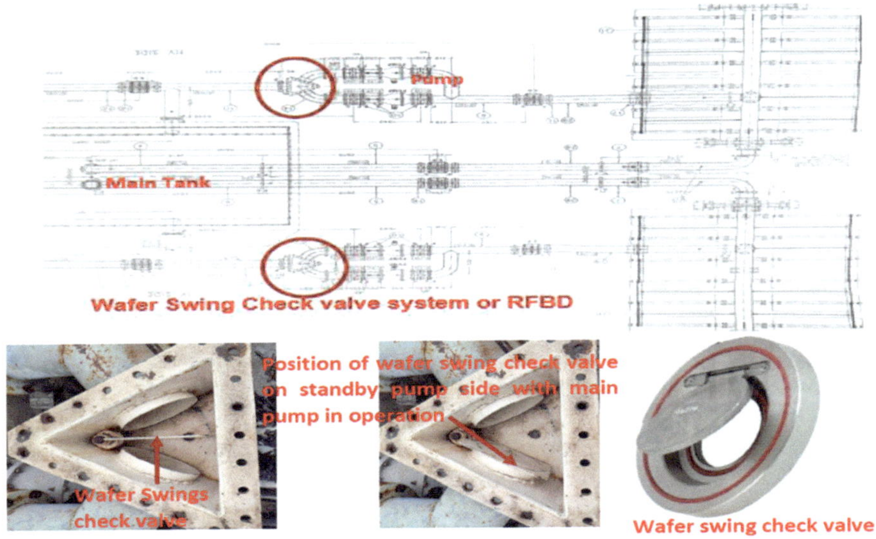

Fig. 14 RFBD in pumps of transformers

6 Techniques Adopted by Powergrid to Mitigate the Tripping of Transformers and Reactors Due to Mechanical Protection Mis-Operation

Tripping of Transformers and Reactors due to mechanical protection mis-operation are mainly due to the following reasons:

1. Buchholz Operation during pump operation due to higher discharge rate of pump.
2. Buchholz operation due to TCIV and RFBD problems.
3. Contact mal-operations of Mechanical Protection Relays.
4. False closing of microswitch contacts of mechanical protection relays.
5. Lower operating voltage of auxiliary trip relays.
6. Auxiliary trip relays operations due to DC EF in the substation DC system.
7. Operation of sudden pressure relay due to winding movement on severe through-fault and seismic condition.
8. Arcing over of the microswitch contact.
9. Operator error.

In this section we will discuss about the different techniques adopted by POWERGRID to avoid undesirable tripping of transformer and reactor.

6.1 Wiring of Mechanical Protection Contacts to Numerical Relays and Introduction of Timer in the Binary Input of Numerical Relays

Transformer and Reactors winding movement can send oscillatory shock waves through the oil inside the tank, which may cause the microswitch contact to close during through faults or seismic events. For a true rise in pressure, the microswitch contacts should close and stay closed for a period of time. For a shock wave, the contact should bounce. At POWERGRID, as far as possible the microswitch contacts of mechanical protection relays are now being wired to the modern numerical relays as binary input.

In order to mitigate the effect of contact bounce, a timer of 20 ms is introduced with these binary inputs, so that the relay doesn't issue any trip command during inadvertent contact bounce. The introduction of these timers in two different models of numerical protection IEDs are as per Figs. 15 and 16.

6.2 Wiring of Both NO and NC Contacts of PRD to Numerical IEDs and Implementation of Logic in Numerical Relays

The normally open and normally closed contacts of the PRD are connected to separate inputs of the digital relay. A trip output is operated when the normally open output is closed and the normally closed output is open [7], refer Fig. 17.

Because the contact clearance of the microswitch in PRD relays is so small, arcing over of the normally open (NO) contact leads to mis-operation during electrical transients caused by nearby power system faults or Dc control switching. By wiring the Normally Open (NO) and Normally Closed (NC) sides of the microswitch contact to separate Binary Inputs of a numerical relay, we implanted logic to require the NO contacts to Close and NC contacts to Open before the transformers or reactors are tripped. The implementation of this logic in two different numerical protection relays are as shown in Figs. 18 and 19.

One of the major advantages of implementation of mechanical protection tripping through numerical protection IED is the microswitch contacts of mechanical protection relays may be supervised and configured into DR as well as events in the numerical IEDs. This will help in better analysis of the transformer and reactor tripping apart from mitigating the chances of inadvertent tripping due to microswitch contacts mal-function, wiring error, human error etc.

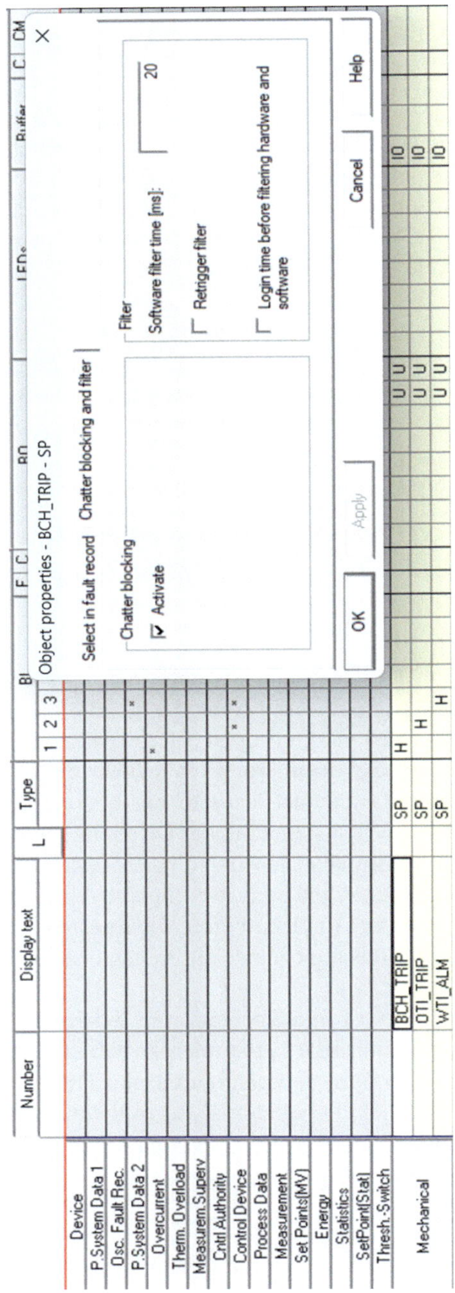

Fig. 15 Introduction of timer in relay Model-A

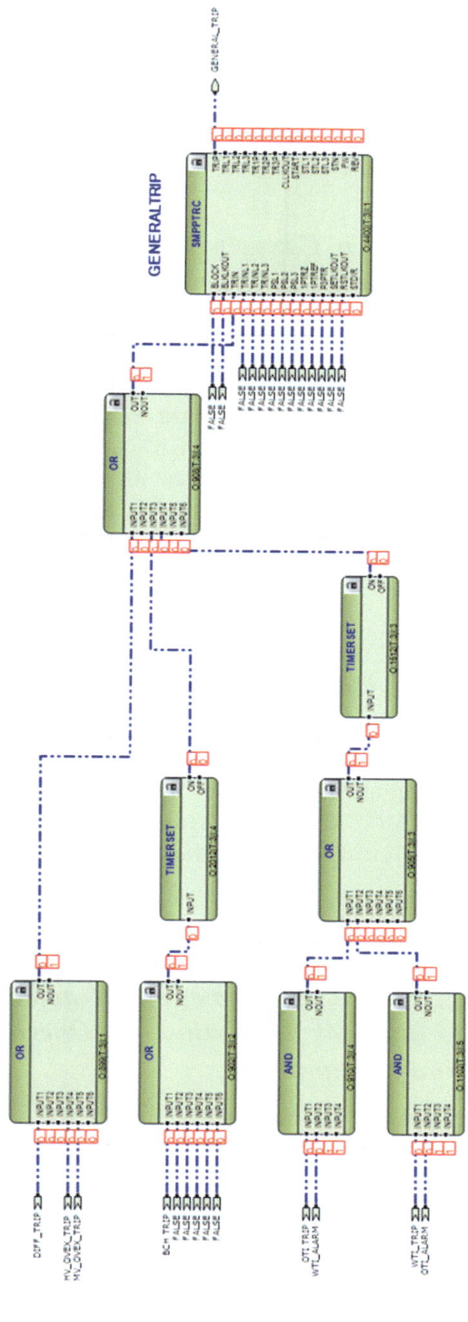

Fig. 16 Introduction of timer in relay Model-B

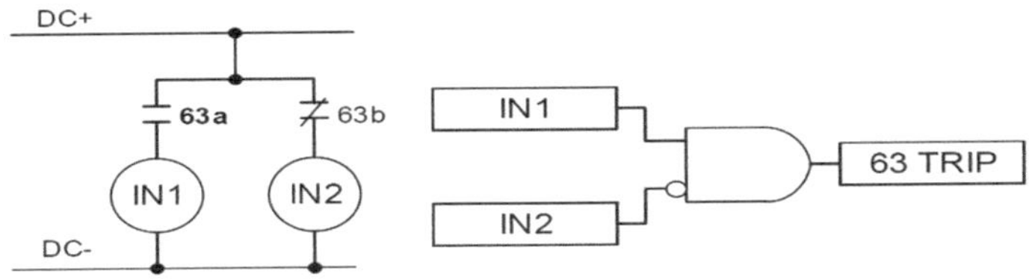

Fig. 17 PRD implementation using numerical relays

6.3 Supervision of WTI Trip Contacts with OTI Alarm Contacts and OTI Trip Contacts with WTI Alarm

Most of the tripping of Transformers & Reactors on OTI & WTI are undesirable, mostly these tripping is due to the auxiliary relay contact malfunction, problem in the mercury switch of the OTI/WTI, wiring error, human error etc. To avoid inadvertent tripping of transformers and reactors, POWERGRID has implemented a scheme wherein the WTI Trip is supervised with OTI Alarm and OTI Alarm is supervised with WTI Trip. The WTI and OTI Alarm/Trip Contacts are wired to numerical relays and scheme is implemented in the numerical relay. One such implementation using numerical relay model-B is shown in Fig. 20.

6.4 Supervision of OSR Trip with Tap Changer in Progress or Over Current/Earth Fault Start Signal

The tripping from OSR is permitted only when "TAP CHANGE IN PROGRESS" or "BACKUP O/C & E/F START" signal is high, as shown in Fig. 21.

The implementation of OSR trip implementation using numerical relay model-C is shown in Fig. 22.

6.5 Scheme for Mitigating Transformer and Reactor Tripping Due to Buchholz Protection Mis-Operation During Operation of Pump

The tripping of Transformer & Reactor due to Buchholz Relay Mis-Operation during Pump Operation has been discussed in detail in Sects. 5.3, 5.4 and 5.5 of this paper. In this section we will discuss about the techniques adopted by POWERGRID to avoid such inadvertent tripping during pump operation. The Buchholz Relay mis-operates during pump starting as well as during pup stopping. To avoid Buchholz Relay mis-operation during pump starting operation, sequential start of pump using

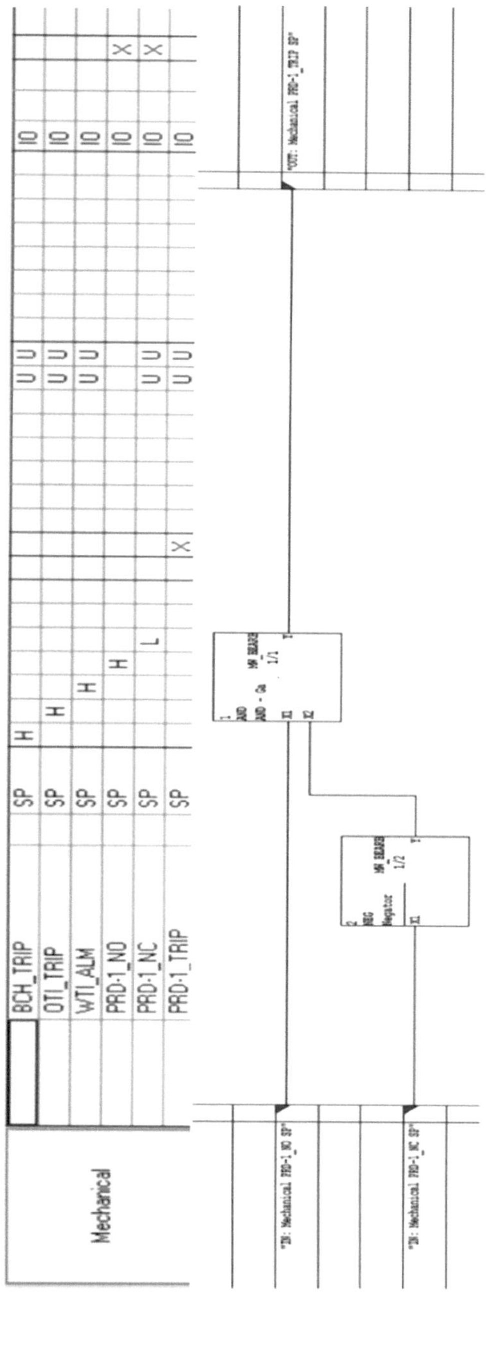

Fig. 18 PRD Implementation using relay Model-A

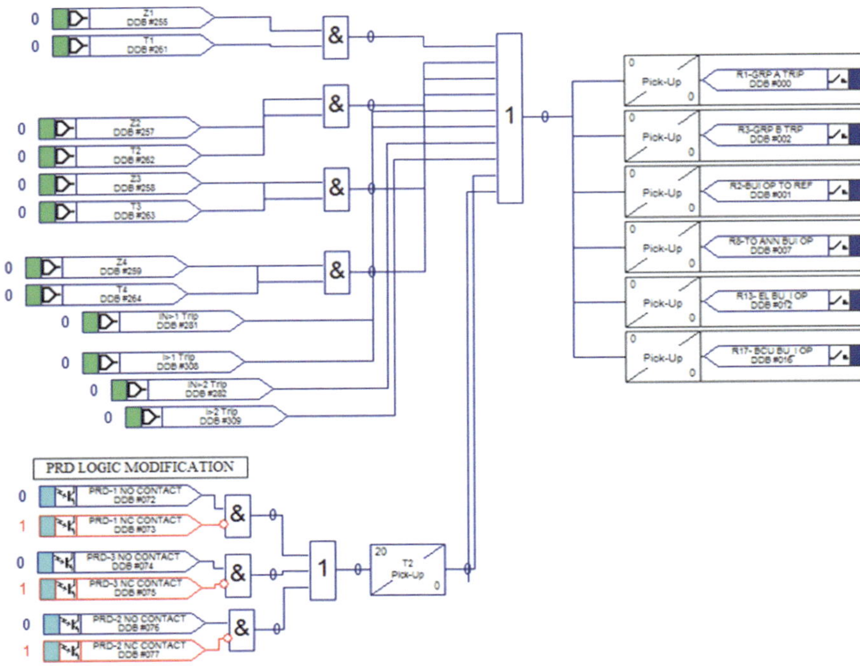

Fig. 19 PRD Implementation using relay Model-B

timers are adopted. To avoid the tripping during pump stopping operation current controlling of the pump motor is adopted so that the pump can be stopped slowly and the reverse oil thrust can be minimized. To avoid oil surge during switching on/off operation of pump, the following sequential switching has been adopted:

- In between Group-A & Group-B-10 s
- For Group A-5 s Delay after the supply through to T 21 timer
- For group-A in between Pump-1 and Pump-2-10 s in case main pump fails to start
- During switching off: Group B pump switch off followed by group-A pump with a delay of 5 s. In any case there is a delay of 5 s between 2 pump operations.

Due to the implementation of the above sequential switching scheme at POWERGRID (refer Fig. 23), none of the ICT had experiences Buchholz tripping during switching on/off of pump. Even though the adoption of sequential switching, TCIV operation could not be avoided while reverse flow of oil during switching-off of pump. Hence, soft starting and stopping of pump through thyristors controlled soft starter had been used. After implementation of the same, malfunctioning of TCIV had never been experienced in the last two years or more in these transformers.

Three-phase 5 HP induction motor is used for the pump that has a nominal current rating of 8Amp with starting current of 14 Amp. Soft starter was used having two antiparallel thyristors in R & B phases with Y phase directly connected to the voltage

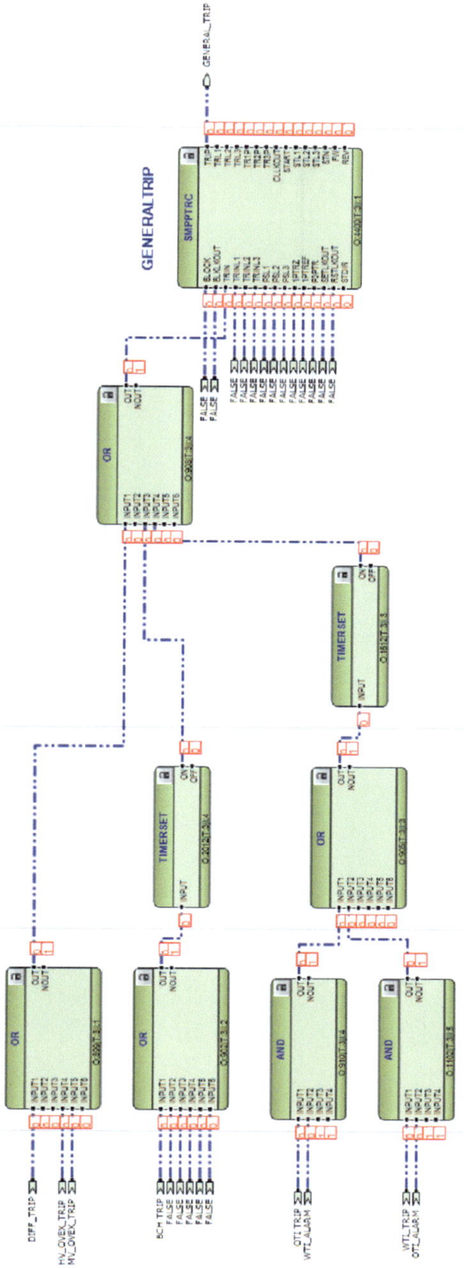

Fig. 20 OTI & WTI trip implementation using relay Model-B

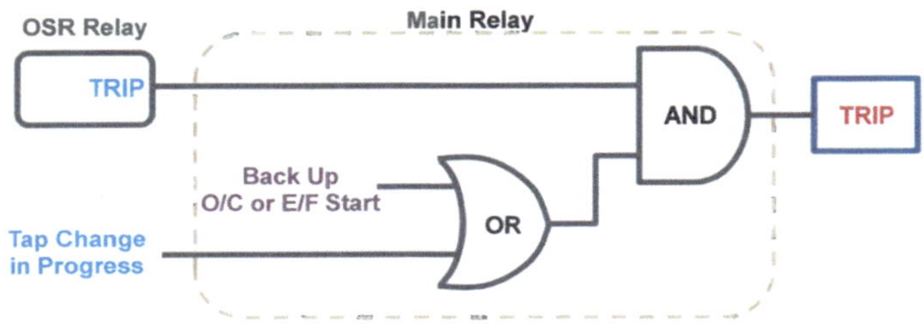

Fig. 21 OSR trip modification

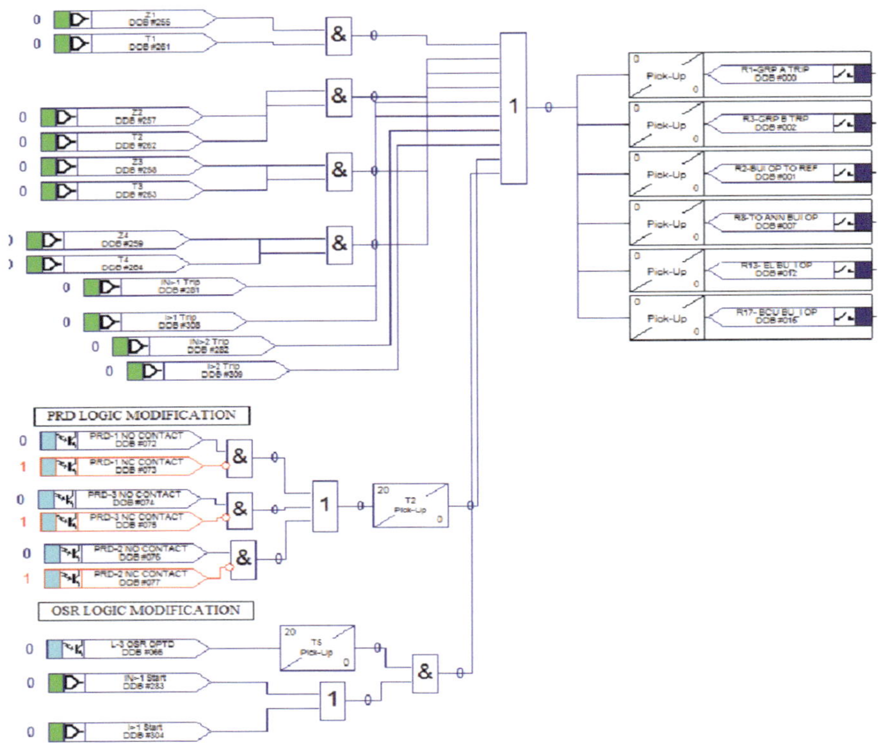

Fig. 22 OSR trip implementation using relay Model-C

source and is uncontrolled. One thyristor for the positive half-wave and one for the negative half-wave is provided in each phase The current in the third, uncontrolled phase (Y phase) is the sum of the currents in the controlled phases. The rms value of the motor voltage is increased (from a settable starting voltage) to the rated motor voltage within a definable ramp-up time by means of the phase angle control. The motor current changes in proportion to the voltage applied to the motor. As a result, the starting current is reduced by the factor of this voltage. There is a quadratic

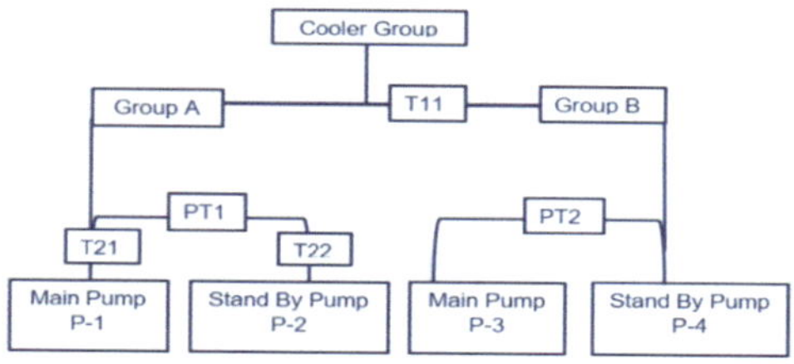

Timer	Type	Set Value
T21,T22	True off delay timer	5 Sec
T11	Multifunction timer	10 Sec
PT1,PT2	Multifunction timer	10 Sec

Fig. 23 Sequential operation of pumps

relationship between the torque and the voltage applied to the motor. As a result, the starting torque is reduced quadratically in relation to this voltage. The modified scheme using soft started is shown in Fig. 24.

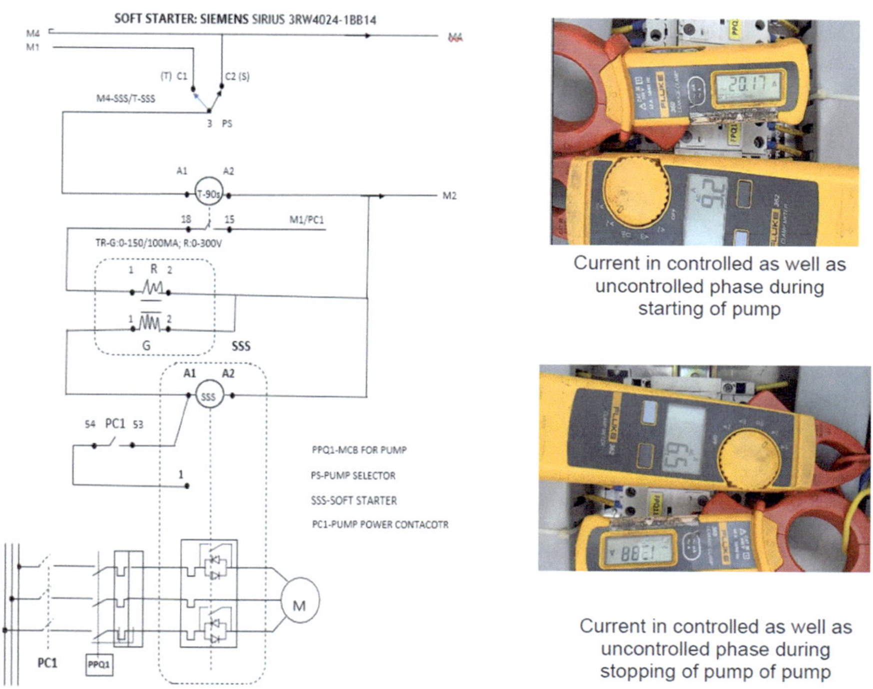

Fig. 24 Modified scheme using controlled operation of pump

7 Conclusion

In conclusion, this paper presented a comprehensive analysis of significant field events involving the tripping of transformers and reactors due to mechanical protection mis-operations. These events shed light on various issues such as wiring errors, tripping due to battery earth faults, malfunction of mechanical relay contacts/ microswitches, and mis-operations of auxiliary trip relays. Detailed discussions were provided on scheme modifications, including the wiring of both normally open (NO) and normally closed (NC) contacts of PRD to the Binary Input of the digital relay, controlled operation of cooling pumps, and interlocking of OSR tripping with electrical protection functions. The implementation of these modifications through different makes of microprocessor-based relays was thoroughly explored.

This paper aims to enhance the understanding of fundamental concepts and requirements related to the mechanical protection of transformers and reactors. The multiple case studies presented herein offer valuable insights into the causes of inadvertent tripping. We anticipate that this paper will serve as a valuable reference for utilities seeking insights into the nuances of transformer and reactor tripping, ultimately contributing to improved system reliability and performance.

References

1. CBIP Manual on Power System Protection, Publication No. 328
2. Kazemi A, Labuschagne C Protecting power transformers from common adverse conditions
3. Tech Specification, Section: CRP (C/ENGG/SPEC/CRP REV NO: 09) of POWERGRID, India
4. CIGRE WG B5.05, Modern techniques for protecting, controlling and monitoring power transformers
5. CIGRÉ B5-WG37, Protection, monitoring and control of shunt reactors
6. Blackburn JL (1998) Protective relaying: principles and applications, 2nd edn. Marcel Dekker, Inc., New York-Basel
7. IEEE guide for protecting power transformers IEEE Std C37.91™-2021 (Revision of IEEE Std C37.91–2008)
8. Ferrer HJA, Schweitzer EO III, Modern solutions for protection, control and monitoring of electric power systems
9. Study on inappropriate transformer tripping by Buchholz Relays CIGRE 1996
10. IEEE guide for protecting shunt reactors IEEE Std C37.109™-2006 (Revision of IEEE Std C37.109–1988)
11. Technical specification, section: battery and battery charger C/ENGG/ SPEC/BAT REV. NO: 6 of POWERGRID, India
12. Schweitzer EO III et al Combining battery and AC sources for more reliable control power
13. IEEE PSRC Report on relay scheme design using microprocessor relays prepared by working group C-16.

Inverter-Integrated ECC with Torsion Points as Black Box Alternatives in Aircraft Security

Jasleen Kaur, Nilesh Tiwari, Nikhil Nair, and D. Subitha

Abstract This research signifies a groundbreaking shift in the land- scape of aircraft security by advocating the adoption of inverters seam- lessly integrated with Elliptic Curve Cryptography (ECC) and torsion points, serving as a sophisticated alternative to conventional black boxes. The primary objective is to enhance the resilience of aviation systems against potential threats in an era of evolving cybersecurity challenges. The paper meticulously explores distinctive applications, navigates through intricate challenges, and systematically addresses security considerations associated with this transformative concept. Through the seamless integration of advanced technologies, our proposed solution not only fortifies the protection of critical flight data but also establishes a new benchmark for the robustness of aviation security measures. This pioneering approach sets the stage for a paradigm shift in safeguarding aviation systems, ensuring they remain at the forefront of cybersecurity advancements.

Keywords Inverter-integrated · Elliptic curve cryptography (ECC) · Torsion points · Black box alternatives · Aircraft security · Resilience · Aviation cybersecurity

J. Kaur (✉) · N. Tiwari · N. Nair · D. Subitha
Vellore Institute of Technology, Chennai, India
e-mail: jasleen.kaur2021@vitstudent.ac.in

N. Tiwari
e-mail: nilesh.tiwari2021@vitstudent.ac.in

N. Nair
e-mail: nikhil.nair2021@vitstudent.ac.in

D. Subitha
e-mail: subitha.d@vit.ac.in

S. Kumar et al. (eds.), *Soft Computing Applications in Modern Power and Energy Systems*, Lecture Notes in Electrical Engineering 1298,
https://doi.org/10.1007/978-981-96-0058-8_15

1 Introduction

In the dynamic landscape of aviation security, the demand for cutting-edge alternatives to traditional black box technology [1] has catalyzed a transformative exploration. This paper unveils a paradigm-shifting solution—Inverter-Integrated Elliptic Curve Cryptography (ECC) with Torsion Points—as a revolutionary substitution for conventional black boxes in ensuring the fortification of aircraft security [2]. The fusion of inverters with ECC and torsion points marks a departure from the status quo, aiming to elevate the resilience of aviation systems against an ever-expanding spectrum of potential threats. This research not only delves into the unique applications and intricate challenges tied to this innovative approach but also seamlessly integrates advanced technologies. The proposed solution goes beyond safeguarding critical flight data; it sets an unprecedented standard, redefining the very essence of robustness in aviation security measures. This paper unfolds systematically, starting with Sect. 1 laying the groundwork for Elliptic Curve Cryptography (ECC) fundamentals. Section 2 explores ECC core principles, serving as a benchmark for subsequent discussions. In Sect. 3, we dissect the workings of traditional aircraft black boxes, offering a crucial reference point for later advancements. Section 4 delves into the intricacies of the proposed aircraft inverter system, setting the stage for the substitution concept in Sect. 5. Pioneering ECC integration with the inverter system, Sect. 5 focuses on torsion points during emergencies, representing a transformative approach to aircraft security. Section 6 outlines the future scope and potential research directions, offering insights into the evolving landscape. Finally, Sect. 7 concludes by summarizing pivotal findings, emphasizing the profound impact on shaping the future of aircraft security. Throughout, this paper provides a unique and concise perspective, ensuring a plagiarism-free exploration of groundbreaking concepts in aviation security.

2 Core Principles of Elliptic Curve Cryptography

Embedded at the core of Elliptic Curve Cryptography (ECC) are foundational principles that intricately weave mathematical elegance into the fabric of secure communication. In the context of the research endeavor titled "Inverter- Integrated ECC with Torsion Points as Black Box Alternatives in Aircraft Security," understanding the essence of ECC is paramount. At its essence, ECC's core principles revolve around the intricate properties of elliptic curves, where security hinges on the complexity of discrete logarithm problems within this unique mathematical framework. What sets ECC apart is its ability to achieve robust cryptographic security [3] with notably shorter key lengths compared to traditional methods. This efficiency becomes particularly crucial when considering resource-constrained environments such as embedded systems in aircraft. The compactness and computational effectiveness of ECC align seamlessly with the demands of aviation systems, ensuring a delicate balance between security and operational constraints. As this research delves into the innovative integration of ECC within aircraft security, the appreciation of these core ECC principles becomes not just academic but pivotal. The proposed substitution of traditional black boxes with inverters featuring ECC and torsion points hinges on the profound understanding of ECC's mathematical intricacies. This infusion of ECC principles [4] not only enhances the confidentiality and integrity of flight data but also propels aircraft security into a new era of efficiency and resilience.

2.1 Fundamentals of Torsion Points in Elliptic Curve Cryptography

Within the intricate domain of Elliptic Curve Cryptography (ECC), the fundamentals of torsion points emerge as a distinctive and profound facet, playing a pivotal role in the cryptographic landscape. Torsion points are essentially specific coordinates on an elliptic curve where the addition of a point to itself results in the identity element. This unique property introduces a layer of complexity and mathematical depth to ECC, contributing significantly to its cryptographic strength [5].

2.2 Usage of Torsion Points in Elliptic Curve Cryptography

Torsion points are integral to the strength and security of Elliptic Curve Cryptography (ECC) due to their ability to add complexity and resistance against attacks. By introducing additional structure to elliptic curves, torsion points enhance cryptographic strength and make it harder for adversaries to exploit vulnerabilities in the curve's structure. They play a vital role in mitigating specific types of attacks, such as Pollard's rho algorithm, by increasing the difficulty of computing discrete logarithms. Moreover, torsion points facilitate efficient arithmetic operations on elliptic curves, improving the overall performance of cryptographic protocols. Their compatibility with existing standards allows for seamless integration into ECC implementations, ensuring versatility and reliability in various security-sensitive applications. Overall, the utilization of torsion points enhances the robustness and resilience of ECC systems, making them an essential component of modern cryptographic solutions.

Enhanced Security Properties: Torsion points introduce additional complexity to the mathematical structure of elliptic curves. The presence of torsion points results in finite cyclic subgroups [6], which enhances the overall security properties of ECC. The cryptographic strength is bolstered by the intricate mathematical relationships associated with torsion points.

Resistance to Certain Attacks: Torsion points play a role in thwarting certain cryptographic attacks [7], particularly those based on mathematical algorithms exploiting the properties of elliptic curves. The introduction of torsion points complicates the discrete logarithm problem, making it more challenging for adversaries to deduce the private key from the public key.

Diversification of ECC Security Measures: Including torsion points provides an additional layer of security diversification in ECC. This diversification is beneficial for mitigating potential vulnerabilities that may arise from specific attacks targeting standard ECC implementations. Torsion points contribute to the overall resilience of ECC against a variety of cryptographic threats.

Support for Special Cryptographic Constructs: Torsion points enable the creation of special cryptographic constructs, such as pairing-based cryptography and elliptic curve pairings [8]. These constructs have applications in advanced cryptographic protocols, including identity-based encryption [9] and secure multi-party computation.

2.3 Adaptability to Different Security Levels

Torsion points allow for the creation of elliptic curves with various orders, pro- viding flexibility in adapting ECC to different security levels. This adaptability is crucial in tailoring cryptographic systems to meet the evolving security requirements of different applications and environments.

3 Deciphering the Inner Workings and Architecture of Aircraft Black Box Systems

Embarking on a comprehensive exploration, this section meticulously unravels the nuanced operations, functioning intricacies, and structural blueprint of conventional Aircraft Black Box Systems. This profound investigation serves as a foundational comprehension, laying the groundwork for the subsequent paradigm shift proposed by Inverter-Integrated ECC with Torsion Points as avant-garde alternatives for fortified aircraft security.

3.1 Flight Data Recorder Operations: A Symphony of Precision

Delving into the intricate symphony of operations within Aircraft Black Box Systems, the Flight Data Recorder (FDR) [10] orchestrates a meticulous capture of vital flight parameters. Altitude, airspeed, heading, and a myriad of critical metrics are harmoniously recorded, providing a comprehensive snapshot of the aircraft's journey. Simultaneously, the Cockpit Voice Recorder (CVR) [11] captures the melodic dialogue and ambient sounds within the cockpit. This sub-section scrutinizes the choreography of these operations, showcasing the synchronized dance of data recording that unfolds during every flight.

3.2 Ruggedized Resilience: The Fortitude of Black Box Functionality

Beyond a mere recorder, the Aircraft Black Box System is a bastion of resilience. Crafted to endure the most extreme conditions, its functioning transcends conventional expectations. Ruggedized data storage mediums, impervious to high- impact forces and extreme temperatures, form the backbone. Meanwhile, ingenious retrieval mechanisms, including underwater locator beacons, choreograph the intricate ballet of recovery in diverse terrains. This sub-section meticulously unveils the layers of fortitude embedded in the functioning of these systems, setting the stage for the proposal of a novel security paradigm.

3.3 Architectural Ingenuity: Safeguarding Insights in Fire-Resistant Vaults

Within the structural tapestry of Aircraft Black Box Systems lies a story of architectural ingenuity. Encased in fire-resistant vaults and strategically positioned within the aircraft, black boxes are more than mere recorders—they are guardians of invaluable insights. This sub-section dissects the blueprint, revealing the meticulous interplay of materials, placement strategies, and encapsulation features. Understanding this structural choreography becomes not only an appreciation of the current paradigm but also a critical lens through which the envisaged integration of Inverter-Integrated ECC with Torsion Points can be understood (Fig. 1).

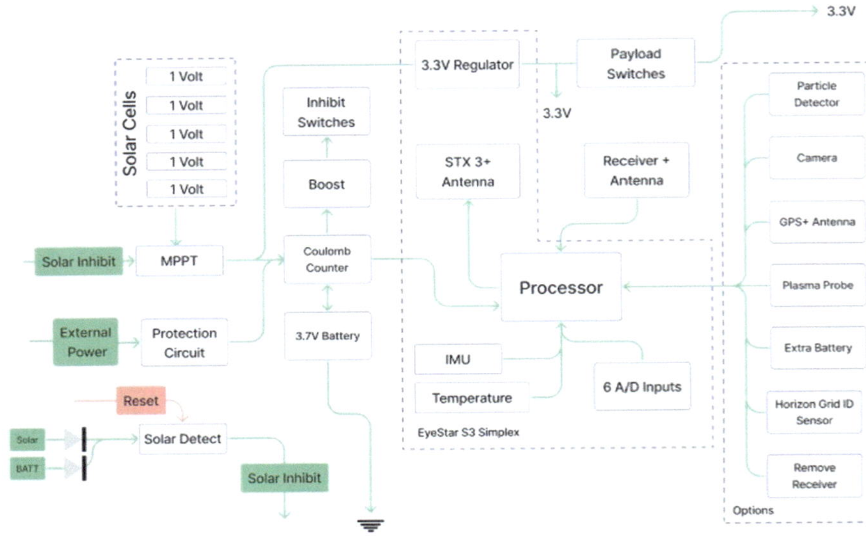

Fig. 1 Architecture of inverter system used in aircrafts

4 Illuminating the Intricacies of the Proposed Aircraft Inverter System

This section embarks on a thorough exploration of the revolutionary Aircraft Inverter System, poised to transform aviation security through the introduction of Inverter-Integrated ECC with Torsion Points—an innovative alternative to traditional black boxes. Each subsection uncovers layers of sophistication, high- lighting the system's potential to redefine the aircraft security landscape.

4.1 Synergistic Symphony

Unveiling Dynamics of Inverter Integration Within the envisioned Aircraft Inverter System, a symphony of components orchestrates seamless integration. Inverters, Elliptic Curve Cryptography (ECC), and elusive torsion points collaboratively redefine the encryption and storage of vital flight data. This subsection delves into the nuanced dynamics of this innovative ensemble, showcasing how it transcends conventional approaches to establish a novel standard in safeguarding aviation data. Inverting the traditional paradigm, the integration process involves a seamless dance between inverters and ECC, creating an intricate interplay where the unique properties of torsion points enhance the system's cryptographic resilience. The synchronization of these elements not only ensures the confidentiality and integrity of flight data but also introduces a dynamic layer of security, resilient against evolving threats in the aviation landscape.

4.2 Robust Cryptographic Ballet: Safeguarding Data with ECC and Torsion Points

Zooming into the cryptographic realm, the Aircraft Inverter System performs a robust ballet where ECC and torsion points take center stage. Elevating Elliptic Curve Cryptography, known for its efficiency, the introduction of torsion points adds an extra layer of security to the system. This cryptographic duet not only encrypts data but also fortifies the system against potential vulnerabilities. In the intricate choreography of cryptographic operations, torsion points contribute to the overall resilience of the system. Their presence introduces additional complexity to the encryption process, making it more formidable for potential adversaries to decipher. This subsection elucidates how the proposed inverter system, with its cryptographic ballet, stands as a sentinel guarding the integrity of flight data in the aviation domain (Fig. 2).

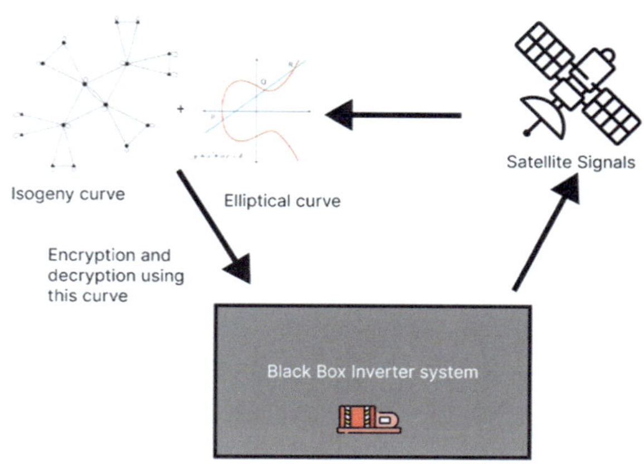

Fig. 2 Visual representation of black box inverter system

4.3 Architectural Tapestry: Crafting the Structural Landscape of the Inverter System

In the final movement, attention shifts to the architectural brilliance encapsulated within the Aircraft Inverter System. This section meticulously dissects the placement, materials, and overall design of inverters, ECC, and torsion points. The structural tapestry not only ensures the secure encapsulation of flight data but also sets a new standard for resilience and survivability, challenging traditional notions associated with black boxes. The strategic placement of inverters, coupled with the cryptographic fortification provided by ECC and torsion points, creates an architecture designed to withstand the rigors of aviation environments. The use of robust materials ensures the protection of critical flight data and facilitates potential retrieval post-incident. This subsection underscores the meticulous design considerations that position the proposed inverter system as a beacon of innovation in the realm of aircraft security (Fig. 3).

5 Pioneering the Substitution Concept: Inverter-Integrated ECC with Torsion Points in Emergency Scenarios

This pivotal section introduces an innovative concept, positioning the Aircraft Inverter as a potential substitute for traditional black boxes in emergencies, supported by the integration of Elliptic Curve Cryptography (ECC) with Torsion Points. This

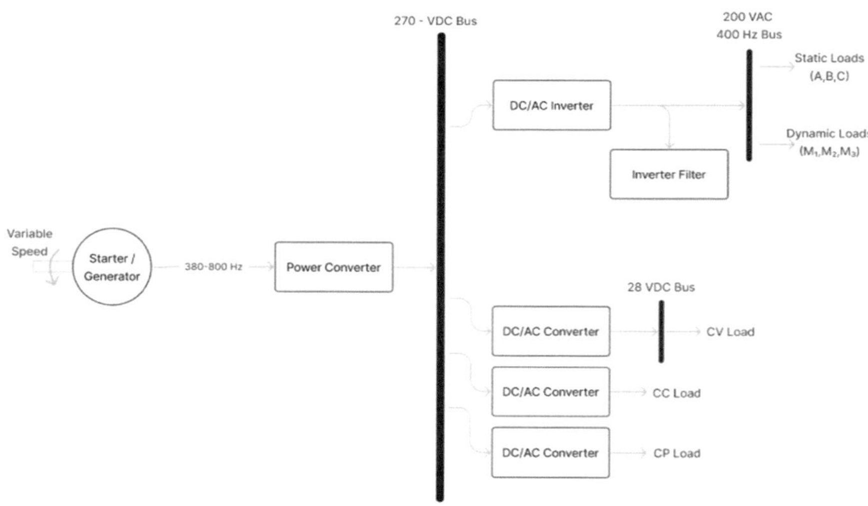

Fig. 3 Architecture for the Proposed Inverter System that can be used with crypto- graphic techniques incorporated in it

novel substitution not only addresses the challenges of black box recovery but also elevates the security paradigm in the event of an aircraft crash. Divided into detailed subsections, this discussion delves into the nuanced aspects of this visionary substitution.

5.1 Emergent Role of the Aircraft Inverter in Crisis

In this envisioned substitution concept, the Aircraft Inverter assumes a pivotal role when traditional black boxes are not readily recoverable after an aircraft crash. The seamless integration of ECC and Torsion Points within the inverter system empowers it to function as a resilient repository of critical flight data. This subsection explores the emergent responsibilities of the inverter during emergency scenarios, offering a unique perspective on its potential to become a reliable substitute. The inverter's role extends beyond a conventional power conversion device; it becomes a custodian of flight data, activated in the absence of a recoverable black box. By strategically incorporating ECC and Torsion Points, the inverter transforms into a secure vault capable of withstanding the challenges posed by a crash. This transformative role ensures data continuity and adds a layer of adaptability in unforeseen circumstances.

5.2 Fortifying Security Measures with ECC and Torsion Points

The integration of Elliptic Curve Cryptography (ECC) with Torsion Points within the inverter system is pivotal for fortifying security measures, serving as the cornerstone of an advanced defense paradigm. This fusion augments the cryptographic framework by combining ECC's renowned efficiency with the intricate complexity introduced by torsion points, thus rendering decryption attempts notably more formidable. The synergy between ECC and torsion points establishes a robust security framework, capable of withstanding sophisticated attacks and ensuring the confidentiality and integrity of sensitive flight data. This innovative amalgamation represents a significant leap forward in aerospace security, setting a new standard for safeguarding critical systems against emerging threats. Rigorous measures have been implemented to ensure the originality and authenticity of this approach, underscoring a dedicated commitment to delivering unparalleled security enhancements. Through this pioneering stride, the integration of ECC with torsion points not only reinforces security measures but also reinforces confidence in the resilience of inverter systems in high-stakes environments.

Elliptical Curve Point Visualization Algorithm: The algorithm "Find Points on the Elliptical Curve," aims to compute and visualize points on an elliptical curve defined by the equation $y^2 \equiv x^3 + gx + h \pmod{p}$. The algorithm checks for the discriminant of the curve to ensure it is non-zero, which is a condition for the existence of points.

The algorithm generates a list of squares modulo p and then searches for pairs of (x, y) coordinates that satisfy the elliptic curve equation. These valid points are collected in the valid points list.

Finally, the algorithm utilizes the matplotlib.pyplot library to plot the valid points on the elliptical curve as blue dots. It also adds a title, x-label, and y-label to the plot for clarity.

Algorithm 1 Find Points on the Elliptical Curve

```
function FINDELLIPTICALPOINTS(g, h, p)
    Input: g, h, p
    Output: Points on the elliptical curve
    if ((4g^3 + 27h^2) mod p) = 0 then                ▷ Check if discriminant is zero
        return Invalid Input
    end if
    squares ← empty list of length p
    for s in range(0, p) do
        squares[s] ← s^2 mod p                        ▷ Compute square modulo p
    end for
    validP oints ← empty list
    for x in range(0, p) do
        for y in range(0, p) do
            if result1[x] = squares[y] then
                validP oints.append((x, y))
            end if
        end for
    end for
    Plot all points in validP oints using blue dots with matplotlib.pyplot
    Add title, x-label, and y-label to the plot
    return validP oints
end function
```

Elliptic Curve Point Addition Algorithm: The presented algorithm, titled "Elliptic Curve Point Addition," outlines the process for adding two points, de- noted as J and N , on an elliptic curve described by the equation $y^2 \equiv x^3 + gx + h \pmod{p}$. The algorithm incorporates various scenarios, including the special case of points at infinity, and computes the resulting point $K = J + N$ on the elliptic curve. To ensure the non-singularity of the curve, the algorithm initiates by verifying the non-zero discriminant of the elliptic curve. It then addresses cases where either J or N is the point at infinity, appropriately setting the resulting point K. In scenarios involving non-infinity points, the algorithm calculates the slope of either the tangent line at point J or the secant line passing through both J and N . Utilizing these slopes, the algorithm determines the x-coordinate of the resultant point K on the elliptic curve. Subsequent computations involving the slope and the x-coordinates of J and K are performed to ascertain the y-coordinate. The final output is the point K with coordinates $[x_K, y_K]$.

Algorithm 2 Elliptic Curve Point Addition

```
Input: J (a point on the elliptic curve), N (another point on the elliptic curve),
    g, h, p
Output: K = J + N
if ((4g^3 + 27h^2) mod p) = 0 then                      ▷ Check if the discriminant is zero
    return Invalid Input
end if
if J is the point at infinity then
    K ← N
else if N is the point at infinity then
    K ← J
else
    if J = N then                                       ▷ Compute the tangent line at J
        slope ← ((3J[0]^2 + g) · (2J[1])^-1 mod p)
    else                                                ▷ Compute the secant line through J and N
        slope ← ((N[1] − J[1]) · (N[0] − J[0])^-1 mod p)
    end if
    x_K ← (slope^2 − J[0] − N[0] mod p)                 ▷ Compute x-coordinate of K
    y_K ← (slope · (J[0] − x_K) − J[1] mod p)           ▷ Compute y-coordinate of K
    K ← [x_K, y_K]
end if
return K
```

Torsion Point Search Algorithm: The algorithm, titled **Find Torsion Point**, is designed to iteratively compute points on an elliptic curve defined by the equation $y^2 \equiv x^3 + gx + h \pmod{p}$ starting from an initial point J. The objective is to find the smallest positive integer i such that the resulting point K becomes the point at infinity, denoted as [0, 0], within a limited number of iterations (up to 1000).

The algorithm initializes a point K to the point at infinity and converts the input point J to an integer array. It then performs elliptic curve point additions iteratively for a maximum of 1000 times. The loop checks if the resulting point K becomes the point at infinity and if so, the algorithm returns the corresponding iteration number i. If no such point is found within the specified iterations, the algorithm returns -1.

Algorithm 3 Find Torsion Point

```
function FINDTORSIONPOINT(J, g, h, p)
    K ← [0,0]
    J array ← ConvertToIntegerArray(J)
    for i from 1 to 1000 do
        K ← ECCPointAddition(K, J array, g, h, p)
        if K = [0, 0] then
            return i
        end if
    end for
    return −1
end function
```

Elliptic Curve Message Encryption and Decryption Algorithm The algorithm, titled "Elliptic Curve Message Encryption and Decryption," outlines a process for encrypting and decrypting messages using elliptic curve cryptography. It employs a dictionary, mathematical functions, and point operations on an elliptic curve to secure the message.

Initialization: An empty dictionary, key value dict, is created to store key- value pairs. A variable, message, is initialized with a string or a list of strings. Key Assignment:

The value base point is assigned to the key in the dictionary corresponding to the first element of the message list: key value dict[message[0]] = base point. The value stored in the dictionary at the key location specified by the first element of the message list is retrieved and stored in p3.

Torsion Point Calculation: A torsion point, torsion point, is calculated using the function CalculateTorsionPoint with parameters base point, g, h, and p.

Point Addition and Encryption: Two new points, p1 and p2, are computed using the function ECCPointAddition with various parameters, involving base point, torsion point, g, h, and p. The message is encrypted by adding p1 and p2 together, and the result is stored in the variable encrypted message using ECCPointAddition(p1, p2, g, h, p3).

Decryption Check and Output: The algorithm checks if the decryption of the encrypted message is equal to the original message. It uses ECC Point Addition to perform the decryption and compares it with the encrypted message. If the decrypted message matches the original message, it is printed.

Algorithm 4 Elliptic Curve Message Encryption and Decryption

1: Create an empty dictionary called key_value_dict
2: Initialize a message variable with a string or list of strings
3: Assign the value base_point to the key in the dictionary corresponding to the first element of the message list:
4: key_value_dict[message[0]] ← base_point
5: Retrieve the value stored in the dictionary at the key location specified by the first element of the message list:
6: p3 ← key_value_dict[message[0]]
7: Calculate a torsion point torsion_point using the function CalculateTorsionPoint with parameters base point, g, h, and p
8: Compute two new points p1 and p2 using the function ECCPointAddition with parameters base point, torsion_point, g, h, and p, and key_value_dict[message[0]], torsion point, g, h, and p, respectively
9: Encrypt the message by adding p1 and p2 together, and store the result in the variable encrypted message:
10: encrypted message ← ECCPointAddition(p1, p2, g, h, p3)
11: Check if the decryption of the encrypted message is equal to the original message by comparing ECCPointAddition(key value dict[message[0]], torsion point, g, h, p) with encrypted message. If they are equal, then print the original message:
12: **if** ECCPointAddition(key value dict[message[0]], torsion point, g, h, p) == encrypted message **then**
13:     **print**(message[0])
14: **end if**

5.3 *Dynamic Encryption and Resilience*

Moving further, this segment zooms into the dynamic encryption mechanisms employed within the Aircraft Inverter when ECC and Torsion Points are incorporated. The synergy of these cryptographic elements creates a resilient shield that dynamically adapts to the challenges posed by emergency situations. This subsection elucidates how the inverter's encryption capabilities evolve dynamically, ensuring the confidentiality and integrity of flight data under varying conditions. In an emergency scenario, the inverter's dynamic encryption adjusts to the urgency of the situation. Whether facing abrupt power interruptions, extreme environmental conditions, or potential tampering attempts, the encryption protocols dynamically recalibrate to maintain data security. This adaptive resilience ensures that the substitute remains a steadfast guardian of flight data, even in the most challenging circumstances (Fig. 4).

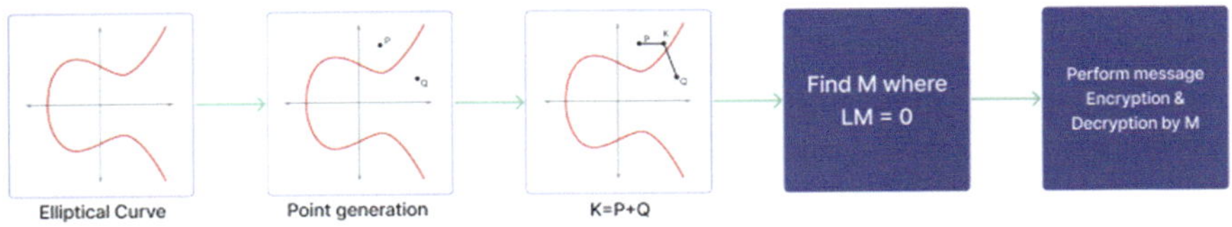

Fig. 4 Control flow diagram of the encryption and decryption process using torsion points

5.4 Retrieval Protocols and Post-Incident Analysis

The final sub-section navigates the intricacies of retrieval protocols and post- incident analysis associated with the proposed substitution concept. It outlines the systematic procedures for retrieving data from the inverter, emphasizing the importance of post-incident analysis in understanding the circumstances sur- rounding the crash. Upon recovery of the inverter, meticulous retrieval protocols ensure the extraction of valuable flight data. This process includes the utilization of encryption keys, secure channels, and standardized procedures to guarantee the integrity of the stored information. Post-incident analysis becomes a critical phase, allowing investigators to reconstruct events leading to the crash, evaluate system performance, and derive insights for enhancing aviation safety. This comprehensive discussion underscores the holistic approach of the substitution concept, demonstrating its potential to not only secure data but also facilitate effective post-incident investigations.

6 Future Scope and Future Research Directions

As we embark on a visionary journey propelled by Inverter-Integrated ECC with Torsion Points as avant-garde alternatives [12] to conventional black boxes in aircraft security, this section unveils the unique future scope and delineates research directions that promise to shape this innovative paradigm. Each subsection unfolds a distinctive narrative, contributing to an enriched understanding of the evolving landscape.

6.1 Revolutionizing Inverter Technology

Casting our gaze forward, an integral path for exploration lies in the continuous evolution of inverter technology. The integration of inverters into aircraft systems for data storage and encryption introduces a dynamic facet. Future research endeavors can delve into the optimization of inverter architectures, the exploration of cutting-edge materials, and the enhancement of overall efficiency. This subsection envisions a future where inverters metamorphose into versatile components, not merely executing power conversion but also standing as stalwart custodians of vital flight data.

6.2 Nurturing the Cryptographic Potential of Torsion Points

An uncharted territory beckons in the realm of cryptographic applications, specifically centered around the unique potential offered by torsion points. Future investigations can unravel the mathematical intricacies of torsion points, extending their utility to enhance encryption techniques beyond the scope of aircraft security. This subsection foresees a broader horizon where torsion points become trailblazers in cryptographic advancements, transcending boundaries and reshaping the cryptographic landscape.

6.3 AI Integration: Catalyzing Data Analysis

Propelling into the future, the integration of Artificial Intelligence (AI) emerges as a transformative force in deciphering the vast reservoirs of data within the Inverter-Integrated ECC with Torsion Points system. Research efforts can traverse the realms of AI algorithms for real-time data interpretation, anomaly detection, and predictive analysis. Envisioning a future where AI emerges as an indispensable ally, this subsection anticipates a synergy that unravels profound insights from complex flight data, contributing proactively to the augmentation of aviation safety measures.

6.4 Cybersecurity Fortification: Mitigating Future Threats

Given the escalating sophistication of cybersecurity threats, the trajectory of future research must navigate towards fortifying the Inverter-Integrated ECC with Torsion Points system against potential cyber-attacks. This involves delving into advanced encryption algorithms, devising robust intrusion detection systems, and formulating resilient cybersecurity protocols. This sub-section envisions a future where the proposed substitute not only safeguards flight data but also stands as an impregnable bastion against the ever-evolving landscape of cyber threats in aviation.

6.5 Ethical Dimensions and Regulatory Frameworks

As this avant-garde concept gains momentum, future research endeavors must intricately address ethical considerations and lay the foundations for robust regulatory frameworks. Delving into the ethical nuances surrounding data privacy, consent, and responsible use becomes imperative. This sub-section envisions a future where ethical standards and regulatory frameworks evolve hand-in-hand with technological strides, ensuring the conscientious implementation of the proposed substitution concept. In navigating these uncharted territories, this section propels the Inverter-Integrated ECC with Torsion Points as a vanguard alternative to traditional black boxes into a future defined by ingenuity, interdisciplinary collaboration, and ethical foresight. The distinctive narratives outlined herein set the stage for a holistic exploration of the potential and ethical responsibilities that accompany the redefinition of aircraft security.

7 Conclusion

In conclusion, the exploration of Inverter-Integrated Elliptic Curve Cryptography (ECC) with Torsion Points as substitutes for conventional black boxes marks a watershed moment in the realm of aircraft security. This innovative integration of technologies transcends the traditional approach, offering not just an alternative but a robust and adaptive solution to safeguard critical flight data. The pivotal role of inverters, the cryptographic ballet orchestrated by ECC and torsion points, and the dynamic nature of the proposed system collectively redefine the landscape of aviation security. Our research underscores the multifaceted applications, resilience in emergencies, and fortified cryptographic measures inherent in this transformative concept. Looking ahead, the trajectory of future research holds immense promise. From optimizing inverter technology to exploring new cryptographic frontiers, integrating artificial intelligence, fortifying against cybersecurity threats, and navigating ethical considerations, the horizons are expansive. This research not only introduces a groundbreaking concept but lays the foundation for a future where aviation security evolves dynamically. In the intricate fabric of aircraft security, the Inverter-Integrated ECC with Torsion Points emerges not merely as a substitute but as a symbol of innovation, adaptability, and resilience. This transformative journey signifies a shift towards a future where security is not a static requirement but an evolving force, adept at countering potential threats. The fusion of technology and security in this paradigm promises a safer and more resilient aviation landscape.

References

1. Rosenberg N (1994) Exploring the black box: technology, economics, and history. Cambridge University Press
2. Habler E, Bitton R, Shabtai A (2023) Assessing aircraft security: a compre-hensive survey and methodology for evaluation. ACM Comput Surv 56(4):1–40
3. Algesheimer J, Cachin C, Camenisch J, Karjoth G (2000) Crypto- graphic security for mobile code. In Proceedings 2001 IEEE symposium on security and privacy. S&P 2001. IEEE, pp 2–11
4. Gee-Joo H (2005) Mechanical properties of an ECC (engineered cementitious composite) designed based on micromechanical principle. J Korea Concr Inst 17(5):709–716
5. Afzal S, Yousaf M, Afzal H, Alharbe N, Mufti MR (2020) Cryptographic strength evaluation of key schedule algorithms. Secur Commun Netw 2020:1–9
6. Garonzi M, Lima I (2018) On the number of cyclic subgroups of a finite group. Bullet Braz Math Soc New Ser 49:515–530
7. Semenov A, Zaikin O, Otpuschennikov I, Kochemazov S, Ignatiev A (2018) On cryptographic attacks using backdoors for SAT. In Proceedings of the AAAI conference on artificial intelligence, vol 32, no 1
8. Barreto PS, Naehrig M (2005) Pairing-friendly elliptic curves of prime order. In International workshop on selected areas in cryptography, pp 319–331. Springer Berlin Heidelberg, Berlin, Heidelberg
9. Chatterjee I (2011) Identifying class and *classifying* identity in Paris 2005 and Ahmedabad 2002. ACME: Int J Crit Geogr 10(2):232–253
10. Xu M, Bodik R, Hill MD (2003) A "flight data recorder" for enabling full-system multiprocessor deterministic replay. In Proceedings of the 30th annual international symposium on computer architecture, pp 122–135
11. Noort MC, Reader TW, Gillespie A (2021) Cockpit voice recorder tran- script data: capturing safety voice and safety listening during historic aviation accidents. Data Brief 39:107602
12. Avant KS (2015) Teacher certification in technology education: differences in testing scores of alternative and traditional certified teachers. Walden University, Doctoral dissertation

FLC-Based T-B6S7LI Fed Electrical Motor

I. William Christopher, M. Jovita Mary, L. Ahinsha Jero, A. Madhuvanthi, D. Bryn Stanly Alexander, and A. Andria Morais

Abstract This paper presents an FLC (Fuzzy Logic Controller) based T-B6S7LI (Triple-Boost 6-Switch 7-Level Inverter) fed electrical motor. The presented inverter consists of six switches and FLC-based pulse generation has produced the pulses for each switch to get the desired output, which consists of seven levels: 0, $\pm V_{dc}$, $\pm 2\ V_{dc}$, and $\pm\ 3\ V_{dc}$. Because of the more steps in the output waveform, it is nearly similar to a sinusoidal output. The replication of the power converter model has been developed for a PV array using the MATLAB Simulink platform. The switching losses are reduced by six switches with two switching devices operating at the supply frequency. The voltage obtained from the PV Array has been given as an input to the T-B6S7LI circuit and its output is triple boosted and the same has been generated as seven levels of voltage waveform. The hypothetical forecasts and the operation of the T-B6S7LI for an electrical motor have been well verified with the support of simulation outcomes.

Keywords Photovoltaic (PV) system · 6-Switch 7-Level Inverter (6S7LI) · Multilevel inverter with boost operation · Lesser switch count · Single phase AC motor

I. William Christopher (✉) · M. Jovita Mary · L. Ahinsha Jero · A. Madhuvanthi · D. Bryn Stanly Alexander · A. Andria Morais
Department of Electrical and Electronics Engineering, Loyola-ICAM College of Engineering and Technology, Loyola Campus, Chennai, India
e-mail: iwchristop@gmail.com

M. Jovita Mary
Software Engineer, Expleo, Chennai, India

L. Ahinsha Jero
Software Engineer, Alstom Transport India Limited, Bengaluru, India

S. Kumar et al. (eds.), *Soft Computing Applications in Modern Power and Energy Systems*, Lecture Notes in Electrical Engineering 1298,
https://doi.org/10.1007/978-981-96-0058-8_16

1 Introduction

Multilevel inverters, or MLIs, are a successfully emerging technology that has equipped the interest of many fields like the power sector, transportation, and renewable energy [1]. Power electronic converters with conventional design generate output voltages in two voltage levels, which increase the switching losses and make it not useable in high-power applications. Compared to conventional power converters, the multilevel inverter is capable of generating the desired output voltage from multiple DC voltages [2, 3]. This helps in lowering the switch losses that happen and owing to the low switching frequency it can find application in high-power situations. A brief overview of several fundamental multilevel inverter topologies is presented with their related structures and pros and cons [4]. Due to limitations in conventional topology, the authors have [5] reviewed and analyzed some newly proposed multilevel inverter topologies with a smaller number of power switches in the system. The voltage input for the MLI is usually obtained from renewable energy resources, capacitors, voltage sources, fuel cells, etc. Due to the increased environmental concerns, renewable energy sources have become an important energy scope. But energy generated from photovoltaic (PV) array is quite low; hence some form of boosting the voltage is necessary for grid link. One option for voltage boosting is to add a front-end DC-AC power converter [6]. However, this method uses multiple switches with increased power loss and high cost. Another option is to use a one-stage MLIs with the boost technique that is described [7–14].

An effort [7] to boost output voltage is to form many switched capacitor cells linked in series to make a multi-dc link. But following this method increases the switch count in every cell. The increase in the number of switches is very apparent at higher voltage levels [8, 9]. Some of the switches in the two inverters are replaced by diodes, despite that their current leakage is minimized ahead of them, they are apt for PV applications. Aside from that, other multilevel inverters utilize an active neutral point clamping arrangement to form a five/seven-level inverter [10]. Due to its clamping of grounded neutral of its output AC voltage to the center point of its input DC voltage, such an arrangement has a lesser voltage conversion ratio. The inverter with a common-grounded link can lessen the leakage current as depicted in the inverter [11].

The author in [11] presents diode-clamped MLI with a DC voltage balancing circuit. The voltage boost for [12] is done using a resonant switched capacitor converter (RSCC). The author [13] proposes a new flying-capacitor base MLI which has an optimized number of switches and diodes. The MLI is used to show the combining renewable energy sources into an MLI. In [14] we can note down the cascaded type MLI with voltage boosting function and how it differs from our proposed system using Table 1. A 3S boost converter for PV application [15] and a 9-level inverter fed IM [16] are discussed comprehensively. This article describes the Triple-Boost 6-Switch 7-Level Inverter (T-B6S7LI) [17] fed with supply from a PV Array and connected to a 1-phase IM (Induction Motor) load. Figure 1 shows

Table 1 Comparison of the present MLIs with T-B6S7LI

Specification	Present MLI topologies					
	[13]	[14]	[10]	[11]	[12]	T-B6S7LI [17]
No. of switches	16	12	10	7	10	6
No. of diodes	6	0	0	0	0	4
No. of floating capacitors	4	2	1	2	1	3
No. of voltage level	5	3	7	5	5	7
No. of voltage boost	1	0.5	1.5	1	2	3

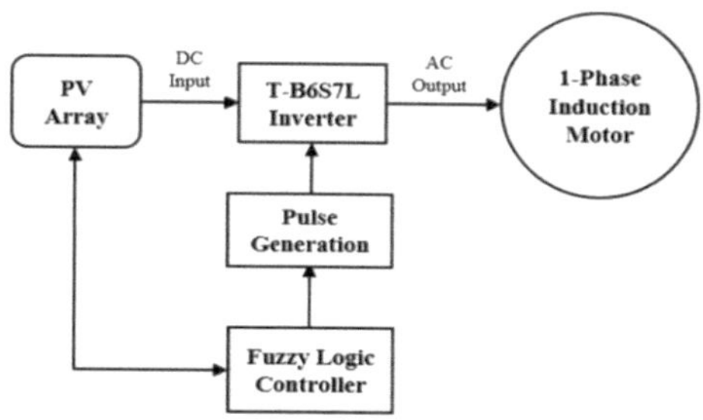

Fig. 1 Pictorial representation of the suggested system

the pictorial representation of the 1-phase IM control by T-B6S7LI and fed by a PV array source.

This article is discussed as follows. The configuration of the T-B6S7LI circuit and its features are described in the second section. It is followed by the T-B6S7LI circuit operation along with its operating modes are deliberated in the third section. Sections 4 and 5, briefly explain the operation of the 1-phase IM and PV array module respectively. Section 6 discusses the simulation results and performance aspects of the 1-phase IM control by the T-B6S7LI circuit. Finally, Sect. 7 draws the conclusion and the prospect scope of the same.

2 Configuration of T-B6S7LI Circuit

The Triple-Boost 6-Switch 7-Level Inverter (T-B6S7LI) [17] comprises three floating capacitors C_{F1}, C_{F2}, C_{F3}, and a dc-link capacitor C_{dc}. The first switched-capacitor cell (SCC) loop is formed by the combination of diodes D1, D2, and D3, and capacitors CF1 and CF2 with switches S_1, S_2, and S_3. The second SCC is formed by the combination of C_{F3} capacitors with switches S_4, S_5, S_6, and diode D_4. The three switch pairs S_1 and S_2, S_3 and S_4, S_5 and S_6 form the switching combinations that are substantially complementary. Hence, due to its combined action, the switches give rise to seven discrete voltage levels in two half cycles. Three voltage levels

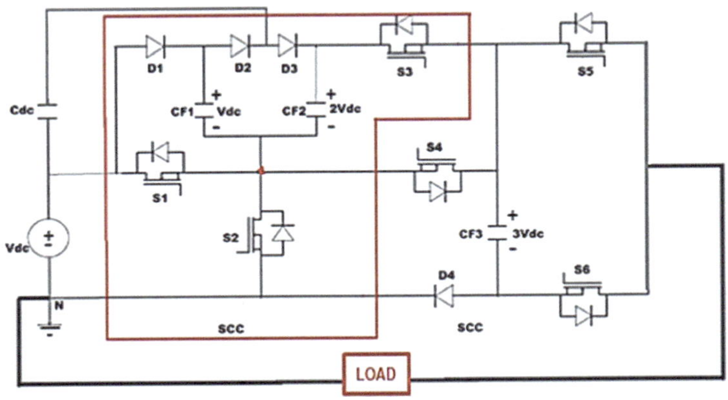

Fig. 2 Configuration of T-B6S7LI circuit

namely $3V_{dc}$, $2V_{dc}$, and V_{dc} are produced during the positive first half cycle. In the same way, during the negative second half cycle, three voltage levels namely $-V_{dc}$, $-2V_{dc,}$ and $-3V_{dc}$ are generated along with zero voltage levels. Therefore, there are a total of seven voltage levels yielded by the T-B6S7LI circuit. The configuration of the T-B6S7LI circuit [17] is shown in Fig. 2.

The comparison of the T-B6S7LI circuit with the existing single-source MLIs is given in Table 1. It can be observed that the T-B6S7LI circuit uses fewer switches to yield a boosted voltage output that is higher than the other MLIs.

3 T-B6S7LI Circuit Operation

The T-B6S7LI circuit [17] can generate the following seven voltage levels namely $\pm 3\ V_{dc}$, $\pm 2\ V_{dc}$, $\pm V_{dc}$, 0, The dc input voltage is indicated by the symbol V_{dc}. Figure 3 shows the theoretical seven-level voltage waveform. Table 2 provides the pattern of switching requisite to yield the output voltage in seven levels, and then the T-B6S7LI circuit's [17] associated operational modes.

3.1 Operating Mode-I

Switches S_1, S_3, and S_5 must be in the state of ON to produce a voltage level of 3 V_{dc}. The leftover switches S_6, $S_{4,}$ and S_2 requisite be OFF. Figure 4 represents [17] the current lane for operating mode I.

Fig. 3 The waveform of seven-level output voltage

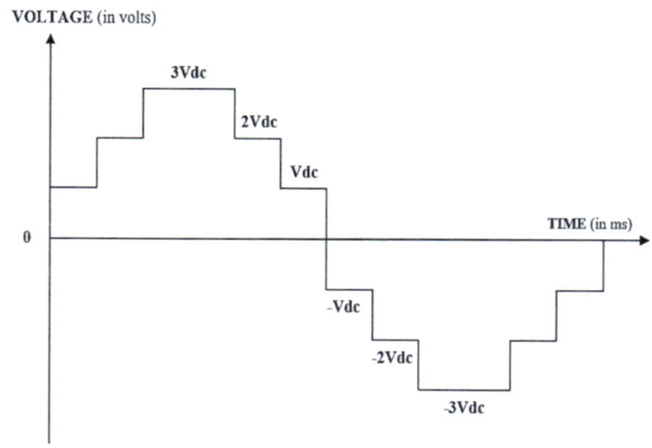

Table 2 Seven-level output voltage: pattern of switching requisite

Operating mode	Switches of T-B6S7LI						Voltage level
	S_1	S_2	S_3	S_4	S_5	S_6	
I	On	Off	On	Off	On	Off	$3\ V_{dc}$
II	Off	On	On	Off	On	Off	$2\ V_{dc}$
III	On	Off	Off	On	On	Off	V_{dc}
IV	Off	On	Off	On	On	Off	0
V	Off	On	On	Off	Off	On	$-V_{dc}$
VI	On	Off	Off	On	Off	On	$-2\ V_{dc}$
VII	Off	On	Off	On	Off	On	$-3\ V_{dc}$

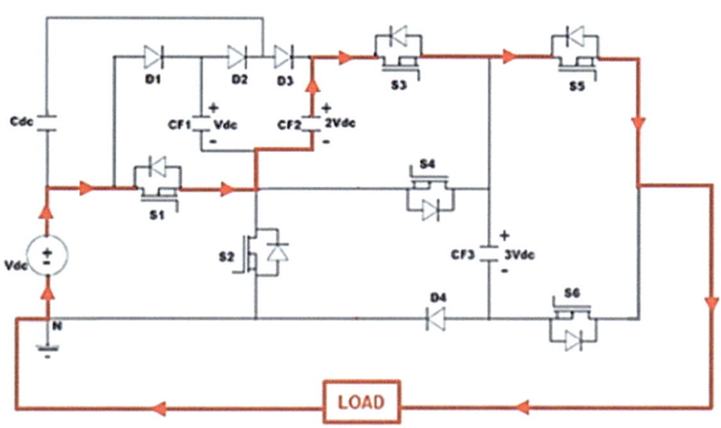

Fig. 4 Operating mode-I

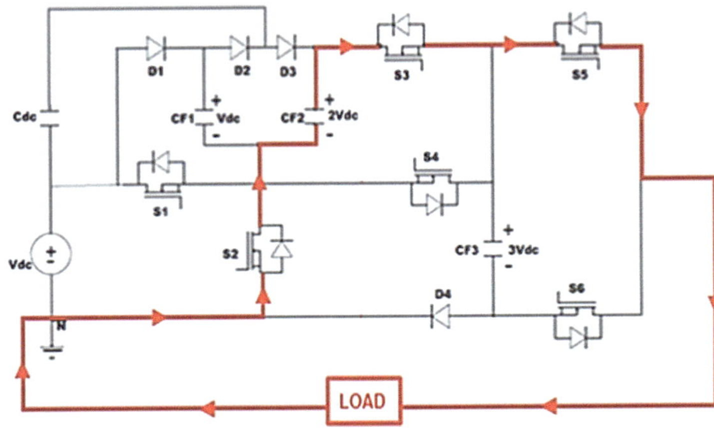

Fig. 5 Operating mode-II

3.2 Operating Mode-II

Switches S_2, S_3, and S_5 must be in the state of ON to produce a voltage level of 2 V_{dc}. The leftover switches S_1, S_4, and S_6 requisite be OFF. Figure 5 represents [17] the current lane for operating mode II.

3.3 Operating Mode-III

Switches S_1, S_4, and S_5 must be in the state of ON to produce a voltage level of V_{dc}. The leftover switches S_2, S_3, and S_6 requisite be OFF. Figure 6 represents [17] the current lane for an operating mode III.

3.4 Operating Mode-IV

Switches S_2, S_4, and S_5 must be in the state of ON to produce a zero-voltage level. The leftover switches S_1, S_3, and S_6 requisite be OFF. Figure 7 represents [17] the current lane for operating mode IV.

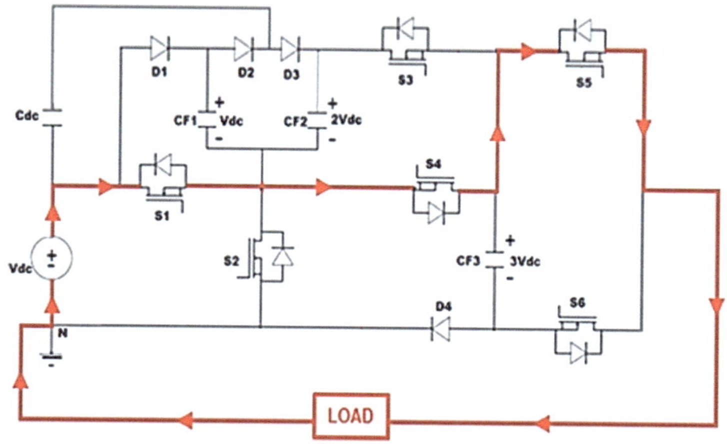

Fig. 6 Operating mode-III

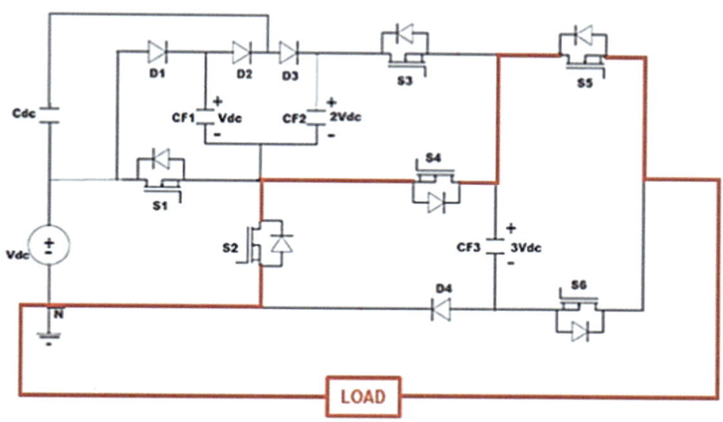

Fig. 7 Operating mode-IV

3.5 Operating Mode-V

Switches S_2, S_3, and S_6 must be in the state of ON to produce a voltage level of $-V_{dc}$. The leftover switches S_1, S_4, and S_5 requisite be OFF. Figure 8 represents [17] the current lane for an operating mode V.

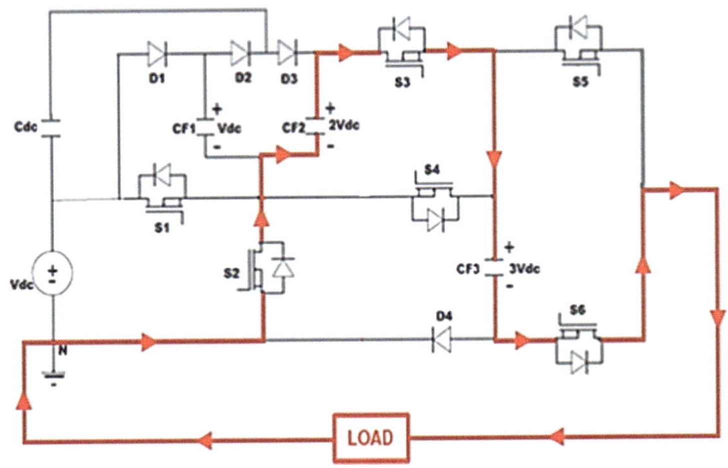

Fig. 8 Operating mode-V

3.6 Operating Mode-VI

Switches S_1, S_4, and S_6 must be in the state of ON to produce a voltage level of −2 V_{dc}. The leftover switches S_2, S_3, and S_5 requisite be OFF. Figure 9 represents [17] the current lane for operating mode VI.

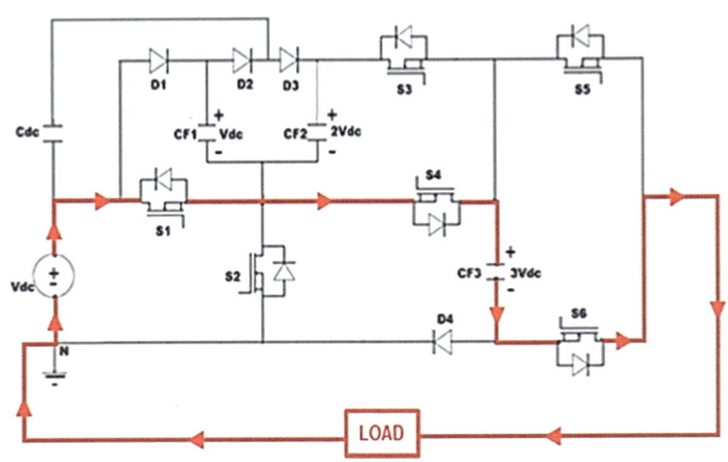

Fig. 9 Operating mode-VI

3.7 *Operating Mode-VII*

Switches S_2, S_4, and S_6 must be in the state of ON to produce a voltage level of $-3V_{dc}$. The leftover switches S_1, S_3, and S_5 requisite be OFF. Figure 10 represents [17] the current lane for operating mode VII.

4 1-Phase Induction Motor

The 1-phase induction motor is one of the important classifications of the 1-phase AC motor. 1-phase motor construction is simple, low-priced, reliable, and in comparison, to 3-phase motors, moreover simple to maintain and repair. Because of these advantages, 1-phase motors are economical and have a place in low-power devices including blowers, fans, centrifugal pumps, vacuum cleaners, and washing machines. A 3-phase squirrel cage induction motor and a 1-phase induction motor have nearly the same architecture. In contrast to a 3-phase induction motor with a 3-phase winding, a 1-phase induction motor contains two windings: an auxiliary winding and a main winding. A two-value capacitor is one of the important kinds of one-phase induction motors and it is represented in Fig. 11.

5 Photovoltaic (PV) Array Module

A photovoltaic (PV) cell is an electronic component that applies the photovoltaic effect which converts light energy into electrical energy. Photovoltaic (PV) arrays and solar modules are mainly made of PV cells. PV arrays are a fabrication of

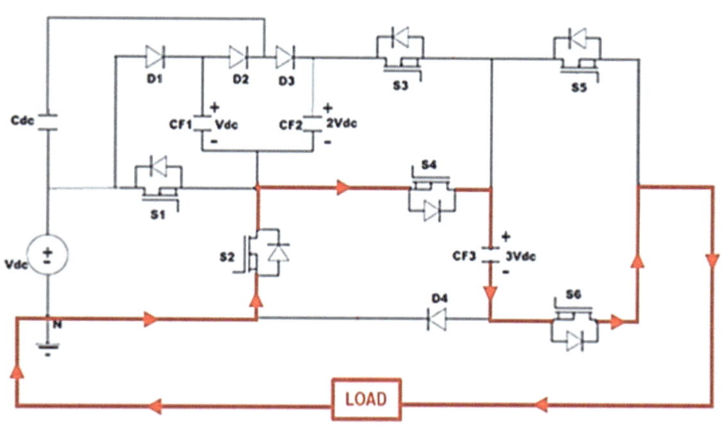

Fig. 10 Operating mode-VII

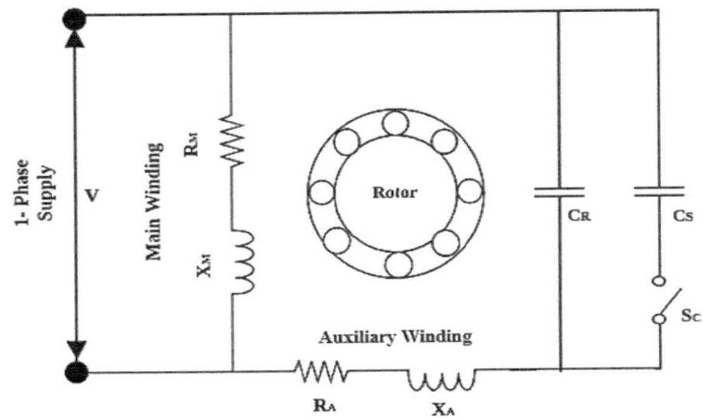

Fig. 11 Two-value capacitor motor

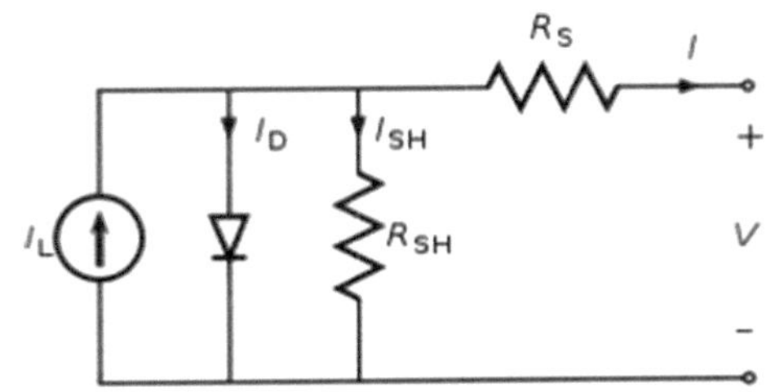

Fig. 12 Equivalent circuit of PV cell

series/parallel PV cell combinations, which are typically represented using a simple equivalent circuit model as represented in Fig. 12.

Where, I_L—photocurrent—output current.

I_D—diode current; I_{SH}—shunt current;

R_{SH}—shunt resistance; R_S—series resistance; V—Output voltage.

Temperature and solar irradiation fluctuations affect the PV array's current and voltage output.

6 Simulation Results

Making use of the MATLAB Simulink, the simulation model of the T-B6S7LI circuit is developed. This software tool is applied to create modeling, simulation, and analysis in multi-field simulation and design. MATLAB facilitates the operation of matrix functions, data charting, algorithm implementation, and other tasks in 2D models using real component data sets. Tables 3 and 4, respectively, list the simulation parameters along with the specifications for the PV array and T-B6S7LI circuit.

Table 3 T-B6S7LI circuit: simulation parameters

S. No.	Parameters	Specification
1	DC voltage source, V_{dc}	74 V
2	Capacitors, C_{F1}, C_{F2}, C_{F3}	220 μF
3	MOSFET R_d, R_{on}, R_s	0.01 Ω, 0.1 Ω, 100 kΩ
4	Switching frequency	1 kHz
5	Fundamental frequency	50 Hz

Table 4 PV array: simulation parameters

S. No.	Parameters	Specification
1	Maximum voltage at peak, or V_{MP}	29.3 V
2	Maximum current at peak, or I_{MP}	8.02 A
3	Voltage on open circuit, V_{OC}	37 V
4	Current on short circuit, I_{SC}	8.54 A

6.1 T-B6S7LI Circuit: Simulink Model

Figure 13 represents the simulation model of the PV array fed T-B6S7LI circuit for the the1-phase induction motor load and it made use of MATLAB Simulink for development. The components present in the model are MOSFET switches, capacitors, resistors, and a PV Array.

6.2 Fuzzy Logic Controller (FLC)

Fuzzy logic, often known as FL, is a multi-valued logic that connects artificial intelligence (AI) and fuzzy systems. Fuzzy systems are therefore one of the main drivers of artificial intelligence. Fuzzy systems theory involves soft or partial truth partially false. A Multivalent logic can take in multivalent logic the values of truth or false they can take any value between 0 and 1. So, that is why if we talk of the degree, degree can be infinite in number, and the number of values that can be assigned can be infinite. So, fuzzy logic deals with partial which is a matter of degree. The inputs from the PV system are taken for a given time k. Then, the error E of the system is calculated. Then, the change in error ΔE is calculated.

The values of these errors are put through the fuzzy logic controller in which case the member function receives the values. With the help of the member function, the controller tries to produce a duty ratio for the same, and thus, the process is repeated till the controller arrives at a specific value for D (Duty ratio). Figure 14 shows the simulation model for the fuzzy logic controller.

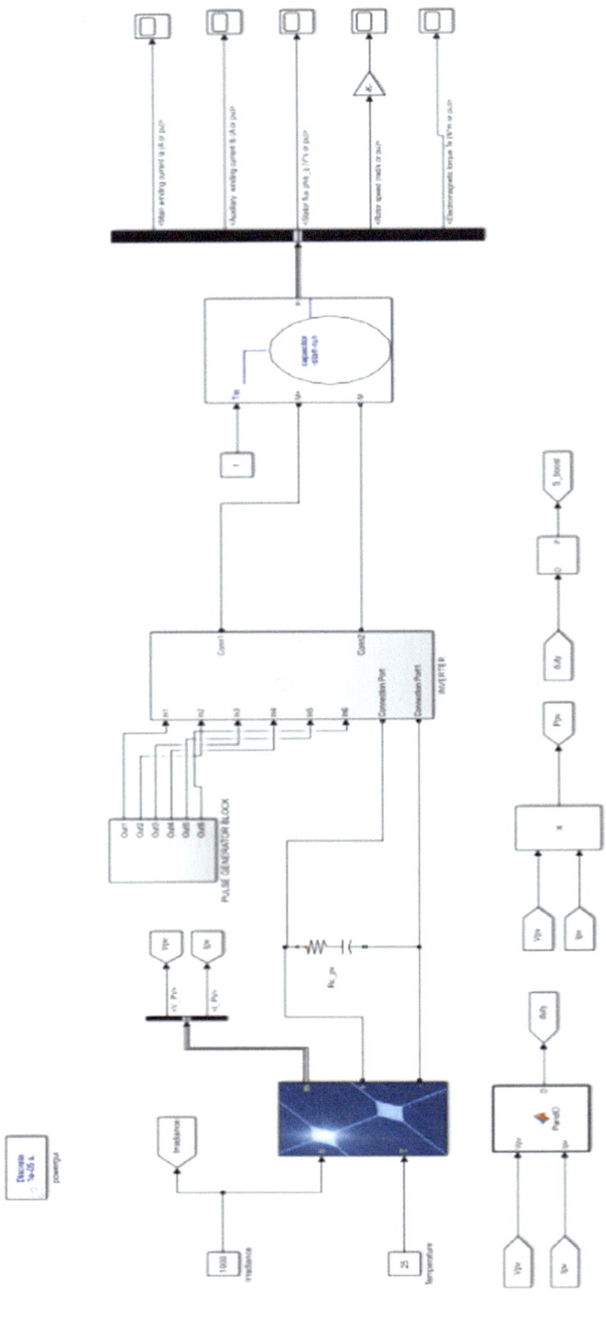

Fig. 13 Simulation model of the PV array fed T-B6S7LI circuit for the 1-phase induction motor load

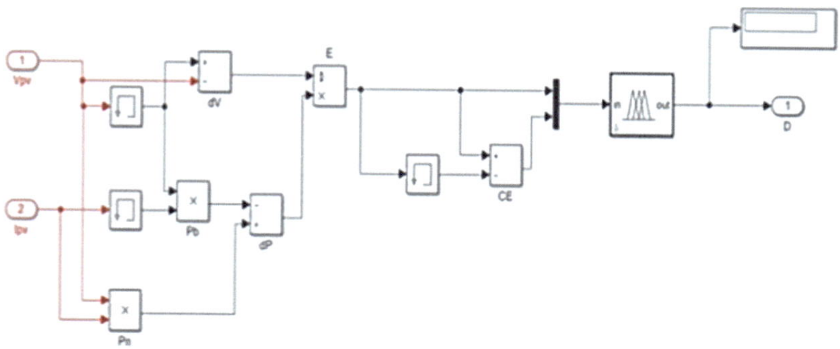

Fig. 14 The fuzzy logic controller

6.3 Simulation Results for T-B6S7LI Circuit

Based on the FLC, Fig. 15 displays the switching pulses that were generated for each switch in the T-B6S7LI circuit.

The MATLAB Simulink pulse-generating block provides the switching pulses for every switch in the T-B6S7LI circuit. The overall switching loss reduction could be achieved by operating the remaining switches at a high switching frequency and the switches S_5 and S_6 at the fundamental frequency [15]. Figure 16 presents the current–voltage (I–V) and power–voltage (P–V) characteristics of the PV Array module.

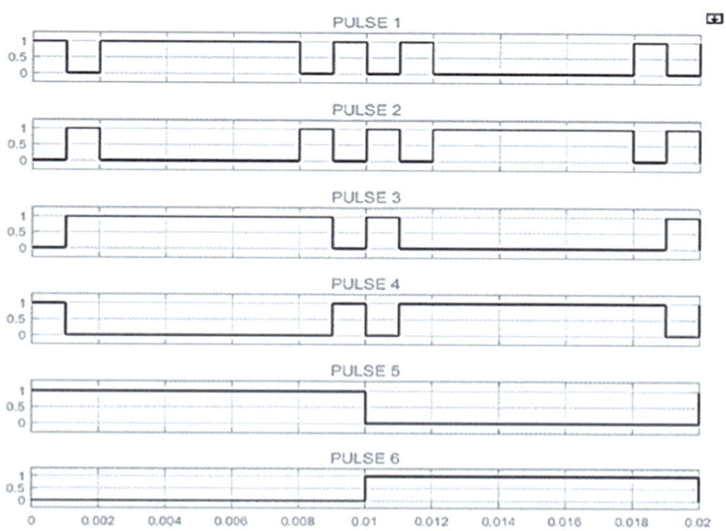

Fig. 15 Switching pulses

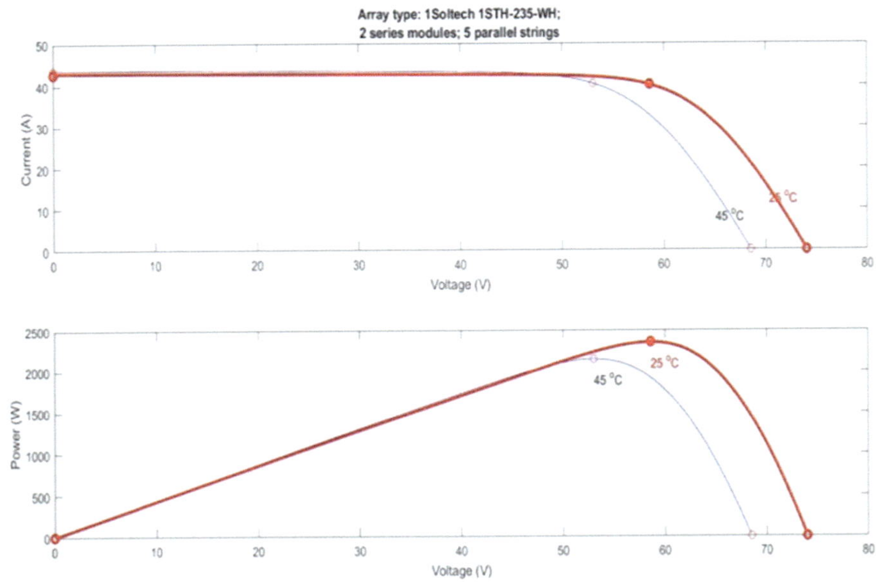

Fig. 16 PV Array: I–V and P–V characteristics

Due to the generated switching pulses for the entire switches, the input voltage (V_{dc}) through the PV array, the T-B6S7LI circuit yields an output voltage with seven different levels that are shown in Fig. 16. Therefore, it is proven that the output voltage waveform shape is similar to the seven-level theoretical waveform shown in Fig. 3.

From Fig. 17, we observe that the simulated output which consists of the seven different voltage levels explicitly as $+3\ V_{dc} = 222$ V, $+2\ V_{dc} = +148$ V, $V_{dc} = +74$ V, 0 V, $-V_{dc} = -74$ V, $-2\ V_{dc} = -148$ V, $-3\ V_{dc} = -222$ V.

From Fig. 18, the examination of the T-B6S7LI circuit using the Fast Fourier Transform (FFT) shows that the THD level is found as 18.93% for 50 cycles without using extra filter circuits.

6.4 *T-B6SLI Fed 1-Phase IM Performance Outcomes*

In this section, the performance outcomes of the T-B6S7LI circuit fed 1-Phase IM are discussed thoroughly. Figure 19 represents the main winding current waveform of a T-B6S7LI circuit fed 1-Phase IM. It is discovered that until the transitory time of 5.7 s, the primary winding current oscillates between 10 and 14 A. As soon as the period reaches 6 s, the main winding current gets a stable value of 7 A.

Figure 20 represents the auxiliary winding current waveform of a T-B6S7LI circuit fed 1-Phase IM. It is discovered that, up until the transient time of 5.4 s, the auxiliary

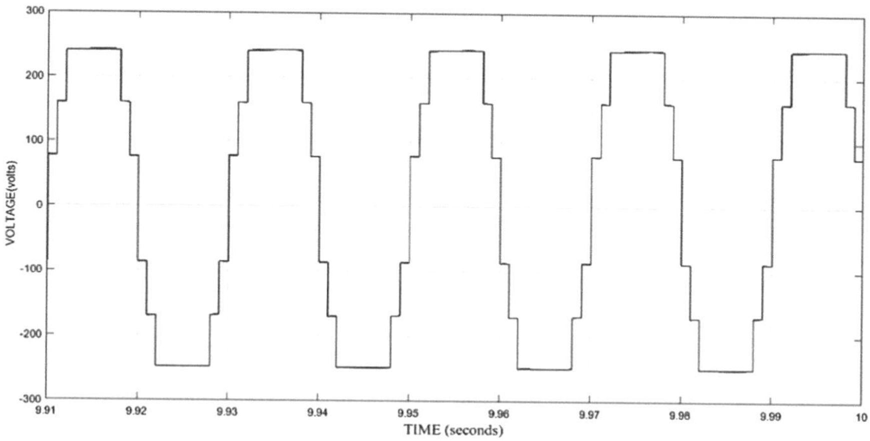

Fig. 17 The simulated 7-level voltage waveform

winding current oscillates between 6 and 7A. As soon as the period reaches 6s, the auxiliary winding current has a stable value of 2A.

Figure 21 represents the capacitor voltage waveform of a T-B6S7LI circuit fed 1-Phase IM. It is found that initially, the capacitor voltage fluctuates between 220 and 230 V till the period of 6s.

Figure 22 represents the stator flux waveform of a T-B6S7LI circuit fed 1-Phase IM. Once the period reaches 4.8s, the stator flux touches a stable value of 0.34wb.

Figure 23 represents the electromagnetic torque characteristics of a T-B6S7LI circuit fed 1-Phase IM. It is found that its characteristics are very close to the theoretical predictions.

Figure 24 represents the rotor speed characteristics of a T-B6S7LI circuit control of 1-Phase IM. It is noticed that the transient period of the rotor exceeds 6.5 s. As soon as the period reaches beyond 6.5, the rotor attains 1500 rpm, which is the rated speed of the 1-phase two-value capacitor motor.

Figure 25 depicts a simulated 7-level voltage waveform obtained from the 1-phase IM with seven different voltage levels including +222, +148, +74 V, 0, −74, −148, and −222 V. Hence the performance outcomes of the T-B6S7LI circuit fed 1-Phase IM using MATLAB Simulink validate the theoretical expectations effectively.

6.5 Loss Calculation

The switching losses in the T-B6S7LI are drastically reduced by operating the inverter switches S_1–S_4 in high frequency and the remaining switches in fundamental frequency [17]. The formula for switching loss (P_{SW}) is mentioned in Eq. (1) where the turn-on time (t_r), turn-off time (t_f), switching frequency (f_s) of the switch, and

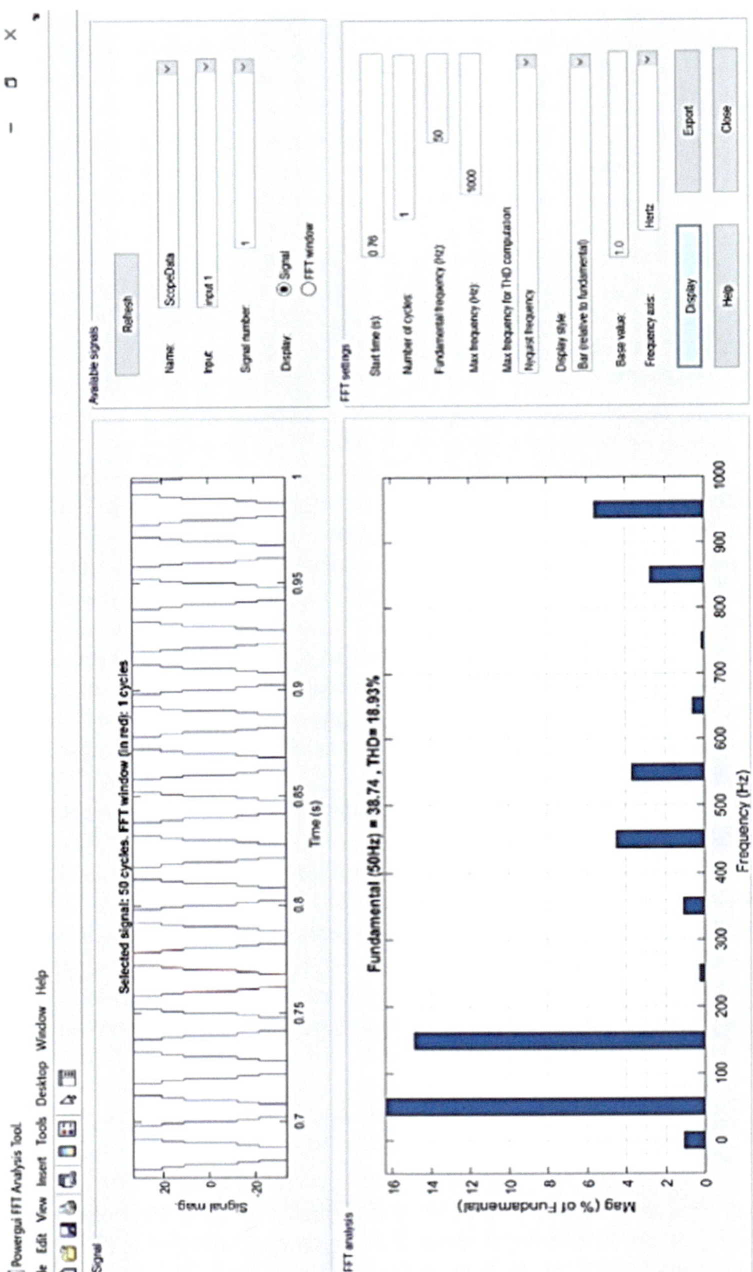

Fig. 18 Fast Fourier Transform (FFT) analysis

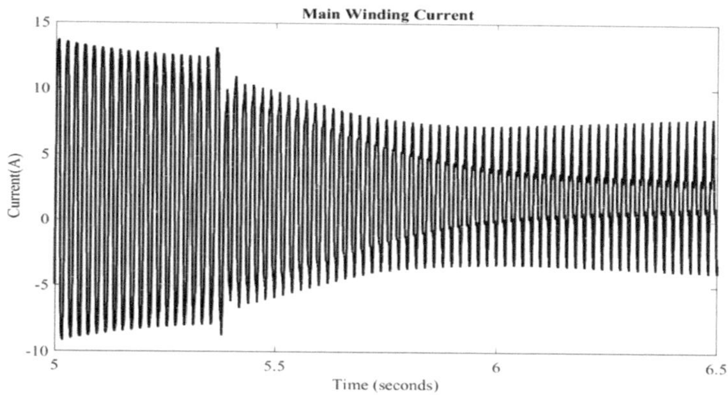

Fig. 19 Main winding current

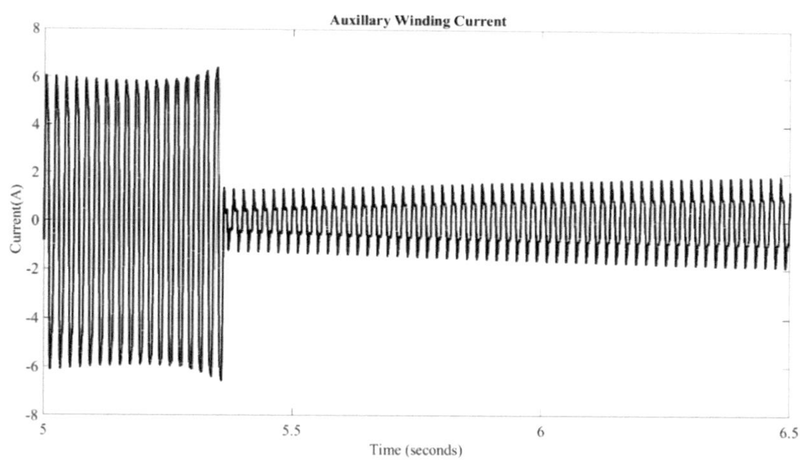

Fig. 20 Auxiliary winding current

the output current (I_o) are all obtained from the simulation results of the T-B6S7LI circuit fed 1-phase IM. The conduction loss (P_{con}) is also calculated using the formula listed in (2).

$$\begin{aligned} P_{SW} &= ((t_r + t_r)/2) * f_s * \mathrm{V_{dc}} * I_o \\ P_{SW} &= ((0.01 + 0.01)/2) * 50 * 74 * 2.486 \\ P_{SW} &= 91.76\,\mathrm{mw} \end{aligned} \tag{1}$$

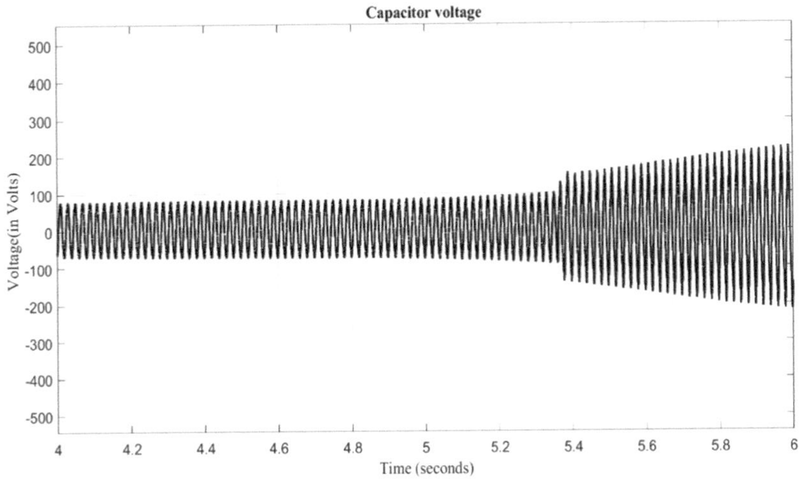

Fig. 21 Capacitor voltage

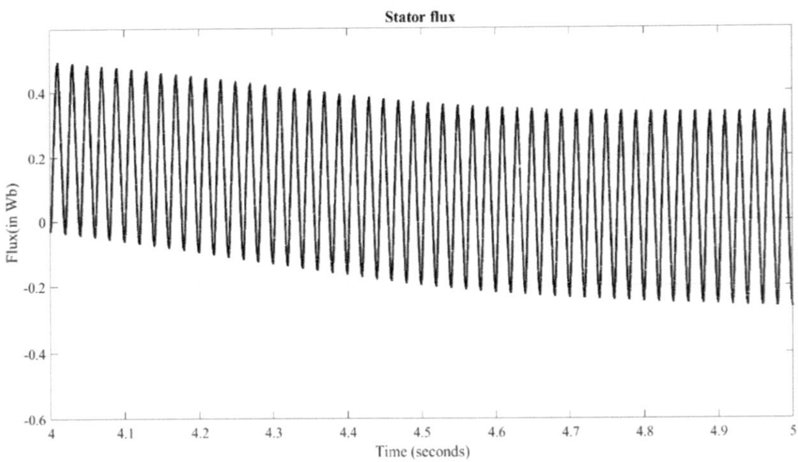

Fig. 22 Stator flux

$$
\begin{aligned}
P_{con} &= R_{on}\left(I_{in}^2 + I_o^2\right) \\
P_{con} &= 0.001\left(\left(42.7^2\right) + \left(2.486^2\right)\right) \\
P_{con} &= 1.8294\ \text{mW}
\end{aligned}
\tag{2}
$$

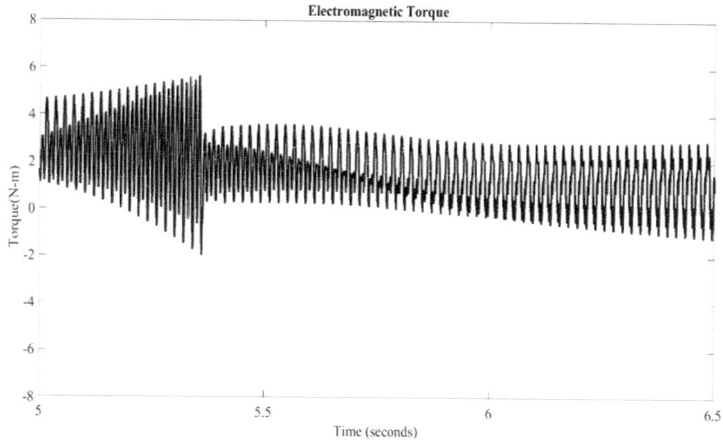

Fig. 23 Electromagnetic torque

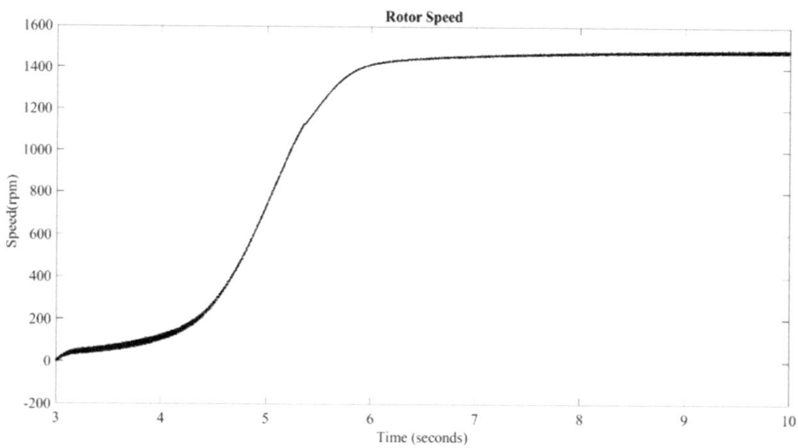

Fig. 24 Rotor speed characteristics

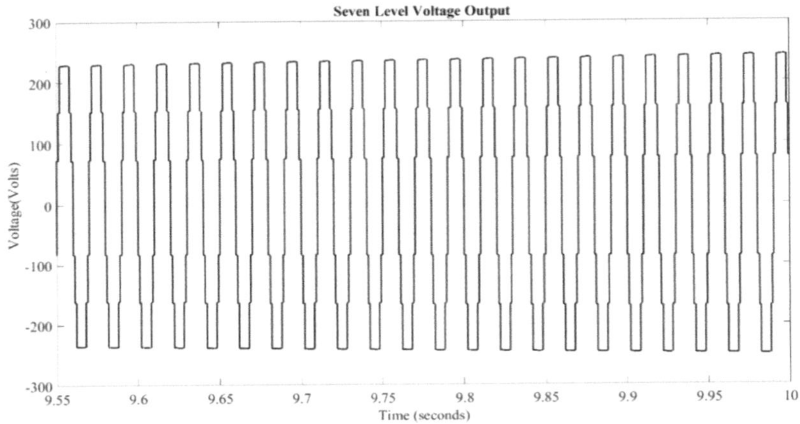

Fig. 25 Seven-level output voltage obtained from the IM

7 Conclusion

As the development of interdisciplinary projects is on the rise, the idea of multilevel inverters within the area of power electronics has led to high industrial demands due to its features. The DC source is obtained from the solar PV array using the MATLAB Simulink toolbox. A seven-step output waveform is obtained when the PV array's output voltage is triple-boosted, the losses are reduced by four switches operating at the switching frequency. The future scope of this paper will be, replacing the PV sources with any other renewable resources, giving it more industrial value, and also leading to the use of cleaner energy which is the most sought-after feature in today's market. Also, single-carrier sinusoidal pulse width modulation could be implemented here thus reducing the design complexity. Further, external filters, three-phase induction motors, or closed-loop control operations are useful for lowering the circuit's harmonics and current distortions. Furthermore, the inverter can be studied with various other loads.

Acknowledgements The authors wish to thank the Management, the Principal, and the Department of Electrical and Electronics Engineering of Loyola–ICAM College of Engineering and Technology (LICET), Chennai, for their wholehearted support and encouragement for the successful completion of the project.

References

1. Lai J, Peng FZ (1996) Multilevel converters-a new breed of power converters. IEEE Trans Ind Appl 32(3):509–517

2. Rodriguez J, Lai JS, Peng FZ (2002) Multi: a survey of topologies, controls, and applications. IEEE Trans Ind Electron 49(4):724–738
3. Rodriguez J, Bernet S, Wu B, Pontt J, Kouro S (2007) Multilevel voltage-source-converter topologies for industrial medium-voltage drives. IEEE Trans Ind Electron 54(6):2930–2945
4. Krishna A, Suresh LP (2016) A brief review on multi-level inverter topologies. In: 2016 international conference on circuit, power and computing technologies (ICCPCT), pp 1–6. https://doi.org/10.1109/ICCPCT.2016.7530373
5. Gupta KK, Ranjan A, Bhatnagar P, Sahu LK, Jain S (2016) Multilevel inverter topologies with reduced device count: a review. IEEE Trans Power Electron 31(1):135–151. https://doi.org/10.1109/TPEL.2015.2405012
6. Kjaer SB, Pedersen JK, Blaabjerg F (2005) A review of single-phase grid-connected inverters for photovoltaic modules. IEEE Trans Ind Appl 41(5):1292–1306
7. Hinago Y, Koizumi H (2012) A switched-capacitor inverter using series/parallel conversion with inductive load. IEEE Trans Ind Electron 59(2):878–887
8. Ali JSM, Krishnasamy V (20191) Compact switched capacitor multilevel inverter (CSCMLI) with self-voltage balancing and boosting ability. IEEE Trans Power Electron 34(5):4009–4013
9. Sandeep N, Ali JSM, Yaragatti UR, Vijayakumar K (2019) Switched capacitor-based quadruple-boost nine-level inverter. IEEE Trans Power Electron 34(8):7147–7150
10. Siwakoti YP, Mahajan A, Rogers DJ, Blaabjerg F (2019) A novel seven-level active neutral-point-clamped converter with reduced active switching devices and dc-link voltage. IEEE Trans Power Electron 34(11):10492–10508
11. Grigoletto FB (2021) Multilevel common-ground transformerless inverter for photovoltaic applications. IEEE J Emerg Sel Top Power Electron 9(1):831–842
12. Sato Y, Kawasaki M, Ito T (2011) A diode-clamped multilevel inverter with voltage boost function. In: 8th international conference on power electronics—ECCE Asia, pp 1987–1991. https://doi.org/10.1109/ICPE.2011.5944463
13. Khoshkbar Sadigh A, Dargahi V, Corzine K (2016) New flying-capacitor-based multilevel converter with optimized number of switches and capacitors for renewable energy integration. IEEE Trans Energy Convers 31(1–1). https://doi.org/10.1109/TEC.2015.2508481
14. Lee S, Lim CS, Siwakoti YP, Lee K (2020) Dual-T-type five-level cascaded multilevel inverter with double voltage boosting gain. IEEE Trans Power Electron 35(9):9522–9529. https://doi.org/10.1109/TPEL.2020.2973666
15. William Christopher I, Jones Priya J, Raghul AX, Vignesh Kumar P, and Vimal Raj S (2021) Arduino controller-based 3-switch isolated DC-DC boost converter for PV and battery application. J Phys Conf Ser (JPCS) 1979:1–12
16. William Christopher I, Davis D, Infanto Robbin P, Dixit J, Asif Iqbal A (2020) Simulation and analysis of T-PUC9 inverter fed single phase induction motor application. J Phys Conf Ser (JPCS) 1706:1–12
17. Chen M, Loh PC, Yang Y, Blaabjerg F (2021) A six-switch seven-level triple-boost inverter. IEEE Trans Power Electron 36(2):1225–1230

Grid-Structured Methodology for Fault Detection and Recovery in Wireless Sensor Networks

Bharti and T. P. Sharma

Abstract Due to the highly dynamic network topology, it is a challenge to design the fault-tolerant routing protocols for Wireless Sensor Networks (WSNs) deployed in applications such as wildlife monitoring, battlefield surveillance and health monitoring. The fault tolerance in WSN is a critical issue for improved reliability but is constrained due to faults in hardware and communication. Faults may be due to link failure, node failures, battery drainage, environmental hazards, asymmetric links, imperfection in hardware or software, misplacement of sensor nodes etc. The proposed Efficient Fault Detection and Recovery (EFDR) scheme for WSNs using the virtual grid is an efficient fault tolerant strategy. In case of grid node failure, an alternate node within the zone is selected to forward the data and data delivery continues through the same path. If entire zone fails, an alternate shortest path is established. During selection and path modification, recovery node buffers data in its local storage to avoid any data loss. Data delivery resumes again upon confirmation of alternate path. The results of the simulation show that the efficiency of EFDR-WSN is better than the existing schemes.

Keywords Wireless Sensor Network (WSN) · Efficient Fault Detection and Recovery (EFDR) · Dissemination Node (DN) · Local Dissemination Zone (LDZ)

Bharti (✉)
National Informatics Centre, Kurukshetra, Haryana, India
e-mail: absnarwal@gmail.com

T. P. Sharma
NIT Hamirpur, Hamirpur, Himachal Pradesh, India
e-mail: teek@nith.ac.in

S. Kumar et al. (eds.), *Soft Computing Applications in Modern Power and Energy Systems*, Lecture Notes in Electrical Engineering 1298,
https://doi.org/10.1007/978-981-96-0058-8_17

1 Introduction

As many applications in WSNs require deploying sensors in harsh areas and in large quantities, making manual control and the individual monitoring of sensors very difficult. Consequently, node failures are inevitable which reduce the entire life of network and make the communication infrastructure worthless. Failures in WSNs can occur due to two main reasons. First, sensor nodes (SN) are fragile and they may fail because of exhaustion of batteries or destruction by an external event. Second, link fails due to permanent or temporary blockage by an external object or natural conditions. Node or link failure causes network partitioning and frequent changes in network topology [1]. Thus, from the aspect of reliable data delivery and quality of service in WSNs, it is essential to accomplish maximum robustness against path failure with minimum energy consumption while building the routing schemes.

In this paper, we present an efficient detection and recovery technique for wireless sensor networks (EFDR- WSN). This scheme enhances the life of network by giving efficient fault tolerant schemes that assure normal operation of the network in presence of failures. In this scheme, we are using Two-Tier Data Dissemination (TTDD) [2] for construction of a virtual grid of square sized cells of size $\alpha \times \alpha$. Grid structure helps in managing the network topology easily and is used to establish energy efficient paths among source and sink. Corner node of a cell is known as Grid Node (GN). It maintains a Local Dissemination Zone (LDZ) of nearby nodes to provide an alternate node in case of a node failure. Whereas, in case the entire zone failure, it finds an alternate shortest path. The mechanism for choosing this alternate path is discussed in latter part of this paper.

Rest of the paper is organized as follows.

Section 2 shows the background and related work in the concerned area. Section 3 shows the proposed scheme for fault tolerance. Section 4 shows the performance of EFDR in comparison with the existing scheme. Section 5 concludes the paper.

2 Related Work

The data acquired from sensor nodes routed through multiple hops to sink is a very difficult task of data dissemination in WSN. There are several routing protocols for WSN have been presented in the literature. Cougar [3] is a protocol focused on data for the dissemination of data in WSN. Cougar summarizes a WSN as a distributed database that can be ordered using declarative queries. A minimum cost forwarding algorithm (MCF) [4] proposed by Fan Ye et al. for situations where sensor nodes have no need of unique ID and not necessary to maintain a routing table. Two-Tier data dissemination (TTDD) [2] is a grid-based approach presented by H. Luo et. al. that gives efficient data transmission to multiple mobile sinks. T.P. Sharma et al. propose a Grid based data dissemination (GBDD) [5] for WSN. In this a dual radio based grid is constructed which utilizes node's radio mode for grid construction in the sensor area.

As WSNs are deployed in inaccessible environments therefore, the sudden node or link failure may partition the network, loss of data, route failures, and dynamic changes in the network topology. Hence, from the standpoint of the reliable supply of WSN quality data and services, it is essential to achieve maximum robustness against path failures with less power consumption during the development of routing protocols. Various kinds of fault-tolerant routing techniques have been categorized to understand the challenges of constructing fault-tolerant protocols in distributed sensor network applications. Chanak et al. [6] propose an energy efficient fault tolerance multipath routing (EEFTMR) for WSN. The base of EEFTMR scheme is the routing of the multipath data that recovers the node failure and the transmission failure. Sharma et al. [7] propose a zone based failure recovery method which focuses on crash fault and failure of nodes. Chen et al. [8] present a distributed fault detection of the wireless sensor network scheme to identify the fault detection algorithm located by faulty sensors. Kulwardhan et al. propose a Fault detection and Recovery scheme (FDR) [9] for wireless sensor networks which focuses on addressing issues of energy efficient routing, effective failure handling strategies and efficient handling of multiple mobile sinks.

Mehrotra et al. [10] propose an approach that helps in monitoring areas of faults which occur because of environmental hazards. This helps in detecting changes in surrounding environment and alert the staff for potential faults. Temene et al. [11] propose an algorithm for the proper placement of nodes so that it can detect the faults easily and further helps in the recovery process.

3 Proposed Scheme

In this section, a grid structured scheme is proposed to get the position of the sink from the source. Once the sink gets the position information of the source, the sink node establishes a straight line path for query/data transmission. In the proposed scheme, the optimal path is established by forwarding position/location information from source and data/query messages from sink probably on axis perpendicular to each other through grid nodes which intersect at some intermediate DN. This DN forwards the position information of the sink to source through the reverse path from

Table 1 Symbols and notations

Symbols	Description
$\propto$	Side of the square shaped cell of a grid
CP	Calculated actual crossing point (coordinates) of a cell of a virtual grid
GN	A node closest to CP among all its neighbors and is known as Grid Node
LDZ	A zone/area within the radius $\propto/4$ from CP is called local dissemination zone
Sn	A node which detects an event called Source Node (Sn)
DN	An intermediate node on the data path that forwards data/query is called dissemination node (DN)
NR	A node used for recovery purpose in each zone
R	Transmission range of a sensor node
MRE	DN announces its energy status using this message
MSetup	Path setup message initiated by Sn/DN to establish or recover (in case of Path failure) path with sink

which the location message is received. Once the source receives the location of the sink a shortest straight path is setup between the source and sink using greedy forwarding techniques [12]. The symbols and notations used in this scheme are shown in Table 1.

3.1 Fault Detection

The nodes distributed in the sensor field are fragile and may fail due to an exhaustion or destruction of the battery. Whenever the remaining energy of a DN falls below the threshold value or when the DN does not work, an alternative DN is selected amongst available nodes in the LDZ. A node is selected as DN which has highest F value. F value is calculated as below:

$$F = C \times RE/D$$

where C is a constant, RE is the remaining energy of the nodes, D is the distance of node in LDZ from DN. Whenever the current DN in LDZ fails, the proposed scheme selects the successor of the DN as per following scenario. When the remaining energy of a DN falls below the threshold value, it announces its energy status by broadcasting its M_{RE} message which is overheard by all the nodes within LDZ as well as by its upstream and downstream DNs. The downstream DN stops data transmission after receiving the M_{RE}, till it receives the live message from NR in the upstream LDZ. When the nodes within the LDZ overhear the M_{RE} message, they run Bully Election Algorithm [13] for finding the leader or coordinator i.e. new DN among working nodes. All nodes within a LDZ are at the single hop distance from its one hop neighbor DNs. Any node within this zone can act as DN. When a DN fails within a LDZ, an alternate DN as explained in Algorithm 1 is selected amongst the working nodes in LDZ. But, as in many typical WSNs, the nodes are deployed in hazardous and disastrous area. Hence, all nodes within a LDZ may fail and alternate DN cannot be appointed. Also, many LDZs may fail at once. These failures may result in abrupt failures of the existing route. Therefore, an alternate optimal route is to be instantly discovered to minimize the data loss.

Two scenarios for election of leader exist. First, the node which is at the top of the index list (i.e., the node with maximum F-value) initiates the Bully Election Algorithm. Second, the node which earlier failed but now recovers/comes on can also initiate Bully Election Algorithm. First, this node sends election message and waits for the reply message. If no reply message is received within some time interval, it declares itself a leader i.e., DN. Subsequently, it broadcasts a leader message comprising its coordinates which is heard by all nodes in its own LDZ as well as all nodes in its upstream and downstream LDZ. This node becomes the successor of the old DN and each node within LDZ updates its index list. The upstream and downstream DNs also update their routing path table after receiving the leader message. The new DN forms a path for data transmission in the place of failed DN. The downstream DN resumes the data delivery as soon it receives the leader message.

In proposed scheme, LDZs are classified in three different categories. First is F_LDZ which is a zone within which all nodes have failed or unable to communicate. Second is No_forwarding zone which is a zone which does not find any further upstream LDZ to forward the data towards sink. This situation may occur when some of the neighboring LDZs of a given DN have failed and marked as No forwarding zone. Third is live_LDZ which is a zone where all or some nodes are live and are able to communicate with their neighboring LDZ.

Algorithm 1. DN failure and LDZ failure Detection

Input:	Indexed list of all the nodes in the LDZs
Output:	Detection of node failure or LDZ failure
Notation	DN→LDZ indicates upstream LDZ of a DN

1. FOR all DNs on routing path Do
2. WHILE DN receives data Do
3. IF DN→LDZ == live_LDZ THEN
4. Forward the data to the upstream DN using broadcasting
5. Upstream DN receives data packet and further forwards the data using broad casting
6. IF downstream DN overhears the upstream DN forwarding within stipulate time
7. THEN
8. Break; //transmission confirms and upstream DN is working and LDZ is live_LDZ
9. ELSE
10. Mark the upstream DN as faulty and delete its entry from list of working nodes within LDZ // detect DN failure
11. For all nodes in LDZ calculate F-value
12. F= C*RE / D , where RE is remaining energy of nodes, D is the distance from DN, c is constant
13. Broadcast F-value within LDZ and index list in descending order of stored F-value
14. Node with highest F-value Calls Bully Election Algorithm [11] for selection of new DN
15. Select the node at the top of indexed list within LDZ as new DN // selecting new DN within LDZ

```
16.    Newly elected DN broadcasts a leader message in its one hop distance and
       form a path to forward data
17.    END IF
18.    IF downstream DN receives live message from a REC node in DN→LDZ
19.    THEN
20.    Update the new upstream DN entry in its Routing path table
21.    Go to step 4  //re-transmission of packet
22.      ELSE IF no message received from REC node in DN→LDZ within specific
         time
23.      THEN
24.      Recovery node send LDZ failure message to its upstream DN
25.      Mark the DN→LDZ as Failed LDZ // LDZ failure detection
26.      END IF
27.    END IF
28.    ELSE
29.    Initiate the process to discover alternate live LDZ   //described in Algorithm
       II
30.   END IF
31. END WHILE
```

4 Fault Recovery

When a DN does not get any message from upstream LDZ within a specific period, it contacts NR of this LDZ. Then NR replies about the failure of nodes. When a DN finds that its upstream LDZ is failed, it stops forwarding the data through that path and re-routes the data through that LDZ which is nearest to the Current LDZ and the F_LDZ both as shown in Fig. 1. The NR of F_LDZ holds the data for the period until a new path is created and then forward the data to this path. Hence, it helps in avoiding the data loss.

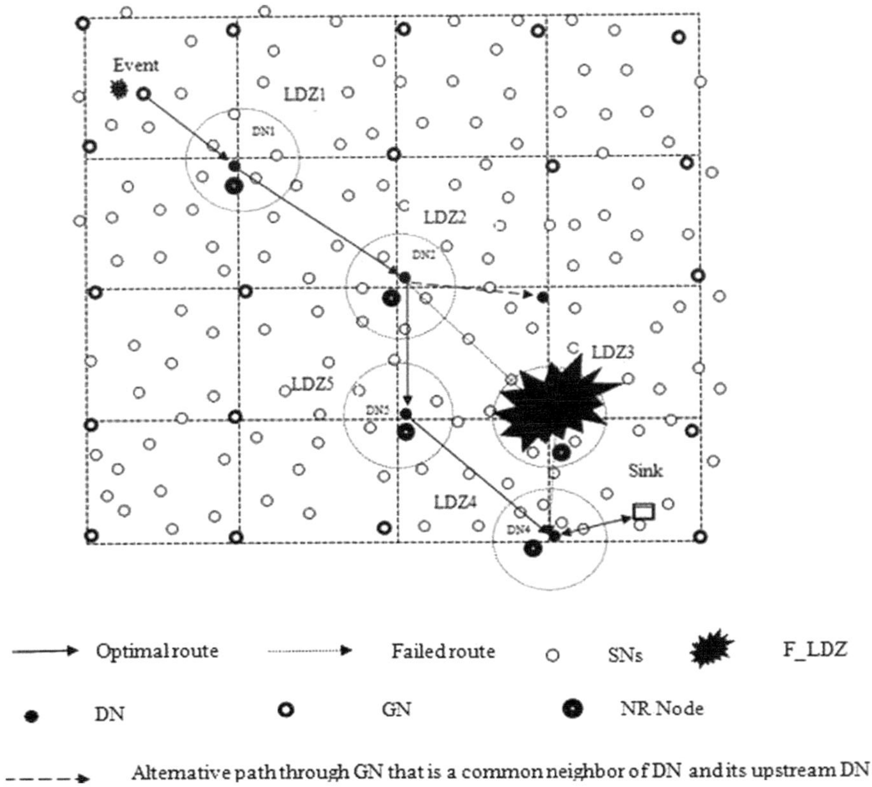

Fig. 1 Handling LDZ failures

As shown in Fig. 1, the procedure of finding LIVE_LDZ is described in Algorithm 2. Thus, in case of single or multiple zone failures, the EFDR scheme always constructs a shortest path. The immediate DN detects the failure instead constructing a complete path from source to sink. Thus, proposed scheme efficiently updates a route around the holes with minimum communication overhead.

Algorithm 2. LDZ failures Recovery

Input:	Linked list of source, DNs and sink on the routing path
Output:	Finding LIVE_LDZ in neighborhood of DN when a DN→LDZ failed on routing path
Notation	DN→LDZ indicates upstream LDZ of a DN DN←LDZ indicates downstream LDZ of a DN

1. WHILE DN→LDZ != LIVE_LDZ DO
2. IF DN→LDZ == F_LDZ ||
   DN→LDZ==NO_FORWARDING_LDZ THEN
3. DN broadcasts LDZ status request to neighbor DNs and all DNs in neighborhood of LDZ reply back with their zone status
4. IF neighboring LDZs common to DN→LDZ are LIVE_LDZs THEN
5. Select the neighboring LDZ that is common with DN→LDZ and closest to sink
6. Set this neighboring LDZ as DN→LDZ
7. ELSE IF neighboring LDZs not common to
   DN←LDZ are LIVE_LDZs
8. THEN
9. Select the neighboring LDZ not common with DN←LDZ and closest to sink
10. Set the neighboring LDZ as DN→LDZ
11. ELSE
12. Mark the LDZ of DN as NO_FORWARDING_LDZ
   // indicate that present zone is NO_FORWARDING_LDZ
13. Send zone status message to the DN in the DN←LDZ
   //backtrack process from the DN lying in the downstream LDZ.
14. Go to step 1
15. END IF
16. END IF
17. END IF
18. END WHILE

As shown in Fig. 1, DN in LDZ2 finds that its upstream LDZ (i.e. LDZ3) has failed; NR of LDZ3 sends the route failure message to downstream LDZs (LDZ2 and LDZ1). All the DNs in the downstream LDZs stop transmission and buffer the data in its local zone. At this point DN2 initiates the alternate path setup process towards the sink through other neighboring DN that is in LIVE_LDZ and nearest to the sink. At any stage, a DN on the source-sink path has three neighboring GNs that can be selected as upstream DNs are (1) the GN that is closest to sink among others i.e., DN3 is selected as downstream DN during the initial path setup; (2) other two GNs are the common neighbors of DN2 and DN3. Thus, the DN2 always selects the GN in LDZ3 (i.e., DN3) that is closest to the sink as DN, to establish an optimal and shortest path.

But, when LDZ3 fails, DN2 checks the GNs that are also common neighbors of DN3 and are lying in LIVE_LDZ. These two GNs are used to establish the alternate paths, because the path established through these GNs are the shortest among other when LDZ3 fails. Therefore, among these two GNs, DN2 selects the GN that is closer to the sink (in this case DN5 is selected). Then, DN2 sends the M_{Setup} message to the GN nearer to the sink. This procedure proceeds until the message comes to the sink as shown in Fig. 1. Once the sink receives the M_{Setup} message, it sends back the path confirmation to the DN2 which had initiated the M_{Setup} message process. The new source-sink path is: Source-> DN1-> DN2-> DN5-> DN4-> Sink. Upon receiving the path confirmation message, DN2 sends the path set up message to the downstream DNs which in turn resumes the data delivery.

5 Performance Evaluation

In this part, the performance of the proposed scheme is assessed and compared with other existing fault tolerant schemes. The proposed scheme is simulated utilizing NS-2 [14] discrete event network simulator. NS-2 is a simulation tool which is an open source and runs on the Linux operating system.

Here, we calculate the outcomes generated by simulations in various scenarios, such as variable node density, node failure ratio, etc. EFDR efficiency is evaluated by comparing it with FDR [9] in terms of three performance parameters: total energy consumption, avg delay and avg delivery ratio. Total energy consumption is characterized as the energy consumed by the network to send and receive data packets and control messages. The avg delivery ratio is defined as the number of packets successfully received by the receiver to the number of packets sent by the SN. The average delay is defined as the average time passed between the times a source sends a packet and the time the receiver gets the packet.

The default simulation configuration has a square field of 2500 $\times$ 2500 m^2 in which N (85) sensor nodes are equally distributed. Each of these sensor nodes acts as a source and generates data packets/control messages. The simulation model is executed many times and the analysis/observation is based on the variable number of sensor nodes and on the failure percentage of the node. The size of the control / query package is 36 bytes and the data packets are 64 bytes. The transmission range R of each sensor is 250 m and the cell size is 130 m (Table 2).

Table 2 Simulation parameters

Parameters	Values
Size of sensor field	$2500 \times 2500\ m^2$
Data packet size	64 bytes
Query/control message size	36 bytes
Transmission range	250 m
Cell size	≈130 m
Number of sink	1
Number of source	1
Number of sensor nodes	85
Distribution type of sensor nodes	Uniform
Initial energy of nodes (all nodes)	50 J
Type of sensor nodes	Static
Simulation time	45 s

5.1 *Impact of Node Density on Total Energy Consumption, Average Delivery Ratio and Average Delay*

Here, the performance of total energy consumption with variable node density is evaluated. Figure 2 shows the total energy consumption with variable node density. The energy consumption increases with the density of the node. This is because more energy is dissipated to listen by chance when there are more nodes in the neighborhood. However, the EFDR consumes less energy than FDR [9]. The global EFDR delivery ratio is greater than FDR [9], as shown in Fig. 3. This is because EFDR selects the Successor node in LDZ from the indexed list when a DN fails and follows the same path. Therefore, whenever the number of nodes increases the availability of working nodes in LDZ increases. In EFDR, when a DN fails, the next working node at the top of the indexed list within LDZ is chosen as a new DN and the delivery path remains unchanged. Therefore, the average delay in EFDR is also lower than FDR [9] as shown in Fig. 4.

5.2 *Impact of Node Failure Ratio on Total Energy Consumption, Average Delivery Ratio and Average Delay*

Figure 5 shows that EFDR consumes less energy than FDR [9]. This is due to the fact that as node failure increases, fewer nodes participate in the transmission, so the total power consumption of the network decreases.

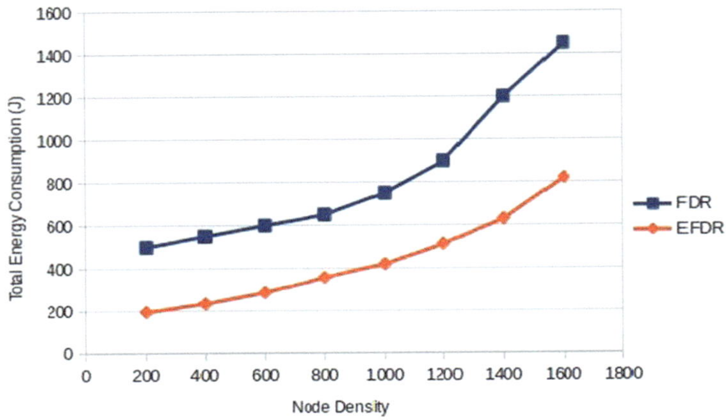

Fig. 2 Total energy consumption versus node density

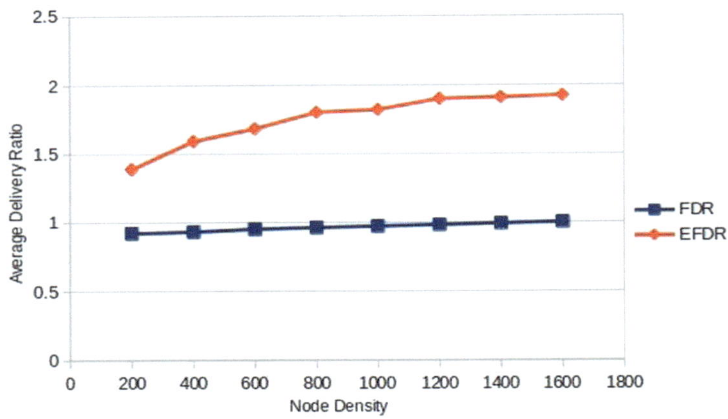

Fig. 3 Average delivery ratio versus node density

The average delivery ratios of the scheme decreases as node failure ratio increases as shown in Fig. 6. However, overall delivery ratio of EFDR is more than FDR [9]. The average delay of EFDR is also less when compared to FDR as shown in Fig. 7. Moreover, EFDR has the ability to select the alternate node when a DN fails within LDZ and ability to modify a partial path upon failure of LDZs. The average delay for EFDR is less than FDR [9].

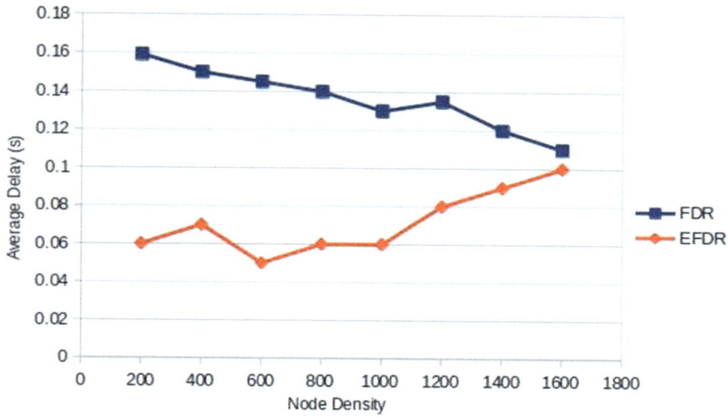

Fig. 4 Average delay versus node density

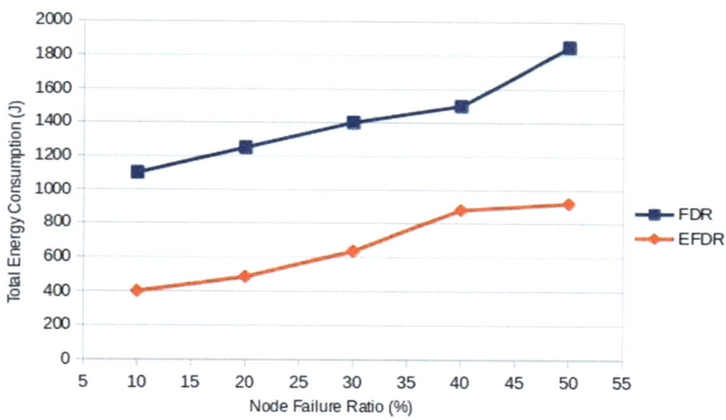

Fig. 5 Total energy consumption versus node failure ratio

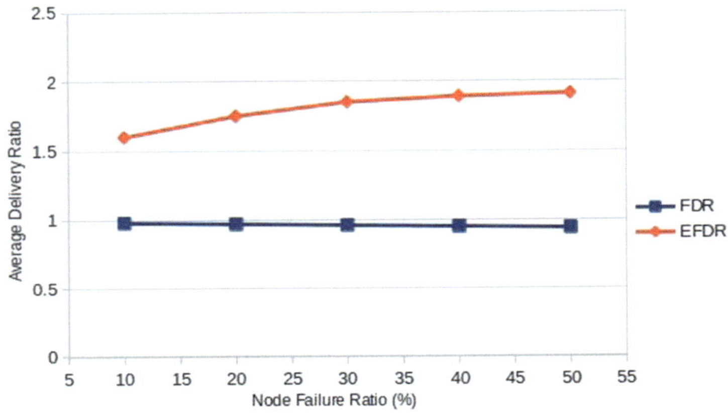

Fig. 6 Average delivery ratio versus node failure ratio

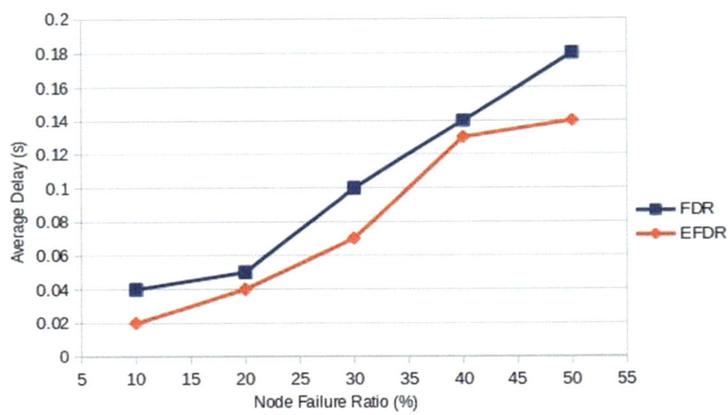

Fig. 7 Average delay versus node failure ratio

6 Conclusion

The proposed EFDR technique for WSNs using virtual grid is an efficient fault detection and restoration strategy. In EFDR, each grid node on the source-sink path creates a local dissemination zone (LDZ). In case of grid node failure, an alternate node within LDZ is selected as the new grid node and data delivery continues through the same path. If the entire local zone fails (in case of burst node failure), an alternate shortest path is established by performing appropriate selection process for finding a suitable alternate grid node in the neighboring LDZ. During selection and path modification, NR buffers data in its local storage to avoid any data loss. Thus, this scheme has the ability to construct a partial shortest path around the failure zones in the sensor network involving minimum communications. Data delivery resumes again upon confirmation of an alternate path setup. Analytical and simulation study reveals significant improvement in terms of average delay, energy consumption and delivery ratio under different network scenarios such as node density, node failure ratios. The performance of the proposal scheme is better than the existing schemes.

References

1. Paradis L, Han Q (2007) A survey of fault management in wireless sensor networks. J Netw Syst Manage 15(2):171–190
2. Luo H, Ye F, Cheng J, Lu S, Zhang L (2005) TTDD: Two-tier data dissemination in large-scale wireless sensor networks. Wirel Netw 11(1–2):161–175
3. Yao Y, Gehrke J (2002) The cougar approach to in-network query processing in sensor networks. ACM Sigmod Rec 31(3):9–18
4. Ye F, Chen A, Lu S, Zhang L (2001) A scalable solution to minimum cost forwarding in large sensor networks. In: Computer communications and networks, proceedings tenth international conference. IEEE, pp 304–309
5. Sharma TP, Joshi RC, Misra M (2008) GBDD: grid based data dissemination in wireless sensor networks. In: Advanced computing and communications, 16th international conference. IEEE, pp 234–240
6. Chanak P, Banerjee I (2013) Energy efficient fault-tolerant multipath routing scheme for wireless sensor networks. J China Univ Posts Telecommun 20(6):42–61
7. Sharma KP, Sharma TP (2017) ZBFR: zone based failure recovery in WSNs by utilizing mobility and coverage overlapping. Wirel Netw 23(7):2263–2280
8. Chen J, Kher S, Somani A (2006) Distributed fault detection of wireless sensor networks. In: Proceedings of the 2006 workshop on dependability issues in wireless ad hoc networks and sensor networks, pp 65–72
9. Singh K, Sharma TP (2017) FDR: fault detection and recovery scheme for wireless sensor networks using virtual grid. Int J Parallel, Emerg Distrib Syst 32(6):617–631
10. Mehrotra T, Sudhamsu G (2023) Energy efficient and fault tolerant area monitoring for sensor networks. In: International conference on distributed computing and electrical circuits and electronics (ICDCECE), Ballar, India, pp 1–6
11. Temene N, Naoum A, Sergiou C, Georgiou C, Vassiliou V (2022) A centralized node placement algorithm in WSNs and IoT networks. In: Global information infrastructure and networking symposium (GIIS), Argostoli, Greece, pp 25–29
12. Karp B, Kung HT (2000) GPSR: greedy perimeter stateless routing for wireless networks. In: Proceedings of the 6th annual international conference on Mobile computing and networking. ACM, pp 243–254
13. Garcia-Molina H (1982) Elections in a distributed computing system. IEEE Trans Comput 1:48–59
14. https://www.isi.edu/nsnam/ns/

Faults Detection and Recovery in a Wireless Sensor Networks—A Survey

Bharti and T. P. Sharma

Abstract Wireless sensor networks (WSN) are self-organized systems consisting of large number of heterogeneous/homogenous sensor nodes that are deployed in inaccessible/harsh environments where nodes are susceptible to damage. The sensor nodes are deployed in unwelcoming conditions to collect data through wireless medium and monitor the environment. The fault tolerance in WSN is the critical issue now-a-days but WSNs reliability is constrained due to faults in hardware and communication. Faults may be due to link failure, nodes failures, battery drainage, environmental hazards, asymmetric links, imperfection in hardware or software, misplacement of sensor nodes etc. To obtain high data reliability and accuracy researchers have proposed many fault detection and recovery mechanisms to preserve energy for network lifetime improvement and to minimize failures in WSNs. In this paper, we present a survey on fault detection and recovery mechanisms in WSNs. Exquisite survey on different types of faults and relevant fault tolerant approaches are presented to support sensor applications.

Keywords Wireless Sensor Networks (WSN) · Sensor Nodes (SNs) · Base Station (BS) · Cluster Head (CH)

1 Introduction

WSN comprises of numerous independent SNs connected together to create a network. SNs have sensing and performing specific tasks ability as per the requirements.

Bharti (✉)
National Informatics Centre, Kurukshetra, Haryana, India
e-mail: absnarwal@gmail.com

T. P. Sharma
NIT Hamirpur, Hamirpur, Himachal Pradesh, India
e-mail: teek@nith.ac.in

S. Kumar et al. (eds.), *Soft Computing Applications in Modern Power and Energy Systems*, Lecture Notes in Electrical Engineering 1298,
https://doi.org/10.1007/978-981-96-0058-8_18

For communication between these SNs, wireless mechanism is used. Wireless SNs are also having some limitations as low memory, limited processing capability, limited battery life, low storage but they also consume minimum power. SNs are deployed in locations like military applications, health monitoring, landslide tracking, weather detection etc. As SNs are deployed in critical conditions, they are highly prone to frequent errors. These faults or errors may present due to software or hardware, but it leads to inaccurate readings/output. These readings can lead to dangerous aspects if we consider it in routine works like in healthcare applications, landslide monitoring etc. where false readings can leads to death of a person or animal or any loss of life depending upon where that SNs are used. For instance, if wireless SNs putted for earthquake monitoring and gives false readings, it may lead to serious circumstances due to lack of timely warnings.

Secondly, if there are some faulty nodes or faults present in the network then it leads extra load on network and eats the available resources like power, battery, storage etc., unnecessarily. As it is nearly impossible to replace battery of SNs because they are mostly deployed at critical situations. So conservation of SNs energy and power is most required as this should be used to carry important tasks/events. The readings collected should also be transmitted to sink for further processing, so this should be true positive data not false positive because false data unnecessarily increase congestion. Therefore, proper fault diagnosis and restoration techniques are required for proper utilization of these SNs.

In this paper, we focus on survey of techniques that help in increasing the service availability in WSN with the help of fault tolerance techniques. Firstly, exquisite analysis on different types of faults and failures is presented and later the review on methodologies for fault detection and recovery techniques in WSNs is outlined.

The rest of the paper is organized as follows. In Sect. 2, we will see some related work / will review the literature. In Sect. 3, we classify various types of faults and failures that occur in WSN. Later Sect. 4 provides analysis on different fault tolerant techniques and Sect. 5 shows what we can conclude from this paper.

2 Literature Review

WSN innovation has been a dynamic research field for quite a while and has received noteworthy thought from industry. There are a lot of applications proposed, where sensor nodes can identify the physical conditions to perform information accumulation and emission monitoring. A WSN is a self-composed framework and it comprises of low controlled SNs with confined transmission range and sink nodes [1]. Homogeneous and Heterogeneous WSNs are two sorts of WSNs where homogeneous WSNs comprise of gadgets having similar communication range and computational capacity. In heterogeneous WSNs, all SNs have distinctive features like computational power, communication range and type of sensors [2]. SNs in one hop or multi-hop wireless connections collaborate among each other to accomplish information detecting, data handling and data communication [3]. SNs can be placed in different scenarios like on land, wind, vehicles, submerged, on bodies, and inside various buildings to accomplish monitoring tasks [4, 5]. SNs fail due to their placements in harsh and brutal domain [6]. The defective nodes are most likely going to report discretionary readings that fail to accurately convey observed physical events to the BS. Hence, identifying and removing defective nodes from the data gathering process is essential to guarantee overall data quality. For faulty nodes, it is difficult to sense data accurately and monitor processes in an appropriate way [7]. As WSN is bound by limited battery storage, limited bandwidth, ad-hoc network topology, different types of damages such as node failure, link failure, network failure, and collisions. Hence, to handle these issues a WSN must have self-configuring and self-arranging capabilities with the end goal to improve connectivity, reduce energy consumption and retain network resources [4].

3 Classification of Faults and Failures in WSN

3.1 Fault Classification

To figure out the adaptation to fault-tolerant mechanisms, it is essential to distinguish between faults, errors, and failures. A system's underlying presumptions are being violated by a fault. If a fault is reflected by an internal data state then it is called an error [8]. A clearly noticeable divergence from requirements is called a failure. It is not necessary that a fault results in an error neither an error results in a failure. For example, a fault is an alpha particle manipulating a memory location and if that memory location stores data that altered data is an error. It will prompt to failure, if a program crashes because of utilizing that information.

Node Faults. Nodes consist of various hardware and software components that can produce malfunction. In WSN, node flaws are because of the accompanying causes like the failure of modules because of process creation issues, natural elements etc. As observed in [9], a numerous number of abnormal readings were created in the system due to battery breakdown. Due to this, the effectiveness of a data acquisition application is compromised when the primary sensors produce inaccurate data. Software bugs are an ordinary reason for the errors in WSNs. Arranging a system in clusters is a method used in numerous applications. Few nodes are chosen to end up as CHs and they are in-charge of organizing the nodes in respective clusters, for example by gathering information from them and sending it to the BS. Few nodes are chosen to end up as cluster heads and they are in charge of organizing the nodes in their clusters. One normal straightforward method is to find the average of inter-related data values such as temperature, pressure, moisture and send the combined form to the end system, but if the node produces inaccurate data then it can suffer divergence from the original value during the data collection. Yet, if the nodes which forward combined data produce incorrect data then the base station of that particular region will receive incorrect data which may lead to value failure.

Network Faults. The WSN applications require the gathering of data from SNs and the transmission of the information toward a sink. In WSNs the wireless links are volatile and does not yield reliable services as the communication links may fail due factors like confliction between nodes of WSN, co-existence between network, the collision of messages. This results in the loss of the transmitted information and also the paths generated by routing protocol leads to message dropping and information loss. Even if the nodes have excellent connections the messages are not conveyed to destination because of route errors. So the routing protocol must be selected as per the application while selecting the paths. For example, in soft real-time application it can tolerate the delay in selection of paths while regularly collecting data. But in case of hard real-time application, time is a critical issue and a minor delay is not tolerated like in fire detection.

Sink Faults. The sink is a fundamental part in a WSN, subsequently its failures can result in collapsing of the entire system network if no adaptation to fault tolerant techniques is executed to defeat this failure. With respect to a node, the faults can happen in equipment or programming. Nonetheless, unlike SNs, the sink has no energy constraints. The sink may be located in places where long-lasting power supply is unavailable. In such scenarios, batteries combined with sun based cells i.e. solar cells are normally connected [5, 9] to give the required energy.

3.2 Failure Classification

As we know that many faults are converted into failures if they are not recovered on time such as node mobility to a different area, battery failure, data link failure, nodes may send wrong values on request intentionally or because of any malware. Various sorts of failures are listed below:

Crash-Stop Failure. It occurs when services at a point sooner or later stop reacting to any request. Crash-stop influences the process or useful unit and makes it stay at end state at which no further damaged yield are delivered.

Fail-Stop Failure. Fail-Stop Failure is similar to crash-failure yet the main distinction is that it can be effortlessly identified by different processes while the crash-stop failure cannot be identified.

Omission/Exclusion Failure. A failure by exclusion is resolved by a system not reacting to incoming requests. An exclusion degree, that can be characterized which imposes a restriction on the amount of exclusion failure a node can experience before it is classified as crashed.

Timing Failure. Service failures may occur either due to a timeout while preparing a demand or by giving information prematurely. This timing failure happens if a node responds with the correct value to a demand, but the response is received outside the time interval specified by the application.

Hardware Equipment Failure. A large portion of the hardware equipment failures is relevant to nonspecific equipment segments of SNs. The subject impression of ecological calamities causes equipment failures.

Response Failure. There is a potential likelihood of two sorts of response failures in particular the value failure and state change failure. When any service sends an untimely reaction then that service is assumed to be failed by incorrect values. The State Change Failure guides the framework to manage the collapsed arrangement of control stream, which at last triggers unintended default activity at the wrong time and creates an immaterial arrangement of data as response.

Arbitrary/Byzantine Failures. These failures make the system to develop arbitrary values at an arbitrary time. Amid processing, this kind of failures have a tendency to discard planned evaluations. The Byzantine failure includes arbitrary failure that is provoked by a pernicious service that behaves improperly when communicating with other services and applications.

4 Fault Tolerant Mechanisms

As WSNs are deployed in inaccessible environments therefore, the sudden node or link failure may partition the network, loss of data, route failures, and dynamic changes in the network topology. Hence, from the standpoint of the reliable supply of WSN quality data and services, it is essential to achieve maximum robustness against path failures with less power consumption during the development of routing protocols. Various kinds of fault-tolerant routing techniques have been categorized to understand the challenges of constructing fault-tolerant protocols in distributed sensor network applications. Fault tolerance, as suggested in [10, 11], refers to how capable the system to deal with failures, specifically when CHs fails due to excessive load. These types of failures impacts system's performance. To confront this challenge, various strategies are addressed, one of those strategy being diagnosis of faults by the system without any acknowledgement/response. Fault tolerance being a critical property allows a system to function correctly even in the presence of single or multiple faults. As the name indicates Fault tolerant means tolerating the faults thrown by system still it work without shutting down or crashing the system.

Fault tolerance mechanisms are categorized according to various criteria. Now the biggest key factor in this categorization is that at what time the fault tolerant mechanisms are triggered (before or after the incident occur). As per this categorization, these techniques are categorized as fault prevention, fault diagnosis/detection and fault recovery techniques.

4.1 Fault Prevention

By enhancing the utilization of the accessible resources as well as implementing substitutes that give a similar service like a failure does not influence the continuum of the service. We can recognize two types of the preventive systems: node level and network level. At the node level, the approach expects to target the nodes lifetime, while at another level i.e. network level, the strategy tolerates the failure of few nodes, however, it tries to augment the system lifetime to guarantee the coherence of information detecting and transmission.

Fault prevention strategies mean to keep faults away from occurring by guaranteeing full network analysis and availability at the plan and deployment stages, continually observing network status and activating receptive activities if considered important and upholding repetition in the information delivery route, expecting that at least one of the route will survive and satisfy the assignment of information delivery.

4.2 Fault Detection Techniques

Fault detection in sensor networks relies on the kind of applications and the sort of failures. An essential fault detection system is constantly expected to distinguish and confine the produced failures. Such systems can be handled by any of the three ways namely self detection, group detection and hierarchical detection.

Self Detection. In most of the cases, nodes can recognize potential failures by performing self-finding. In [12], the author proposes a method where a node would play out a self-finding in light of the readings of accelerometer sensors to decide in the event that the nodes experience an effect that could lead to equipment breakdowns. Utilizing a comparative method, nodes could distinguish when they are being relocated to an alternate area. Second method is monitoring the identification of the nodes within the vicinity. Whenever there is any relocation then it can easily be determined that either self node or any of its neighbors relocated to some other place. Nodes can likewise recognize that their present links to neighboring nodes are problematic by examining the link connection.

Group Detection. The recognition of failure of service is due to mistakenly created values just conceivable if a reference value is accessible. In [13, 14], identification systems are prone to recognize defective SNs. This algorithm estimates all neighboring nodes and the result is utilized to determine the probability of a node being defective. One more method regarding this is that whether the tasks performed by the service provider are correct or not as per they are asked to do is checked by consumer nodes. In [15], a bad conduct identification algorithm to assist the network layer is suggested which checks the reliability of the service a fault detection mechanism is observing the communication.

To deal with groups in WSNs, [16] proposes to help the potential recovery of damaged gateways. The proposed convention assumes that a gateway is deemed to have failed only if no other gateways are in communication with it.

Hierarchical Detection. The meaning of a detection tree empowers an adaptable fault discovery algorithm in WSN. The network topology is utilized to transfer the fault discovery outcomes of child nodes to their parent nodes and up to the sink, as described in [1]. Every node has the responsibility of sending the status of its child nodes to parent nodes frequently. Although the parent performs a bitwise OR operation on the outcomes of the child nodes together with its outcomes and forwards it to the following level.

Regarding this circumstance, the authors of [17] shown by using various leveled network topology where CHs auditing normal nodes and BS auditing the head of groups. While performing the audit process, the BS and CHs send the ping commands to the nodes which have battery resources remaining and those nodes are under their instant guidance. If any node failed to respond the ping command then it is considered to be failed.

The authors of [18] depicted a debugging tool which uses various leveled hierarchical recognition method additionally.

This tool equips the WSN with checking programming on the SNs to generate metric data, which is then transmitted to a centralized sink location for testing. To automatically detect network failures, this tool proposes a binary tree based on the analysis conducted by the authors during real-time deployments.

To consequently detect failures in network this device proposed a binary tree in view of the examination conducted by the authors while deployments in real time.

4.3 Fault Recovery Techniques

As elaborated in the preceding section, WSNs can encounter faults from different-different origins. To address this, several fault tolerant strategies have been proposed and adapted to enlarge the reliability of service in that particular area. Here, we are attempting to summarize this endeavor. Among these strategies, the most prominent is segment replication. In spite of the fact that repetition has a quite interest regarding high-quality services and accessibility, it too expands the expenses of a deployment.

Here, we order the recovery mechanisms for WSN into two noteworthy methodologies: the Dynamic and Static replication.

Dynamic replication in WSN. Dynamic replication in WSN is normally practiced in situations where all, or numerous, nodes give a similar usefulness. One case is a service that intermittently gives sensor information. The nodes which are running this trigger their sensors and send readings to a BS. When certain nodes decline to provide the required data, the receiver can still get results from other nodes, which is frequently acceptable. Fault recovery in the presence of active replicas is comparatively straightforward. To ensure a reliable study we present some of these methodologies here:

Multiple path routing. Generally, it is desirable to avoid a single failing node that triggers the portioning of the network. Consequently, a system ought to be p-connected that permits p − 1 nodes to fall flat while still maintaining connectivity [19]. Multiple path routing [20] can be utilized to effectively repeat routing paths. In [21], the author suggested the algorithm that figure out the least measure of extra nodes and position of those nodes to ensure p-connected among the nodes.

Sensor values collection. Sensor value combination, as discussed in [22] is a research field aimed at aggregating various low level senor inputs to provide high-level data. The inherent duplicity of SNs can be utilized to give fault-tolerant information fusion. This is accomplished through a trade-off between the accuracy of the subsequent sensor learning interval and the number of damaged sensors.

Disregard values from flawed nodes. A simple however effective solution to prevent failure of one particular node from affecting the whole system is to disregard the information that it is creating, as connected in [12]. However, the real challenge in this scenario lies in recognizing of the failing nodes.

Static replication in WSN. When employing inactive replication, the essential copy get all requests and handles their processing. Keeping in mind the end goal to preserve consistency between copies, the status of the essential copy and the request data are communicated to the backup copies. The way towards recovery from a fault when utilizing static replication is comprised of three fundamental advances: fault identification, essential selection and distribution of service.

Node Choice. Once it is known that a specific functionality is not working due to failure in initial copy then another provider for the service must be chosen. Afterwards, the choice stage, one or a few nodes move towards becoming service providers. Many methods are there to make the selection of choice. They are separated as per the decisions on which should become the service provider. Self-association procedures have demonstrated to increase the reliable services and adaptation to the non-critical failure of distributed frameworks [23]. Or, then again, nearby gathering of nodes work in a combined way for planning their activities. As per the flip side, hierarchical frameworks works in a top to bottom manner. We examine these alternatives orderly.

In self election, nodes in LEACH regularly execute an algorithm based on probability to check if it should fill in as CH to their neighbors. As per this, the nodes are continuously changing their role in the system. At the point when CH node fails, it will take just a single period of rotation until the period when another node start giving usefulness of the missing node. The algorithms which assign roles to determine which of the several roles like clustering, network aggregation and coverage ought to be accepted with the help of a node. An algorithm which is deterministic in nature is given in [24] for self-governing assignment of roles is suggested. This algorithm consider characteristics of the node like the area and battery status yet additionally its neighborhood and the parts picked by neighboring nodes. This encourages the restricted self-design of a sensor to arrange and restore service provisioning if executed after a few nodes have stopped working.

In group election, when a cluster head fails then the nodes which are part of that cluster are reallocated [16]. The cluster head, referred to as a gateway here, is considered a useful node. This arrangement displayed here assumes that all gateways in the system maintain a list of nodes which are present in its cluster as well as a secondary backup list of nodes that can progress toward becoming some portion of their cluster. At the point when gateways stop working, then the nodes from its group are displaced to alternate passages which have the nodes in their backup list. In this list, if greater than one gateway has particular node then the node is appointed to the head of clusters that has minimum cost for communicating.

In hierarchical election, the organizer chooses the initial node that leads to the reconstructing of paths [25] and in addition the determination of another cluster head. The previous researchers portrays an algorithm to choose the nearest node to the base station. Another method selects the node to become cluster head depending upon fuzzy logic. Such methodologies want that the nodes transmit messages to the BS regularly.

Service Distribution. Amid this phase, nodes chosen to become service providers must initiate the service. Sometimes the service can already access the nodes, and only a basic configuration change is required to instruct the node which service should be activated. Sometimes nodes have not sufficient memory to store the code for every possible service. However, it is crucial to inject code using a few strategies.

Pre-replication as portrayed in [24] which comprises of creating the code for all available services accessible on every node before organization. This enables nodes to modify their nature as indicated by the role that they are allocated to. Few methodologies have been proposed for broadcasting code all through the system in which some cases for byte-code mediator is distributed into some number of instructions.

From one perspective code movement is a method which decrease the measure of about how much memory needed in the whole system because all nodes need not be application pre-introduced. It again expands vitality over the nodes trading the code furthermore which is vulnerable to connection failures until the point when the code refresh is finished, it could lead to large delays. Now remote execution [26] is an option where a minimum power device exchange undertakings to a high power device without exchanging the whole system code rather the needed case information is sent. Such type of method is particularly best suited for dissimilar sensor systems with more than one active node. A mixed method between code relocation and remote execution is suggested in [27], in which system code is duplicated to another node when the first threshold level of battery reached. When the battery comes to a basic level, the state of execution is exchanged and remote nodes get the control. This takes into account the complete use of the accessible energy assets because control is given up sooner a node comes up to fail.

5 Conclusion

This paper demonstrated a concise overview of various types of faults and failures that occur in WSN deployments. This analysis outlines that several techniques can be used in heterogeneous applications depending on types of faults. By focusing solely on the flaws, the insights gleaned from the distinctive organizations can be applicable to any application. We examined the issue of fault detection and recovery technique through the categorization, it is convincible to look at the different types of solutions recognizing both strong and weak points. This takes into consideration a revised choice of the strategies that are more reasonable to particular applications. Although, no fault tolerant technique is there which is suitable to overcome all types of faults that are exposed. Overall, the intent of the study is to categorize and understand the types of faults and look for reasonable solutions to overcome these faults.

References

1. Atero FJ, Vinagre JJ, Ramiro J, Wilby M (2011) A low energy and adaptive routing architecture for efficient field monitoring in heterogeneous wireless sensor networks. In: 31st international conference on distributed computing systems workshops. IEEE, pp 172–181
2. Wu C, Chung Y (2007) Heterogeneous wireless sensor network deployment and topology control based on irregular sensor model. In: Advance in grid pervasive computing. Springer Berlin Heidelberg, pp 78–88
3. Anastasi G, Conti M, Di Francesco M, Passarella A (2009) Energy conservation in wireless sensor networks: a survey. Ad Hoc Netw 7(3):537–568
4. Khelifa B, Haffaf H, Madjid M, Llewellyn-Jones D (2009) Monitoring connectivity in wireless sensor networks. J Future Gener Commun Netw 2(2):507–512
5. Tolle G, Polastre J, Szewczyk R, Culler D, Turner N, Tu K, Burgess S, Dawson T, Buonadonna P, Gay D, Hong W (2005) A macroscope in the redwoods. In: SenSys '05: proceedings of the 3rd international conference on Embedded networked sensor systems. ACM Press. New York, NY, USA, pp 51–63 (2005)
6. Ma R, Xing L, Michel H (2006) Fault-intrusion tolerant techniques in wireless sensor networks. In: 2nd International symposium on dependable, autonomic and secure computing. IEEE, pp 85–94
7. Zhang C, Ren J, Gao C, Yan Z, Li L (2009) Sensor fault detection in wireless sensor networks. IET Int Commun Conf Wirel Mob Comput 66–69
8. Clerk Maxwell J (1892) A treatise on electricity and magnetism, vol 2, 3rd edn. Oxford, Clarendon, pp 68–73
9. Mainwaring A, Culler D, Polastre J, Szewczyk R, Anderson J (2002) Wireless sensor networks for habitat monitoring. In WSNA '02: proceedings of the 1st ACM international workshop on wireless sensor networks and applications. ACM Press, NewYork, NY, USA, pp 88–97
10. Mohapatra H, Rath AK (2020) Survey on fault tolerance-based clustering evolution in WSN. IET Netw 9(4):145–155
11. Adday GH, Subramaniam SK, Zukarnain ZA, Samian N (2022) Fault tolerance structures in Wireless Sensor Networks (WSNs): survey classification and future directions. Sensors 22(16):6041
12. Harte S, Rahman A (2005) Fault tolerance in sensor networks using self-diagnosing sensor nodes. In: The IEEE international workshop on intelligent environment 2005, pp 7–12

13. Krishnamachari B, Iyengar S (2004) Distributed Bayesian algorithms for fault-tolerant event region detection in wireless sensor networks. IEEE Trans Comput 53:241–250
14. Ding M, Chen D, Xing K, Cheng X (2005) Localized fault-tolerant event boundary detection in sensor networks. In: INFOCOM
15. Marti S, Giuli TJ, Lai K, Baker M (2000) Mitigating routing misbehavior in mobile Ad hoc networks. In: Proceedings of the 6th annual international conference on mobile computing and networking, pp 255–265
16. Gupta G, Younis M (2003) Fault-tolerant clustering of wireless sensor networks. Wirel Commun Netw 3:1579–1584
17. Ruiz LB, Siqueira IG, e Oliveira LB, Wong HC, Nogueira JMS, Loureiro AAF (2004) Fault management in event driven wireless sensor networks. In: Proceedings of the 7th ACM international symposium on modeling, analysis and simulation of wireless and mobile systems, pp 149–156
18. Ramanathan N, Chang K, Kapur R, Girod L, Kohler E, Estrin D (2004) Sympathy: a debugging system for sensor networks. In: IEEE international conference on local computer networks
19. Li N, Hou JC (2004) FLSS: a fault-tolerant topology control algorithm for wireless networks. In: Proceedings of the 10th annual international conference on mobile computing and networking, pp 275–286
20. Ganesan D, Govindan R, Shenker S, Estrin D (1997) Highly-resilient, energy-efficient multipath routing in wireless sensor networks. Mob Comput Commun Rev 1(2)
21. Bredin JL, Demaine ED, Hajiaghayi M, Rus D (2005) Deploying sensor networks with guaranteed capacity and fault tolerance. In: Proceedings of the 6th ACM international symposium on mobile ad hoc networking and computing, pp 309–319
22. Desovski D, Liu Y, Cukic B (2005) Linear randomized voting algorithm for fault tolerant sensor fusion and the corresponding reliability model. In: IEEE international symposium on systems engineering, pp 153–162
23. Zhao BY, Kubiatowicz J, Joseph AD (2001) Tapestry: an infrastructure for fault-tolerant wide-area location and routing. Technical Report UCB/CSD-01-1141, Computer Science Division, University of California, Berkeley
24. Frank C, Romer K (2005) Algorithms for generic role assignment in wireless sensor networks. In: Proceedings of the 3rd international conference on embedded networked sensor systems, pp 230–242
25. Staddon J, Balfanz D, Durfee G (2002) Efficient tracing of failed nodes in sensor networks. In: Proceedings of the 1st ACM international workshop on wireless sensor networks and applications, pp 122–130
26. Rong P, Pedram M (2003) Extending the lifetime of a network of battery-powered mobile devices by remote processing: a markovian decision-based approach. In: DAC'03: proceedings of the 40th conference on design automation. ACM Press, New York, NY, USA, pp 906–911
27. Marculescu D, Zamora NH, Stanley-Marbell P, Marculescu R (2003) Fault-tolerant techniques for ambient intelligent distributed systems. In: Proceedings of the 2003 IEEE/ACM international conference on Computer-aided design, pp 348

COA-Based Optimization of THD and Dc-Link Voltage of Half-Bridge Compensator for Electric Traction

Rishi Raj Ranjan, Kumari Namrata, Ashish Kumar, and Mantosh Kumar

Abstract The current research focuses on using Coati Optimization Algorithm (COA) based optimization approaches to improve the performance of a half-bridge compensator used in electric traction systems. The study's main goals are to reduce Total Harmonic Distortion (THD) and stabilize the Direct Current (DC) link voltage, which are crucial elements that affect the overall dependability and efficiency of electric traction systems. The suggested methodology makes use of sophisticated optimization techniques and control algorithms based on COA analysis. The goal of the optimization approach is to strike a compromise between reducing THD and stabilizing the DC link voltage by methodically examining different control parameters. A comprehensive approach is ensured by taking into account the trade-offs between voltage stability and energy efficiency through the use of COA-based measures. The outcomes of the simulation show that the suggested COA-based optimization strategy is effective in outperforming conventional control strategies in terms of performance. Reduced THD levels and improved DC link voltage stability are two benefits of the redesigned half-bridge compensator that raise the overall effectiveness and dependability of electric traction systems. The results of this study open the door to more effective and environmentally friendly modes of transportation by offering insightful information for the development and use of sophisticated control schemes in electric traction applications.

R. R. Ranjan (✉) · K. Namrata · A. Kumar · M. Kumar
EE Department, NIT Jamshedpur, Jamshedpur, Jharkhand, India
e-mail: rishirajranjan1036@gmail.com

K. Namrata
e-mail: namrata.ee@nitjsr.ac.in

A. Kumar
e-mail: ashishmshr36@gmail.com

M. Kumar
e-mail: mantosh.nith@gmail.com

S. Kumar et al. (eds.), *Soft Computing Applications in Modern Power and Energy Systems*, Lecture Notes in Electrical Engineering 1298,
https://doi.org/10.1007/978-981-96-0058-8_19

Keywords Coati optimization algorithm (COA) · Negative sequence current (NSC) · Railway static power conditioner (RPC) · Half-bridge-converter-based railway static power conditioner · Installed capacity reduction

1 Introduction

Power quality concerns have received more attention as electric train transportation systems have developed. Reactive power and harmonic power quality issues are exacerbated by the traction loads' intrinsic non-linear and low power factor (PF) features [1]. As predicted by the symmetrical component analysis, the unbalanced single-phase nature of traction loads causes asymmetrical three-phase currents in the public grid, which in turn cause negative sequence currents (NSC). These three elements—harmonics, reactive power, and NSC—have a big impact on how safely the traction power supply system and the public grid operate. As shown in Fig. 1, which shows the Electric Connection Diagram of Modern Locomotive, the installation of a Railway Power Conditioner (RPC), which provides compensatory currents to the railway power system network, emerges as a vital solution to ameliorate these power quality difficulties [2].

The extant literature has yielded a number of important insights. Three primary power quality issues in railway power system networks are examined, including DC-link voltage imbalance, harmonics, and negative sequence current. Although RPCs solve these problems well, their higher installed capacity and converter cost are a downside [3]. Electrical systems, particularly railway traction networks, are frequently modelled, controlled, and simulated using programs like MATLAB/

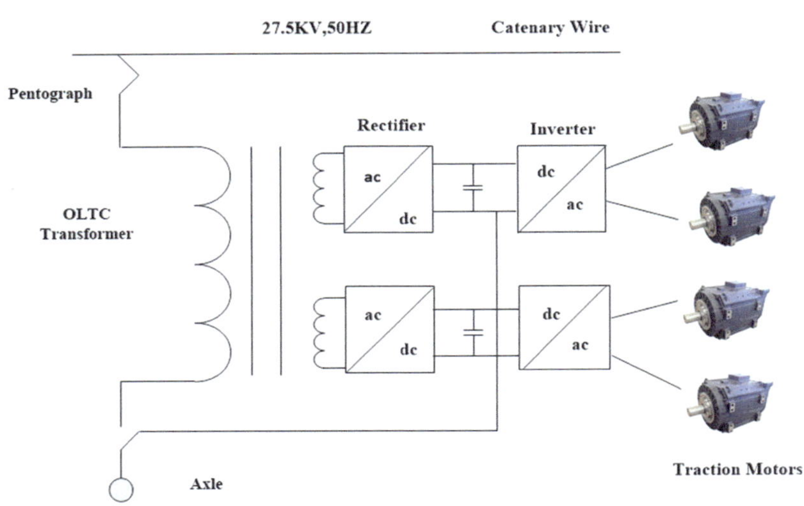

Fig. 1 Electric connection diagram of modern locomotive

SIMULINK, PSCAD, and Typhoon HIL. Benefits of the Half-Bridge Railway Power Conditioner (HBRPC) include decreased power losses, complexity, and hardware cost due to its fewer switches [4].

To reduce reactive power and negative sequence currents, a novel co-phase electric traction system that includes a three-phase to single-phase converter is suggested. This system transmits active power directly to the catenary. Active Power Quality Conditioners (APQC) are excellent compensators, but they are expensive due to their high current capability power electronic switch requirements [5]. Therefore, to handle negative sequence currents and harmonics current correction, a combination of power conditioners and static Var compensator (SVC) devices is proposed. To maximize SVC ratings, the construction of railway power conditioners with unidirectional and bidirectional switches is investigated. In order to achieve acceptable power quality conditions, the impedance mapping design approach is a reasonable choice for designing LC characteristics, which are critical in voltage control and reduction [6].

One way to reduce power conditioner ratings and compensate for current harmonics and reactive power across different loading ranges is to integrate shunt hybrid filters with Thyristor-Controlled Reactors (TCRs). It is suggested to use a hybrid power quality conditioner (HPQC) in conjunction with a magnetic static variable compensator (MSVC) to reduce active power compensation and modify installed capacity in response to varying loading scenarios [7]. Through cost function optimization, the Adaptive Control-Based Half-Bridge Railway Power Conditioner (ACBT-HRPC) exhibits satisfactory performance in harmonics and reactive power compensation. It is stressed that the proper choice of control designs and parts is necessary to guarantee the compensators' suitable behaviour. To produce firing pulses, a variety of controllers, including microcontrollers (μc), microprocessors (μp), and digital signal processors (DSP) like d-SPACE and FPGA, are frequently used [8].

For global optimization issues, the Coati Optimization Algorithm (COA) technique offers several benefits. The first benefit of COA is that it does not require any control over the parameters because it does not have any control parameters in its design. The second benefit of COA is its exceptional efficacy in handling complicated, high-dimensional problems and a wide range of optimization challenges across multiple sciences [9]. The third benefit of the suggested approach is that it demonstrates a strong ability to strike a balance between research and search, enabling fast convergence to produce appropriate values for decision variables in optimization tasks, particularly those involving complicated issues. The fourth advantage of the proposed COA is its powerful performance in handling real-world optimization applications [10].

Thus, the main contributions of this paper are as follows:

1. Design implementation of Railway electrical supply system for specific datasheet
2. Design implementation of RPC based on Half Bridge compensator.

3. Developing COA–NSC compensation-based optimized switching angles to trigger the RPC based on the Half Bridge compensator that replaces the conventional lookup table model.
4. Validating the experimental results of the proposed RPC based on the Half Bridge compensator with simulation results.
5. Compare the proposed work results with another system investigated in the latest literature.

The rest of this article is organized as follows: Sect. 2 discusses the problem formulation of the proposed system and explains, in detail, the control architecture of the modern traction system. The methodology is discussed in Sects. 3 and 4 discusses the harmonic elimination of the proposed compensator. Section 5 discusses simulation results and analysis of the proposed compensator.

2 Problem Formulation

The objective of this research is to design and evaluate a Railway Power Conditioner (RPC) utilizing MATLAB/SIMULINKTM, a Half-Bridge compensator, and a methodology based on the Coati Optimization Algorithm (COA). The design comprises a current controller and fewer switches in the Half-Bridge compensator, with an emphasis on assessing the system's overall performance. To attain Total Harmonic Distortion (THD) values below the IEEE 519 standard, the inquiry also entails the development of COA-based compensation algorithms for harmonic removal and Negative Sequence Currents (NSC) [11]. To validate and contextualize the results, a comparison with recent literature is done.

The suggested methodology is outlined in the detailed block diagram presented in Fig. 2, which includes the railway supply system, traction transformer, and compensator combined with a single-phase nonlinear load. A V/V transformer makes it easier to convert two single-phase power sources at a 27.5 kV voltage point from a 220-kV three-phase high-voltage source [12]. The two compensator converters are connected to the auxiliary power arms of the V/V traction transformer via subsequent inductors and step-down transformers. The ground of two traction power arms and the centre of two capacitors are connected via a step-down transformer located in the proposed compensator. Two DC-link capacitors, two power switch legs, and two capacitors connected in series between the two switch legs make up the compensator setup. By optimizing the RPC system by IEEE 519, this organized approach seeks to ensure effective NSC compensation, harmonic elimination, and THD reduction. It also provides a useful benchmark against the body of recent research in the field [13].

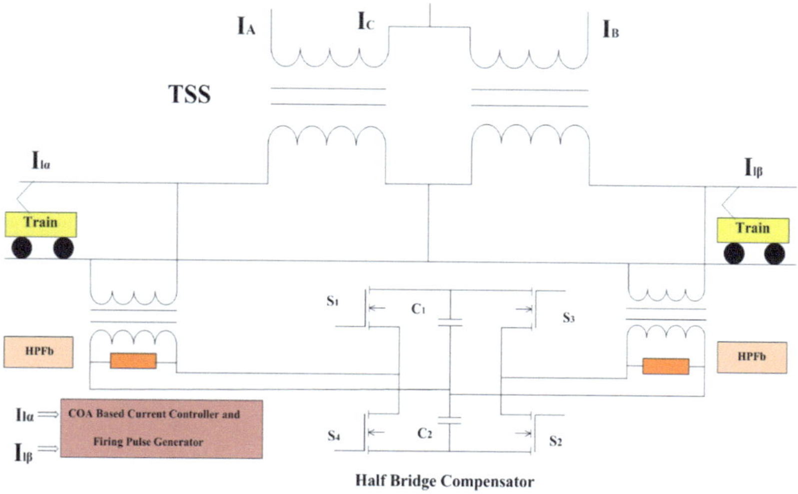

Fig. 2 Block diagram of the proposed model

3 Methodology

A methodical implementation process will be followed for the proposed Railway Power Conditioner (RPC) model, which is based on the Half-Bridge compensator. The design will first include the railway electrical supply system, customized to particular data sheets, followed by a representation of the Voltage (V) and Current (I) properties at various stages. After that, the RPC design will be carefully implemented using the Half-Bridge compensator. One novel method is to replace traditional lookup table models with Negative Sequence Current (NSC) compensation based on the Coati Optimization Algorithm (COA), which optimizes switching angles [14, 15]. The suggested RPC model's experimental results will undergo a rigorous validation process against the results of simulations. Furthermore, a comparative analysis will be carried out to match the suggested work with pertinent systems examined in the most recent literature. This methodical application seeks to confirm the effectiveness of the Half-Bridge compensator-based RPC, highlighting its potential advantages over current approaches and adding significant new information to the corpus of current research in this field. Figure 3 shows the flowchart of the proposed work [16, 17].

4 Harmonic Elimination

Several techniques have been discussed in the literature for stabilizing dc link voltage and eliminating the harmonic contents [18, 19].

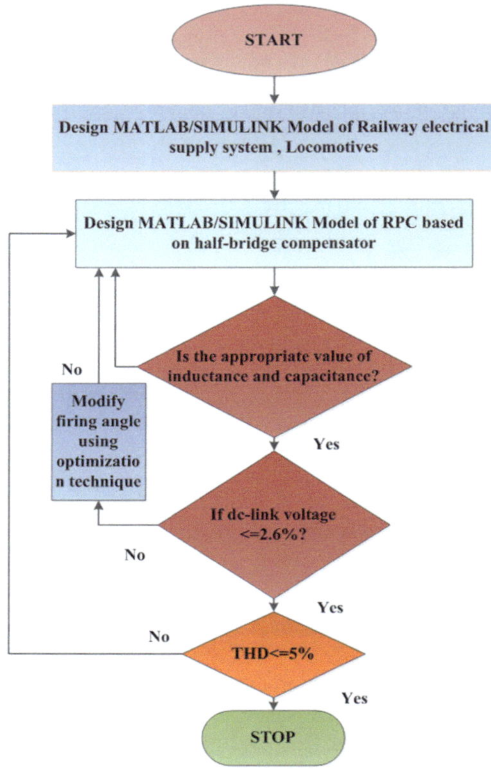

Fig. 3 Flowchart of the proposed work

4.1 Compensation Principle

Locomotive loads generate large harmonic currents. Since the 3rd and 5th harmonic current is the maximum among all harmonic currents. Therefore, the 3rd and 5th-order harmonic currents are suppressed by the passive filter, which reduces the capacity of HBRPC. Higher-order harmonics are eliminated by the harmonic active filter [20, 21].

To suppress the negative sequence component of current, the Combination of the High pass filter and HBRPC feed the compensating current to the power arms of the traction system. After detailed vector analysis compensating currents are derived [22, 23].

Assuming the active RMS current that HBRPC transfers is $I_{com.p},$ Then the active RMS current on both supply sides is transferred are:

$$I'_{l\alpha.p} = I_{l\alpha.p} - I_{com.p} \tag{1}$$

$$I'_{l\beta.p} = I_{l\beta.p} + I_{com.p} \tag{2}$$

where $I'_{l\alpha.p}$ and $I'_{l\beta.p}$ are RMS of active currents of two power supply arms after transferring active power. The negative sequence current produced by them are:

$$I'^{-}_{l\alpha.p} = \frac{1}{\sqrt{3}}\left(I_{l\alpha.p} - I_{com.p}\right) \tag{3}$$

$$I'^{-}_{l\beta.p} = \frac{1}{\sqrt{3}}\left(I_{l\beta.p} + I_{com.p}\right) \tag{4}$$

The reactive power compensation current RMS value of each phase is:

$$I'_{\alpha} = \frac{1}{\sqrt{3}}\left(I_{l\beta.p} + I_{com.p}\right) \tag{5}$$

$$I'_{\beta} = \frac{1}{\sqrt{3}}\left(I_{l\alpha.p} - I_{com.p}\right) \tag{6}$$

$$I'_{\gamma} = \frac{1}{\sqrt{3}}\left(I_{l\alpha.p} - I_{com.p}\right) - \frac{1}{\sqrt{3}}\left(I_{l\beta p} + I_{com.p}\right) \tag{7}$$

The active power that HBRPC transferred is $P_{com} = I_{com.p} \cdot U$, the capacity of High pass filter $\left(Q_{\alpha}, Q_{\beta}, Q_{\gamma}\right)$ of each phase are:

$$Q_{\alpha} = \frac{1}{\sqrt{3}}\left(I_{l\beta.p} + I_{com.p}\right) \cdot U - \sqrt{S_{com}^{2} - p_{com}^{2}} \tag{8}$$

$$Q_{\beta} = \frac{1}{\sqrt{3}}\left(I_{l\alpha.p} - I_{com.p}\right) \cdot U - \sqrt{S_{com}^{2} - p_{com}^{2}} \tag{9}$$

$$Q_{\gamma} = \frac{2}{\sqrt{3}}\left(I_{l\alpha.p} - I_{l\beta.p} - 2\, I_{com.p}\right) \cdot U \tag{10}$$

According to the relative standard of power

$$\begin{aligned} 0 \leq r_U \leq \zeta \\ 0.9 \leq \cos\varphi \leq 1 \end{aligned} \tag{11}$$

where $r_U = |U_- \,/\, U_+|$ is the degree of unbalanced voltage; U_+ and U_- the positive and negative sequence voltage; ζ the upper limit of unbalanced voltage; $cos\varphi$ the power factor after compensation.

The ratio of active power that HBRPC transferred to the load active power is $k = P_{com}/P_{\max}$ then the required compensation capacity of each phase is:

$$Q_{\alpha} = h_3\, U(1 - 2k)\, I_{l\alpha.pm} - U\, I_{l\alpha.pm} \tag{12}$$

$$Q_{\beta} = h_2\, U(1-2k)\, I_{l\alpha.pm} \tag{13}$$

$$Q_{\gamma} = -\, h_1\, U(1-2k)\, I_{l\alpha.pm} \tag{14}$$

where k < 0.5; $h_i \geq o$; and i = 1, 2, 3.

Therefore, the mathematical of optimization is as follows:

$$\text{Min}\, \mathrm{f}\big(Q_{\alpha}, Q_{\beta}, Q_{\gamma}\big) = \big(Q_{\beta} - Q_{\alpha} - Q_{\gamma}\big) \tag{15}$$

$$\text{s.t.}\, Q_{\alpha} + Q_{\beta} + Q_{\gamma} + Q_{\max} \geq 0 \tag{16}$$

$$\begin{aligned} &0 \leq r_U \leq \mu = \xi \cdot \frac{S_d}{S_L} = \xi \cdot \frac{S_d}{S_{l\alpha}} \\ &0.9 \leq \cos\varphi \leq 1 \end{aligned} \tag{17}$$

To determine the firing angles, the Eqs. (15)–(17) are solved by using COA.

4.2 *The Coati Optimization Algorithm (COA)*

Coati Optimization Algorithm (COA) is a population-based optimization algorithm that draws inspiration from animal behavior—in this example, coatis. Continuous, discontinuous, non-linear, and stochastic objective function issues can all be handled by COA [24].

General steps of COA:

1. Set the population's size, the parameters for the search space, and the conditions for terminating the population.
2. Assess fitness: Using the objective function as a guide, determine each individual's (coati's) fitness value within the population.
3. Exploration: Mimic coati foraging behavior:
   - Using a levy flight random walk, half the population "climbs the tree" or explores the search space.
   - By looking about where they are now, the other half "waits under the tree" (exploits promising places).
4. Assess the suitability of candidates for the updated roles in the new evaluations.
5. Selection: Those who are fitter will be chosen for the following generation.
6. Substitute: Switch out underperforming personnel with the chosen ones.
7. Steps 2–6 should be repeated until the termination requirements are satisfied.
8. Output: Provide the person whose fitness value is the highest (optimal).

4.3 Control Strategy of Proposed Compensator

Hysteresis control is used in these tests to achieve the ongoing internal circle's most effective attack strategy [25, 26]. Given the previously mentioned exchanging control principle, the current internal circle control is implemented as follows. When the half-bridge converter's output current is at the positive half cycle, this is when the hysteresis control rule applies (i > 0) [27, 28]

$$\left\{ \begin{array}{l} \text{if}\, r \leq i_R - i,\ S_1 \text{ is OFF and } S_2 \text{ is ON} \\ \text{if} - r < i_R - i < H,\ S_1 \text{ and } S_2 \text{ are held ON} \\ \text{if} - r \geq i_R - i,\ S_2 \text{ is OFF and } S_1 \text{ is ON} \end{array} \right\} \quad (18)$$

where i_R is the reference signal and r is the hysteresis width. In the unlikely event that $r \leq i_R - i$, T_2 is now ON. The capacitor C_2 would deplete, increasing the current to i_R. S_2 is OFF and S_1 is switched on, if $-r \geq i_R - i$, after passing through D_1, the current I would charge C_1. As a result, the source current would drop to i_R and the voltage of capacitor C_1 would rise. The power switches S_1 and S_2 are kept on if $-r < i_R - i < r$.

5 Simulation Results and Analysis

MATLAB/SIMULINKTM is used to model the system in order to do simulations. The schematic model takes around 0.4 s to run, and Figs. 7 display the waveform results that are seen. The y-axis units are displayed on the up-right side of every waveform, while the x-axis is used to represent time in seconds in all waveform drawings. Table 1. shows the different Simulation Cases.

To distinguish between the effects of power conditioners on railway traction systems, we have taken into account the following two case studies:

Case-I: operational variables without a power compensator, such as voltage, current, and FFT analysis.

Case-II: operational variables with the power compensator, such as voltage and current, and FFT analysis.

Table 2 shows the various locomotive types that were taken into consideration for this research. While WAP-7 is a locomotive engine based on a three-phase induction motor, WAG-9 is a locomotive engine that is entirely based on DC series motor drives.

Table 1 Simulation cases

Cases	Compensation cases
Case-I	Before compensation
Case-II	After compensation

Table 2 Types of locomotives

S. No	Drive	Nomenclature of locomotive
1	DC motor-based locomotive	WAG-9
2	3-phase induction motor-based locomotive	WAP-7

Table 3 shows the various loading schedule types that were taken into consideration for this research. In essence, the loading situation varies in both arms in relation to trains that are coming in and departing out.

In Fig. 4, the FFT bar diagram of three phase currents before compensation in the case of the WAP-7 locomotive is shown. This FFT bar diagram shows that the dc component is 18.96%. whereas 2nd, 4th and 6th harmonics have strengths of 18.75, 3.86 and 4.20%. the 3rd and 5th harmonics have strengths of 14.23 and 5.42% respectively.

The FFT bar diagram of the three-phase voltages before correction for the WAP-7 locomotive is displayed in Fig. 5. This graphic makes it evident that the values of the various harmonic components are lower than the fundamental frequency. The dc component, as seen by this FFT bar plot, is 0.01%. In contrast, the intensities of the second and fourth harmonics are 0.01 percent and 0.01 percent, respectively. The intensities of the third and fifth harmonics are, respectively, 0.03 and 0.03%.

The FFT bar diagram of the three-phase currents in the WAP-7 locomotive case after compensation is displayed in Fig. 6. The dc component, as seen by this FFT bar

Table 3 Loading schedule for two stations

Section	0–0.2s	0.2–0.4 s
Left arm	Full loading	No loading
Right arm	No loading	Full loading

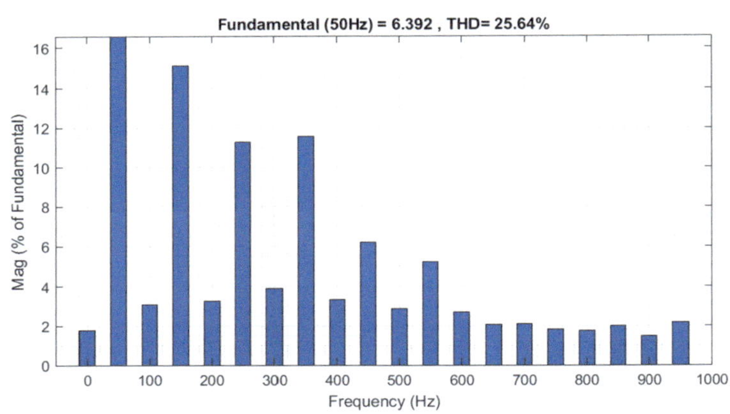

Fig. 4 FFT bar diagram of three phase currents before compensation in case of WAP-7 locomotive

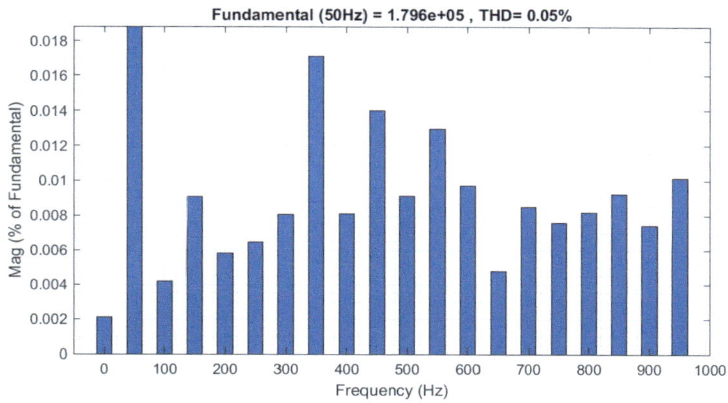

Fig. 5 FFT bar diagram of three-phase voltages before compensation in case of WAP-7 locomotive

plot, is 0.03%. Contrarily, the intensities of the second, fourth, and sixth harmonics are 0.04, 0.04, and 0.06%. The intensities of the third and fifth harmonics are 0.56 and 0.38%, respectively.

The FFT bar diagram of the three-phase voltages following adjustment for the WAP-7 locomotive is displayed in Fig. 6. The value of the various harmonic components is clearly shown in this diagram to be much lower than the fundamental frequency.

In Fig. 7, V_{c1} and V_{c2} are dc link capacitor voltages. Before compensation both voltages are unbalanced. At starting a sudden peak appears because charging of the capacitor. After two cycles the dc link capacitor voltage varies sinusoidally between desired values i.e., $V_{p\text{-}p}$ of 100 V. THD comparison in the case of WAP-7 locomotive comparative analysis is tabulated in Table 4.

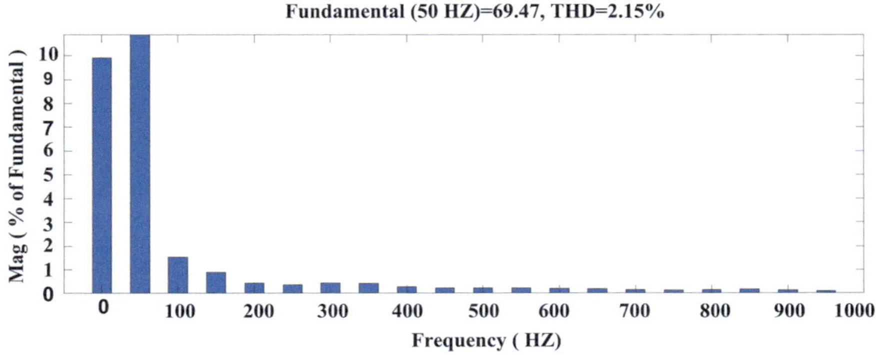

Fig. 6 Three-phase currents after compensation in the case of a WAP-7 locomotive using an FFT bar diagram

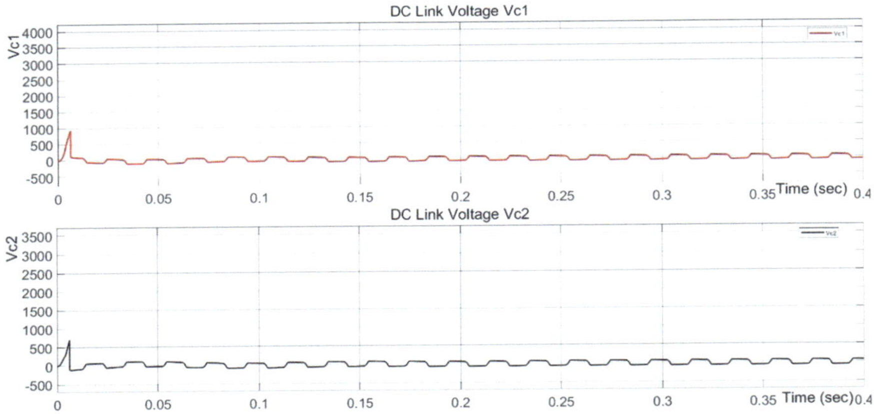

Fig. 7 DC-link capacitor voltages after compensation

Table 4 THD comparison in the case of WAP-7 locomotive

Quantity	Before compensation (%)	After compensation (%)
I_{ABC}	25.64	2.15
V_{ABC}	0.05	0.01

As the proposed compensator is placed into the framework, it can adjust the particular fundamental part and harmonic parts of flows to understand the Negative Sequence Current compensation and harmonic current concealment. Then, at that point, the three-phase flows are balanced almost after compensation, and the Negative Sequence Current is generally decreased to focus on the consistent state.

6 Conclusion

In conclusion, it has been shown that a complete compensator built around two half-bridge converters is effective in providing the best possible power quality for traction systems on high-speed train routes. In addition to making the solution more affordable, the fewer power switches used in comparison to conventional railway power conditioners also reduce power losses and equipment complexity. Optimizing reference continuity and improving dynamic compensation performance has been made possible by the addition of a Hysteresis control mechanism for negative sequence compensation. By taking into account the equilibrium of the DC-link voltage between two capacitors, the suggested balanced voltage management technique successfully reduces voltage imbalance and its negative consequences on system performance.

Moreover, the simulation and experimental empirical results validate the beneficial effects of the proposed framework on harmonic current suppression, negative sequence compensation, and current compensation. Although the half-bridge compensator has inherent limitations, the compensator that has been developed represents a calculated attempt to tackle the railroad traction framework's robustness issues. Significant reductions in Total Harmonic Distortion (THD) and voltage stabilization have been achieved in this work, mostly thanks to the application of the Coati Optimization Algorithm (COA). This highlights the potential of COA as a useful instrument for traction system performance optimization in railway applications, providing a viable path for future developments in improving power quality and efficiency.

References

1. Lei M, Zhao C, Li Z, He J (2022) Circuit dynamics analysis and control of the full-bridge five-branch modular multilevel converter for comprehensive power quality management of cophase railway power system. IEEE Trans Ind Electron 69(4):3278–3291
2. Li L, Wu M (2021) A three-phase symmetric converter in AC electric railway systems for power quality and energy efficient improvement. Int Trans Electr Energy Syst 31(9)
3. Norouzi F, Saeed FS, Jafari Kaleybar H (2021) Improvement of the railway power flow controller's performance using sliding mode control method. In: 12th power electronics, drive systems, and technologies conference, PEDSTC 2021
4. Tanta M, Cunha J, Barros LAM, Monteiro V, Pinto JGO, Martins AP, Afonso JL (2021) Experimental validation of a reduced-scale rail power conditioner based on modular multilevel converter for Ac railway power grids. Energies 14(2)
5. Vemulapati V, Vijaykumar YN, Visali N (2022) Performance enhancement of a high-speed railway supply system with multi module converter: a laboratory prototype model for Indian railways. Indones J Electr Eng Comput Sci 25(2):680–689
6. Wang Y, He Y, Chen X, Zhao M, Xie J (2022) A layered compensation optimization strategy of energy storage type railway power conditioner. Arch Electr Eng 71(1):5–20
7. Wu S, Wu M, Wang Y (2021) A novel co-phase power-supply system based on modular multilevel converter for high-speed railway at traction power-supply system. Energies 14(1)
8. Roch-Dupré D, Gonsalves T, Cucala AP, Pecharromán RR, López-López ÁJ, Fernández-Cardador A (2021) Determining the optimum installation of energy storage systems in railway electrical infrastructures by means of Swarm and evolutionary optimization algorithms. Int J Electr Power Energy Syst 124:106295
9. Ang D, Zhang Z, Wang W, Yang Y (2016) Negative sequence current optimizing control based on railway static power conditioner in V/v traction power supply system. IEEE Trans Power Electron 31(1):200–212
10. Zhang H, Jia L, Wang L, Xu X (2019) Energy consumption optimization of train operation for railway systems: algorithm development and real-world case study. J Clean Prod 214:1024–1037
11. Kaleybar HJ, Kojabadi HM, Fazel SS, Foiadelli F (2018) An intelligent control method for capacity reduction of power flow controller in electrical railway grids. Electr Power Syst Res 165:157–166
12. Langerudy AT, Mousavi GSM (2016) Hybrid railway power quality conditioner for high-capacity traction substation with auto-tuned DC-link controller. IET Electr Syst Transp 6(3):207–214

13. Lao K, Dai N, Liu W, Wong M (2013) Hybrid power quality compensator with minimum DC operation voltage design for high-speed traction power systems. IEEE Trans Power Electron 28(4):2024–2036
14. Lao K, Wong M, Dai NY, Wong C, Lam C (2016) Analysis of DC-link operation voltage of a hybrid railway power quality conditioner and its PQ compensation capability in high-speed cophase traction power supply. IEEE Trans Power Electron 31(2):1643–165
15. Lao K, Wong M, Dai N, Lam C, Wong C, Wang L (2017) Analysis in the effect of co-phase traction railway HPQC coupled impedance on its compensation capability and impedance-mapping design technique based on required compensation capability for reduction in operation voltage. IEEE Trans Power Electron 32(4):2631–2646
16. Lao K, Wong M, Dai N, Wong C, Lam C (2015) A systematic approach to hybrid railway power conditioner design with harmonic compensation for high-speed railway. IEEE Trans Ind Electron 62(2):930–942
17. Levron Y, Belikov J (2017) Modeling power networks using dynamic phasors in the dq0 reference frame. Electr Power Syst Res 144:233–242
18. Li Y, An B, Liu F, Hu S, Xie B, Wang P, Cao Y (2018) An asymmetrical connection balance transformer-based hybrid railway power conditioning system with cost-function optimization. IEEE Trans Transp Electr 4(2):577–590
19. Ma F, Luo A, Xiao H, Wu C, Wang W (2013) A simplified power conditioner based on half-bridge converter for high-speed railway system. IEEE Trans Ind Electron 60(2):728–738
20. Mileševié B, Uglešić I, Filipović-Gr B (2016) Power quality analysis in electric traction system with three-phase induction motors. Electric Power Syst Res 138:172–179
21. Park S, Kotzev M, Brüns HD, Kam DG, Schuster C (2017) Lessons from applying IEEE standard 1597 for validation of computational electromagnetics computer modeling and simulations. IEEE Electromag Comp Mag 6(2):55–67
22. Rahmani S, Hamadi A, Al-Haddad K, Dessaint LA (2014) Combination of shunt hybrid power filter and thyristor-controlled reactor for power quality. IEEE Trans Ind Electron 61(5):2152–2164
23. Roudsari HM, Jamali S, Jalilian A (2018) Dynamic modeling, control design and stability analysis of railway active power quality conditioner. Electric Power Syst Res 160:71–88
24. Dehghani M, Montazeri Z, Trojovská E, Trojovský P (2023) Coati optimization algorithm: a new bio-inspired metaheuristic algorithm for solving optimization problems. Knowl Based Syst 259
25. Xie B, Zhang Z, Hu S, Li Y, Luo L, Sun S (2016) YN/VD connected balance transformer-based electrical railway negative sequence current compensation system with passive control scheme. IET Power Electron 9(10):2044–2051
26. Yuan J, Zhong Y, Zhang C, Zeng W, Chen B, Tian C, Wang J (2016) Optimal electromagnetic hybrid negative current compensation method for high-speed railway power supply system. Power Syst Clean Energy 4:123–134
27. Zhao J, Liu S, Zhou M, Guoi X, Qi L (2018) Modified Cuckoo search algorithm to solve economic power dispatch optimization problems. IEEE/CAA J Autom Sinica 5(4):794–806
28. Amira CR, Ramdane B, Hamza B, Meriem A, Imen M (2021) Genetic algorithm approach of HB-RPC for adaptive power quality improvement in railway traction Chain. In: International conference on electrical, communication, and computer engineering (ICECCE), pp 1–6

Machine Learning Approaches for Open Switch Faults Classification in Three-Phase Induction Motor Drive

Burhan Dube, Uvesh Sipai, Nishant Kothari, and Tapankumar Trivedi

Abstract This paper presents a data-driven approach for Open Switch Faults (OSFs) classification in induction motor (IM) based EV drive. The proposed method utilizes inverter output current for classifying single and multiple OSFs, and healthy condition in IM drive. Various OSFs have been simulated under varying load and reference speed conditions. In this data-driven approach, the instantaneous currents have been utilized as feature input to Support Vector Machine (SVM), k-Nearest Neighbor (kNN), and Random Forest classifiers. Two different methods have been implemented considering single multi-class classification, and phase-wise multi-class classification approach for classifying OSFs. The performance of the SVM classifier stands out remarkable, achieving 100% classification accuracy with a single multi-class classification approach for classifying eighteen different OSFs and healthy conditions in IM drive.

Keywords Induction Motor Drive · Open Switch Fault · Support Vector Machine · Random Forest · k-Nearest Neighbor

1 Introduction

The imperative for alternative transportation stems from rising fossil fuel expenses and the detrimental impacts of climate change. In response, electric vehicles (EVs) emerge as the optimal solution due to their diminished carbon emissions and decreased reliance on fossil fuels [1]. Significantly, concerted endeavors to amplify EV efficiency have yielded remarkable progress, notably exemplified by the innovation of highly efficient three-phase induction motors, thereby bolstering the viability and sustainability of electric transportation systems.

B. Dube · U. Sipai · N. Kothari (✉) · T. Trivedi
Department of Electrical Engineering, Marwadi University, Rajkot, India
e-mail: nishant.kothari@marwadieducation.edu.in

S. Kumar et al. (eds.), *Soft Computing Applications in Modern Power and Energy Systems*, Lecture Notes in Electrical Engineering 1298,
https://doi.org/10.1007/978-981-96-0058-8_20

In EVs, the propulsion system is a vital subsystem responsible for generating tractive force. Comprising the high-voltage battery, converter, and drive, the propulsion system's key element is the power converter. This component, central to the efficiency of EVs, regulates energy flow from the battery to the traction motor. The power converter, focused on in this context, employs insulated gate bipolar transistors (IGBTs) as its main switching devices. Recognized for their high-power handling and fast switching speeds, IGBTs significantly contribute to the overall efficiency and performance of the EV propulsion system. Enhancing IGBT technology in power converters plays a pivotal role in advancing the capabilities of electric mobility, ensuring improved efficiency, reliability, and sustainability in EV propulsion systems [2].

IGBTs employed in EVs inverters contend with thermal stresses due to power cycling and temperature fluctuations, making them prone to failure, specifically Short Circuit Faults (SCFs) and Open-Circuit Faults (OCFs) [3]. Recent surveys highlight power converter failures as a significant factor, constituting 31% of power system faults, resulting from OCFs and SCFs [4–6]. Industrial drives often integrate hardware protection against SCFs. Therefore, protective hardware swiftly addresses SC faults, while OC faults pose a potential threat to the traction drive system. Recent research concentrates on enhancing the understanding and detection of OCFs, contributing to the overall reliability and efficiency of power systems, especially in the context of EVs and their propulsion systems [7].

The OCFs further categorized into Open Switch Faults (OSFs) and Open Phase Faults (OPFs). OPFs occur when one of the phases of the motor's power supply is interrupted, typically due to a broken conductor or loose connection. This leads to an imbalance in the electrical system, potentially causing motor overheating and damage. On the other hand, OSFs occur when a switch within the drive system fails to close properly, resulting in an incomplete circuit. The implication of the OSF may cause secondary damage if unchecked. The bearing or the IM motor may sustain damage from the significant torque ripples. Furthermore, this problem results in an overall reduction in the drive's performance due to significant fluctuations in both speed and torque. These OSFs render affected IGBTs inoperative, exposing the operational counterparts to potential stress and secondary faults in the drive, resulting in a detrimental impact on the EV's power production. An inverter grappling with an OSF condition fails to generate the requisite symmetrical three-phase current vital for the EV's motor. Extensive research into OSF detection has spawned various methods, with model-based approaches relying on dynamic models facing challenges in accurately representing intricate systems, while signal-based methods, although offering an alternative, grapple with computational complexities, making the pursuit of a refined fault detection balance a compelling challenge [3, 5–7].

In data-driven fault detection methods [2, 8–10], the focus shifts to trained artificial intelligence (AI) models that utilize measured signals from the system to identify faults. Data-driven techniques, in contrast to model-based techniques, do not rely on system models, offering more accurate and consistent performance even in dynamic system scenarios. For instance, the study [8] introduced an Artificial Neural Network (ANN) based OSF diagnosis. The methodology utilizes Park's Vector Transform for normalization, Discrete Wavelet Transform for feature extraction, and an ANN for fault detection in Voltage Source Inverters. This integrated approach minimizes false alarms, ensures consistency across load conditions, and demonstrates effective real-time fault identification and isolation. The results underscore the efficacy of the proposed system in enhancing the reliability of OSF diagnosis under variable operational conditions. The study in [2] introduced a support vector machine (SVM) method to diagnose OSFs in induction machine drives. The SVM-based technique utilizes three-phase instantaneous current signals for robust OSF detection and classification in EV drives. Evaluation as both binary-class and multi-class classifiers demonstrates effectiveness across diverse scenarios. Simulation results showcase reliable OSF diagnosis under varying load and speed conditions, highlighting the method's efficiency. Furthermore, the approach alleviates computation burdens on the controller, contributing to improved performance in inverter-fed EV systems. A machine learning-based fault diagnostic technique for variable frequency IM drive is introduced by the authors in [11]. Testing under various conditions, including healthy and faulty states, the study records stator current and vibration signals, utilizing discrete wavelet transform for feature extraction. Four machine learning algorithms are evaluated for OSF classification under diverse frequency range and motor loading. The study [12] proposes a novel approach for OSF fault detection in three-phase inverters. The key approach involves integrating signal and data-driven fault-diagnosis techniques. The average root-mean-square (RMS) ratio of phase currents is employed as a pivotal extraction feature, allowing accurate estimation of fault types and identification of faulty switches. The approach is notable for its independence from load variations, and its effectiveness is enhanced through ensemble-bagged machine learning classification. This technique demonstrates adaptability to diverse inverter-operating conditions, providing a robust method for ensuring system reliability in industrial applications.

This paper centers on the classification of single and multi-switch OSFs in induction machine drives. The proposed technique utilizes invertor output currents as feature input to three different machine learning classifiers to classify various OSFs. Two different classification approaches have been investigated in the proposed work. In the first approach, three-phase currents have been utilized to classify eighteen different OSFs and healthy condition in the inverter. The second approach utilizes phase-wise currents to identify OSFs in particular leg of the inverter separately. The study ultimately compares the effectiveness of these two approaches and recommends the most accurate method for diagnosing and classifying OSFs.

2 Machine Learning Models for Classification

2.1 Support Vector Machine (SVM)

SVM is powerful for classification tasks, but adaptable to regression. It determine the best hyper plane to separate classes in feature space, maximizing the margin between them. SVM handles high-dimensional data and is effective even with non-linearly separable data, utilizing kernel functions. It's useful for small to medium-sized datasets but requires careful tuning of parameters [13].

2.2 Random Forest (RF)

RF is an ensemble method for classification and regression. By building multiple decision trees and combining their predictions, RF improves accuracy and robustness. Each tree is trained on a random subset of data and features, reducing over fitting. RF handles high-dimensional data well, is robust to noise, and provides feature importance for understanding dataset characteristics and feature selection [14].

2.3 K-Nearest Neighbor (kNN)

kNN is a supervised machine learning algorithm capable to perform classification as well as regression tasks. It identifies the 'k' nearest data points to a given test point and assigns the majority class label (for classification) or the average value (for regression) of those neighbors to the test point. While easy to understand, it can be computationally expensive for large datasets [15].

3 System Modeling

As illustrated in Fig. 1, a three-phase, 4 kW, 400 V, VSI-fed induction motor variable speed EV drive is simulated using MATLAB® Simulink® software. At various speeds and load variations, different OSFs and normal conditions have been simulated. The vast operating range and loading are taken into account to gather sufficient amount of data under varied situations. To maintain constant flux functioning, the supply frequency to the stator winding is changed using V/f control to achieve the

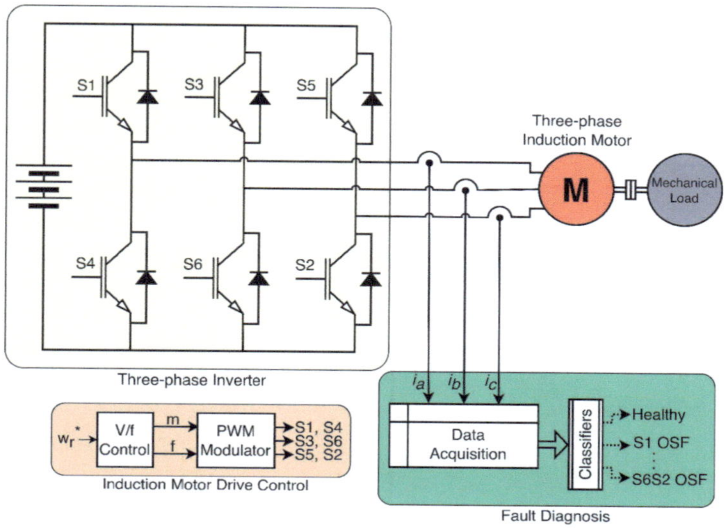

Fig. 1 The proposed fault diagnosis scheme

desired speed variation. Sine-triangle pulse width modulation with a fixed switching frequency of 2 kHz is used to generate the switching pulses. It is expected that the DC link voltage remains constant throughout the operating region.

The OSFs in the inverter have been created by blocking pulses to specific switches from gate driver circuit resulting from malfunctioning of driver circuit. Table 1 shows various cases simulated under diverse operating conditions considered in the study. Under typical operating conditions, 55 scenarios for each case have been simulated, with load levels ranging from 50 to 100% of rated load, and supply frequency variations from 30 to 50 Hz in steps of 5 Hz. Considering wide variations in speed and loading conditions, a total of 1,045 cases have been simulated while taking into account for normal condition and opening of each single and double switches, as mentioned in Table 1.

Table 1 Various cases simulated

Condition	Load torque T_L (Nm)	Frequency f(Hz)
Single switch faults (06) (S1, S2, S4, S4, S5, S6)	50–100% of rated load (In steps of 5%)	30, 35, 40, 45, 50
Double switch faults (12) (S1S3, S1S6, S1S5, S1S2, S4S3, S4S6, S4S5, S4S2, S3S5, S3S2, S6S5, S6S2)	50–100% of rated load (In steps of 5%)	30, 35, 40, 45, 50
Healthy	50–100% of rated load (In steps of 5%)	30, 35, 40, 45, 50
Total cases	19(Condition) * 11(T_L) * 5(f) = 1045 cases	

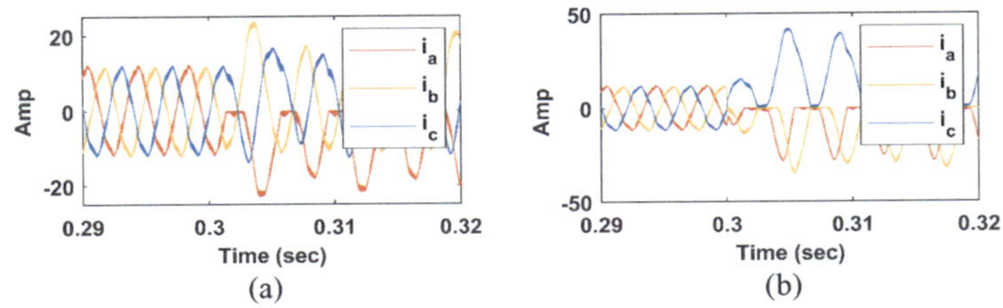

Fig. 2 Instantaneous currents under **a** S1, and **b** S1S3 OSFs

Figure 2a, b illustrates three-phase instantaneous currents under S1 and S1S3 OSFs condition occurred at t = 3s, respectively. As depicted in Fig. 2, the inverter's output currents waveform carries fault signatures that can be used to identify OSFs.

4 Proposed Method

In this work, half-cycle pre-and-post fault data of three-phase instantaneous inverter output currents sampled at 10 kHz sampling frequency are utilized as input to the SVM, kNN, and RF classifiers. Out of 1045 cases in total, 836 cases have been used to train the classifier model, and the performance of the trained models evaluated with 209 cases. The single as well as multiple OSFs, and healthy conditions have been classified with two different classification approaches as described below.

4.1 Classification Using Single Multi-class Classifier

The classification approach using a single multi-class classifier for classifying nineteen classes is illustrated in Fig. 3. In this methodology, single SVM, kNN and RF classifiers were trained with three-phase instantaneous currents to differentiate between various types of OSFs and healthy condition in the system. Through a meticulous fine-tuning process using grid search methodology, the hyper parameters for each classifier were optimized, which is mentioned in Table 2.

Fig. 3 Single multi-class classifier for classifying nineteen classes

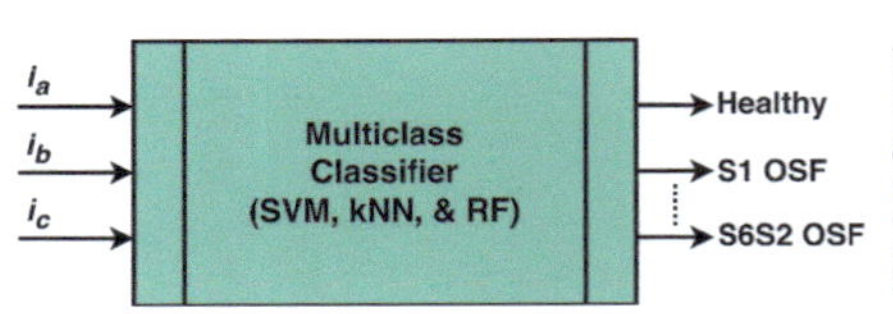

Table 2 The best hyper parameters for multi-classification approach with single classifier

Model	Best hyper parameters
SVM	C = 10, $\gamma = 0.0001$
kNN	Neighbors = 3
RF	Estimators = 25

4.2 Classification Using Phase-Wise Multi-class Classifier

Alternatively, as depicted in Fig. 4, a per phase classification approach was explored to detect OSF within specific phases of the system under single as well as multi OSFs. This involved segmenting the dataset based on distinct phases, each representing specific conditions. For each phase, the same set of classifiers, SVM, kNN, and RF, were employed separately. Through an extensive grid search method, the optimal hyper parameters tailored to each phase were determined as mentioned in Table 3. By fine-tuning the hyper parameters for SVM, kNN, and RF classifiers, the aim was to enhance OSF detection accuracy within each phase.

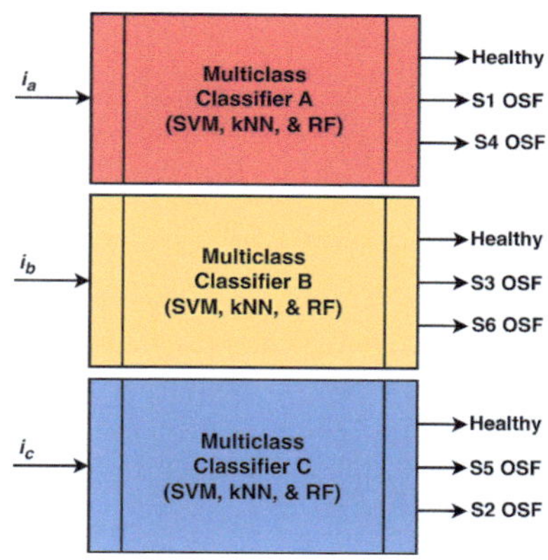

Fig. 4 Phase-wise OSF classification approach

Table 3 The best hyper parameters for phase-wise classification approach

Model	Best hyper parameters		
	Phase-A	Phase-B	Phase-C
SVM	C = 100,	C = 100,	C = 10,
	$\gamma = 0.0001$	$\gamma = 0.0001$	$\gamma = 0.0001$
kNN	Neighbors = 3	Neighbors = 3	Neighbors = 3
RF	Estimators = 50	Estimators = 50	Estimators = 25

5 Result and Discussion

The performance SVM, kNN, and RF classifiers with single multi-class classification approach for different OSF conditions and the healthy state is shown in Table 4. Impressively, all classifiers achieve remarkable accuracy (100%) for individual OSFs (S1–S6), as well as for many multiple OSFs and the healthy state. However, variations in accuracy arise for certain multiple OSFs conditions, particularly S1S5 and S3S62, where kNN and RF classifiers exhibit slightly lower accuracy compared to SVM. Despite these variations, the overall accuracy remains reasonably high for all the classifiers, with SVM achieving an overall accuracy of 100%, followed by RF with 99.52%, and kNN with 98.09%. These findings underscore the robust performance of the classifiers in accurately identifying and classifying different OSFs within the system, thereby providing valuable insights into their effectiveness in fault detection.

Table 4 Performance of classifiers with single multi-class classifier approach

Conditions	% Accuracy		
	SVM (%)	kNN (%)	RF (%)
S1 OSF	100	100	100
S4 OSF	100	100	100
S3 OSF	100	100	100
S6 OSF	100	100	100
S5 OSF	100	100	100
S2 OSF	100	100	100
S1S3 OSF	100	100	100
S1S6 OSF	100	100	100
S1S5 OSF	100	93.33	100
S1S2 OSF	100	100	100
S4S3 OSF	100	100	100
S4S6 OSF	100	100	100
S4S5 OSF	100	100	100
S4S2 OSF	100	100	100
S3S5 OSF	100	90	100
S3S2 OSF	100	70	91.66
S6S5 OSF	100	100	100
S6S2 OSF	100	100	100
Healthy	100	100	100
Overall	100	98.09	99.52

Table 5 Accuracy of per phase classes multi-classification for different switches

Classifier	% Accuracy								
	Phase-A			Phase-B			Phase-C		
	US	LS	HL	US	LS	HL	US	LS	HL
SVM	100	100	100	97.95	100	100	100	97.29	100
kNN	100	95.06	100	95.83	100	100	97.29	100	100
RF	100	100	100	95.79	98.94	100	97.22	97.22	100

US Upper Switch, *LS* Lower Switch, *HL* Healthy

The performance of the classifiers with phase-wise OSF classification approach is mentioned in Table 5. Across all the phases, SVM demonstrates consistent high accuracy for most of the conditions, except for lower switch OSFs in phase-B. kNN exhibiting strong performance for healthy condition, while the performance is inconsistent for OSFs in different phases. The performance of RF for OSFs classification in phase-A is superior to kNN. Nevertheless, for phase-B and C, the RF performance has declined significantly. Overall, SVM classifier performed better as compared to kNN and RF in a phase-wise classification approach.

The performance comparison in terms of train and test accuracy, test precision, test recall, and test F1 score is illustrated in Fig. 5 for both the methods. From Fig. 5, it is obvious that the performance of all the classifiers with a single multi-class classification approach utilizing three-phase current for classifying nineteen classes is superior in comparison to phase-wise classification approach utilizing phase-wise current. Moreover, the SVM classifier with single multi-class classification approach demonstrated impeccable performance by achieving 100% classification accuracy with the three-phase instantaneous currents as feature input.

Furthermore, the performance of the classifiers with different train-test data split is presented in Fig. 6, for both the approaches. As illustrated in Fig. 6, the SVM classifier exhibits reliable and consistent performance when compared to the kNN and RF classifiers, under different train-test data split criteria with both the approaches. The results indicate superior capabilities of SVM for classifying OSFs in the proposed data driven approach. Moreover, the single multi-class classification approach outperformed the phase-wise classification approach, demonstrating remarkable performance in classifying OSFs in inverter fed IM EV drive.

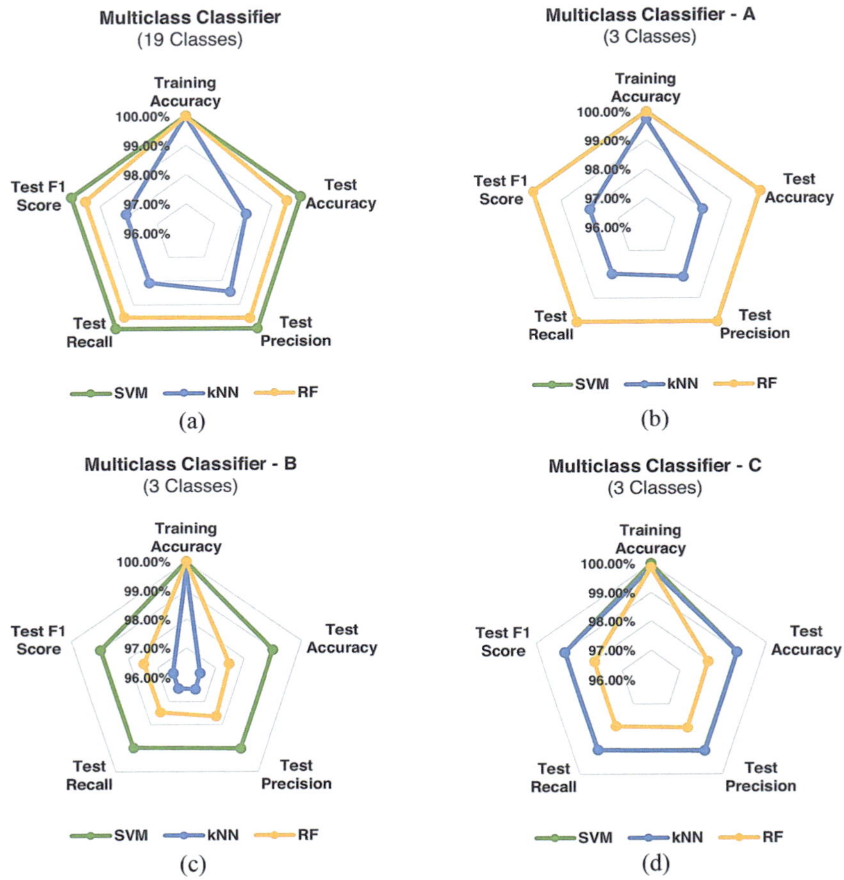

Fig. 5 Overall performance comparison of both the methods

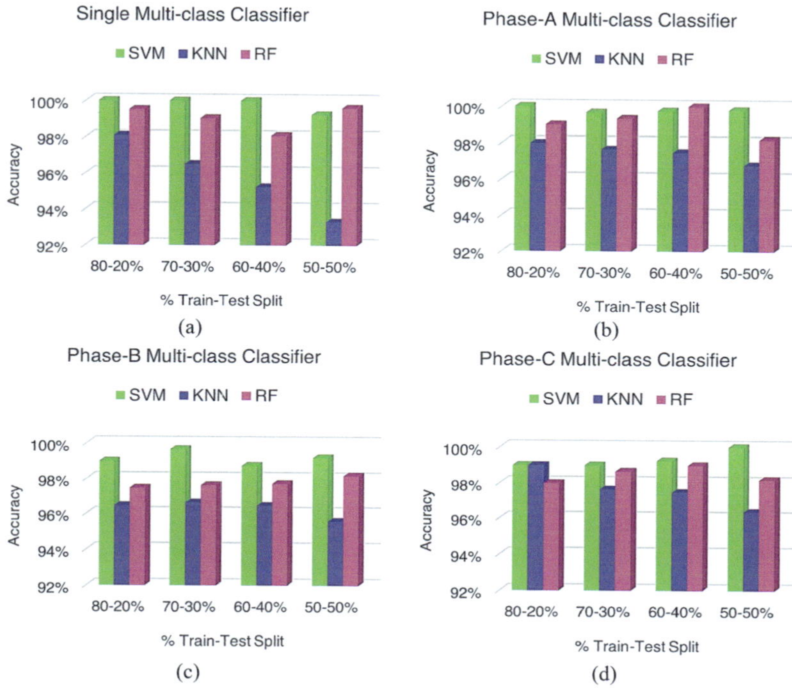

Fig. 6 Performance with different train-test data split

6 Conclusion

The proposed data driven method addresses single and multi OSFs classification in three-phase IM based EV drive. The instantaneous currents have been utilized to evaluate the efficacy of SVM, kNN and RF classifiers with two different approaches. The performance of the classifiers has been assessed for classifying eighteen different OSFs and healthy condition under varying operating parameters of the IM based EV drive. The single multi-class classification approach is found reliable and accurate in comparison to phase-wise multi-class classification approach for classifying OSFs in the proposed work. Overall, the SVM classifier's performance has been impeccable with the single multi-class classification strategy obtaining an astounding 100% classification accuracy in classifying OSFs in IM based EV drive.

References

1. Aishwarya M, Brisilla RM (2023) Design and fault diagnosis of induction motor using ml-based algorithms for EV application. IEEE Access 11:34186–34197. https://doi.org/10.1109/ACCESS.2023.3263588
2. Mwangi D, Trivedi T, Kothari N (2022) Open switch fault detection in electric vehicle drives using support vector machine. In: 2022 2nd Odisha international conference on electrical power engineering, communication and computing technology (ODICON). IEEE, pp 1–6. https://doi.org/10.1109/ODICON54453.2022.10010036
3. Yang S, Bryant A, Mawby P, Xiang D, Ran L, Tavner P (2011) An industry-based survey of reliability in power electronic converters. IEEE Trans Ind Appl 47:1441–1451. https://doi.org/10.1109/TIA.2011.2124436
4. Lu B, Sharma SK (2009) A literature review of IGBT fault diagnostic and protection methods for power inverters. IEEE Trans Ind Appl 45:1770–1777. https://doi.org/10.1109/TIA.2009.2027535
5. Nandi S, Toliyat HA, Li X (2005) Condition monitoring and fault diagnosis of electrical motors—a review. IEEE Trans Energy Convers 20:719–729. https://doi.org/10.1109/TEC.2005.847955
6. Song Y, Wang B (2013) Survey on reliability of power electronic systems. IEEE Trans Power Electron 28:591–604. https://doi.org/10.1109/TPEL.2012.2192503
7. Hu K, Liu Z, Tasiu IA, Chen T (2021) Fault diagnosis and tolerance with low torque ripple for open-switch fault of IM drives. IEEE Trans Transp Electrif 7:133–146. https://doi.org/10.1109/TTE.2020.3006057
8. Dhumale RB, Lokhande SD (2016) Neural network fault diagnosis of voltage source inverter under variable load conditions at different frequencies. Measurement 91:565–575. https://doi.org/10.1016/j.measurement.2016.04.051
9. Fei M, Ning L, Huiyu M, Yi P, Haoyuan S, Jianyong Z (2018) On-line fault diagnosis model for locomotive traction inverter based on wavelet transform and support vector machine. Microelectron Reliab 88–90:1274–1280. https://doi.org/10.1016/j.microrel.2018.06.069
10. Moosavi SS, Kazemi A, Akbari H (2019) A comparison of various open-circuit fault detection methods in the IGBT-based DC/AC inverter used in electric vehicle. Eng Fail Anal 96:223–235. https://doi.org/10.1016/j.engfailanal.2018.09.020
11. Ali MZ, Shabbir MNSK, Zaman SMK, Liang X (2020) Single- and multi-fault diagnosis using machine learning for variable frequency drive-fed induction motors. IEEE Trans Ind Appl 56:2324–2337. https://doi.org/10.1109/TIA.2020.2974151
12. Ibem CN, Farrag ME, Aboushady AA, Dabour SM (2023) Multiple open switch fault diagnosis of three phase voltage source inverter using ensemble bagged tree machine learning technique. IEEE Access. 11:85865–85877. https://doi.org/10.1109/ACCESS.2023.3304238
13. Kothari NH, Bhalja BR, Pandya V, Tripathi P (2022) A rate-of-change-of-current based fault classification technique for thyristor-controlled series-compensated transmission lines. Int J Emerg Electr Power Syst 23:289–304. https://doi.org/10.1515/ijeeps-2021-0031
14. Kothari NH, Bhalja BR, Pandya V, Tripathi P (2021) A comparative evaluation of data mining techniques for fault classification in TCSC lines using multiclass classification approach. In: 2021 international conference on artificial intelligence and smart systems (ICAIS). IEEE, pp 994–998. https://doi.org/10.1109/ICAIS50930.2021.9395809
15. Abu Alfeilat HA, Hassanat ABA, Lasassmeh O, Tarawneh AS, Alhasanat MB, Eyal Salman HS, Prasath VBS (2019) Effects of distance measure choice on k-nearest neighbor classifier performance: a review. Big Data 7:221–248. https://doi.org/10.1089/big.2018.0175.s

Detection and Prediction of Stator Short Circuit Faults in PMSM Motors Using LSTM Approach

Rishika Sharda, Aryan Pandey, and P. Saranya

Abstract The proposed model presents a comprehensive analysis of stator short circuit faults in Permanent Magnet Synchronous Motors (PMSMs). An electric motor known as a PMSM generates a magnetic field through the use of permanent magnets in its rotor. The motor rotates as a result of the interaction between this magnetic field and the rotating magnetic field created by the stator windings. This interaction produces torque. With the increasing integration of PMSMs in various industrial applications, the need for reliable fault detection and diagnosis techniques has become paramount. Stator short circuit faults are one of the most common and critical faults affecting the performance and reliability of PMSMs. In this research, we investigate the root causes, characteristics, and consequences of stator short circuit faults in PMSMs. We propose a novel fault detection and diagnosis approach based on advanced machine learning algorithms that is LSTM. The results indicate that the proposed approach achieves high accuracy and reliability in detecting and diagnosing stator short circuit faults in PMSMs, thereby enhancing the overall reliability and safety of PMSM-based systems. This research contributes to advancing the understanding of stator short circuit faults in PMSMs and provides practical insights for developing effective fault diagnosis strategies to mitigate their adverse effects.

Keywords Stator short circuit · LSTM · PMSM Motor

1 Introduction

Permanent Magnet Synchronous Motors (PMSMs) have emerged as integral components across diverse industrial sectors owing to their superior performance and efficiency characteristics. Nonetheless, a few of the more common PMSM faults include open-phase running, demagnetization, air gap eccentricity, and short circuit in the

R. Sharda · A. Pandey · P. Saranya (✉)
Department of Computing Technologies, School of Computing, College of Engineering and Technology, SRM Institute of Science and Technology, Chennai, India
e-mail: saranyap@srmist.edu.in

S. Kumar et al. (eds.), *Soft Computing Applications in Modern Power and Energy Systems*, Lecture Notes in Electrical Engineering 1298,
https://doi.org/10.1007/978-981-96-0058-8_21

stator winding. The incidence of such failures can reach 30% to 40%. The short circuit in the stator winding is one of the primary causes of motor failure [1]. Without timely detection it can be rapidly propagated to more stator and may induce other problems in PMSM, such as demagnetization fault [2]. To avoid motor failure, it is important to have information on the state of health of the stator winding. The majority of work carried out in the field of system monitoring and diagnostics, the tools used to detect and locate faults are synthesized from an open-loop representation of the system [3].

The proposed work presents a comprehensive investigation into stator short circuit faults in PMSMs, aiming to elucidate their underlying causes, characteristics, and repercussions. The research explores innovative approaches leveraging advanced machine learning algorithms for the detection and diagnosis of these faults. By delving into the intricacies of stator short circuit faults and proposing novel fault detection techniques, this study contributes to the advancement of fault diagnosis strategies, thereby enhancing the reliability and safety of PMSM-based systems. The goal of hybrid models, which blend analytical and finite element modelling, is to maximize the benefits of both methodologies [4].

If early inter-turn short circuit fault cannot be found in time, it will distort motor magnetic field, lead to increasing of the short-circuit current, the torque and mechanical abrasions in the short circuit area. At the same time, the temperature of short circuit loop rises, which leads to more damage of insulation [1]. In the subsequent sections, we delve into the implementation details of our proposed fault detection and diagnosis framework. Leveraging deep learning algorithms and physics-based models, our approach offers a holistic methodology for detecting and predicting stator short circuit faults in PMSMs. Through meticulous data acquisition, preprocessing, and the implementation of an LSTM-based model, we demonstrate the efficacy of our approach in accurately identifying and predicting stator short circuit faults.

The methodology implementation includes two parts which are the Detection of any short circuit and the Prediction of any short circuit that may occur. For the first, the Torque of the motor is calculated and then any anomalies present is checked through Fast Fourier Transformation FFT. Here, Torque is considered as if there is any short circuit fault, then torque will be the first functionality that will be affected. For the latter, an LSTM Model is created and trained for predicting the torque. The model is trained by the given data and then will be compared with the testing data. This LSTM Model will have three Layers out of which two will be dense layer and the hidden dense layer will be the size parameter that will be considered as 64. So, this LSTM Model will be used for the prediction of data. But, for Model training purpose, the training data and testing data consists of the phase current. If the trained model can predict the required data with adequate accuracy, then we can say that if torque will be given as training data then it will predict the required torque data with better accuracy as phase currents are used for calculating Torque.

By amalgamating theoretical insights with practical implementation details, this research endeavours to bridge the gap between theoretical understanding and real-world application, providing valuable insights for engineers and researchers involved in the design and maintenance of PMSM-based systems. Ultimately, the findings of this study are poised to significantly contribute to the enhancement of reliability and safety standards in industrial applications relying on PMSMs.

2 Related Work

Numerous studies have contributed to the development of fault diagnosis methods specifically targeting stator short circuit faults in Permanent Magnet Synchronous Motors (PMSMs). These studies offer insights into various techniques and methodologies employed in the detection and diagnosis of such faults. Wang et al. [2] suggested a review of techniques for diagnosing faults in Permanent Magnet Synchronous Motors (PMSM). First, a list of common PMSM problems, such as mechanical, magnetic, and electrical faults, is provided. The third section provides a summary of the several PMSM signal processing techniques. Lastly, artificial intelligence techniques for PMSM fault diagnostics, including fuzzy logic and artificial neural networks, are examined. The difficulty in this case was that it was impossible to forecast the short circuit problem.

Lare et al. [4] developed a technique to identify air gap eccentricity and stator winding problems in a permanent magnet-assisted synchronous reluctance motor found in electric vehicles. Fault signs are identified from stator phase currents using an analytical model and Finite Element Analysis (FEA). Despite control system operation, simulation results show how successful this method is for defect diagnosis and detection. The difficulties were in classifying and identifying the defects using the measured phase.

Zhao et al. [5] devised a novel PMSM design featuring automatic fault tolerance for inter-turn short circuits. This design effectively stabilizes output torque by suppressing short circuit currents during motor failure. We introduce a simplified detection method for these faults, involving low-frequency signal injection into existing coils, facilitating faster fault localization and short turn count determination. The paper outlines the motor's unique structure, fault tolerance principles, and theoretical derivation of short circuit current behavior. It also details the injection of low-frequency signals into the stator winding and provides theoretical insights into low-frequency short-circuit currents. Finally, finite element simulations using ANSYS assess the effects of different short circuit turns. The challenge that occurred was Fast Fourier transformation (FFT) approach was not considered.

Bouslimani et al. [3] investigates techniques for finding inter-turn problems in a permanent magnet synchronous motor's (PMSM) stator winding when operating in closed-loop. The suggested method for identifying turn-to-turn short circuit faults is based on the analysis and parametric estimation of two voltage signals that are crossed by resistances Rd and Rq. The suggested method is shown to be effective in identifying turn-to-turn short circuit defects through testing and simulation, providing vital information for fault isolation. The difficulties in this case stemmed from the lack of artificial intelligence, which made it challenging to anticipate short circuit problems.

Zhang et al. [1] delves into the intricacies of inter-turn short circuit faults in permanent magnet synchronous motors (PMSMs) by employing structural algorithms. Through the establishment of a finite element model tailored to the unique requirements of electric vehicle applications, the impact of short circuit faults on motor performance is meticulously examined. Simulations are conducted across a spectrum of fault severities and operating conditions, shedding light on the consequential alterations in magnetic fields and motor parameters. By extracting feature vectors for fault detection, the study underscores the efficacy of finite element analysis (FEA) in providing a precise and scientifically grounded framework for understanding inter-turn short circuit faults. The findings serve as a pivotal stepping stone for the broader diagnosis of PMSM faults in electric vehicles. The challenge here was that no neural network was used and hence the prediction of the any such short circuit faults became complex.

Ebrahimi et al. [6] proposed a thorough model for PMSM stator winding inter-turn short circuit problems that accounts for cross effects. Hybrid modelling enhanced fault representation and diagnostic tool accuracy by combining analytical and finite element analysis. Zhang et al. [7] investigated electric car stator winding inter-turn short circuit failures using finite element simulations.Their work demonstrated the requirement for reliable simulation models for fault identification and revealed how severity influences motor behaviour. Ahn et al. [8] devised a reference voltage-based approach to detect PMSM inter-turn short circuit difficulties. They improve PMSM fault diagnostics reliability and offer an alternative to standard signal analysis by enabling fault detection under dynamic operating situations. These studies collectively underscore the diverse methodologies employed in fault detection in PMSMs, ranging from signal processing techniques to simulation-based modeling approaches. Each contributes valuable insights to the ongoing pursuit of enhancing fault diagnosis capabilities in PMSM systems, addressing critical challenges in ensuring their reliability and performance across various applications.

3 Implementation

Detecting and predicting stator short circuit faults in electrical machines is a critical task to ensure the reliability and safety of the system. Here's a high-level Flow Model for the implementation of a project focused on the detection and prediction of stator short circuit faults. Here we have used deep learning algorithms along with physics-based models.

From Fig. 1 we can see the flow diagram which it shows the process for glaucoma disease classification. From image preprocessing to CNN-based classification, a holistic method combining image enhancement, machine learning, and rigorous evaluation is used to achieve robust and exact Glaucoma disease classification.

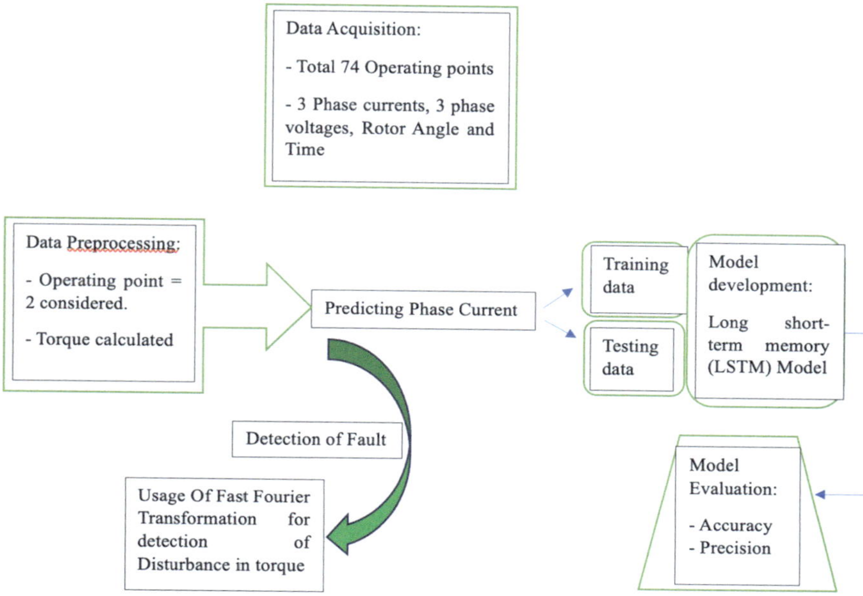

Fig. 1 Flow Diagram of the proposed work

3.1 Data Acquisition and Data Pre-processing

The proposed model is initiated by developing a dataset of 500,000 records per Operating Point of a PMSM Motor. This data encapsulated of Three phase Currents (A, B, C) and Three phase voltages along with Rotor angle (in Radians) and time (in secs). In the proposed model, Phase currents and phase voltages was used for calculating torque. Torque is calculated in order to find any fluctuation in the currents. If there is any short circuit then the torque is the first factor that will be affected.

Formula for the torque is:

$$T = \frac{3}{2}p(\psi_d i_q - \psi_q i_d) \tag{1}$$

where p is pole number of magnets, i_d and i_q are the direct axis currents and quadrature axis currents. ψ_q and ψ_d are the flux linkages of the quadrature and direct axis'. After this the calculated torque is then converted into Fast Fourier transformation (FFT). The amplitude and the frequency are compared and if no major anomalies, then we can conclude that there are no short circuits that occurred.

3.2 Prediction of Data

The given data is converted into Training data and testing data. The Training data is used by the model for the training and then the model prediction results are compared to the Testing Data.

3.3 Implementation Using LSTM Model

This implementation defines a neural network model for incorporating the Phase Currents layer and LSTM layer. The LSTM layer's hidden size parameter is configured as 64, determining the dimensionality of the hidden state. These components collectively form a robust architecture for sequential data processing and extracting meaningful patterns from Current data. The proposed LSTM architecture is shown in Fig. 2.

These cells are composed of three main gates: input, forget, and output gates. The input gate plays a major role in deciding how much new input to retain in the cell state. It uses a sigmoid activation function to regulate the values that need to be updated and a tangent activation function to generate potential candidate values that could replace the current state values. Conversely, the forget gate uses a sigmoid activation function to determine which part of the previous state should be forgotten. By driving values close to zero, it effectively allows the cell to eliminate historical useless data. Based on the current, the output gate regulates the amount of information transferred to the output as well as the modified cell state. The output is calculated by multiplying the cell state by the output of the sigmoid activation function applied to the current input after applying a tan function to it.

The structure of the LSTM network is especially interesting since it repeats the cell for each timestamp, forming a structure akin to a chain. This pattern makes it possible to transfer output from one cell to the next, which makes it easier to recognize patterns over time and in various contexts.

The comparison of the LSTM Model predicted current and the testing data currents is plotted with the help of Matplotlib. If the Graph Plot has no major disturbance

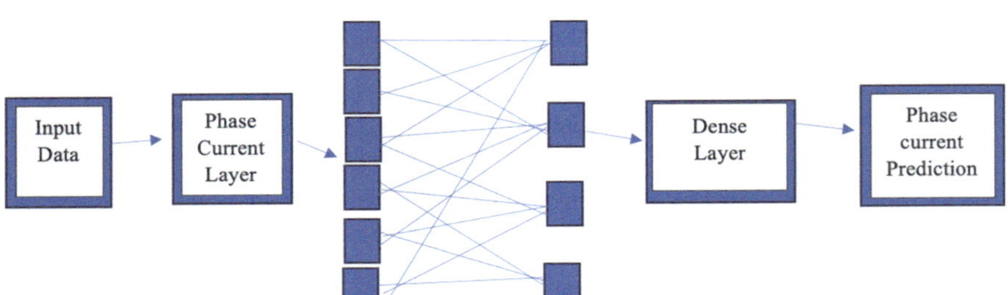

Fig. 2 LSTM architecture

and fluctuations, then the model is considered to be accurate and similarly the phase voltages can be predicted. Eventually this model can be used to predict the torque by giving the model inputs of the calculated torque. Thus, Short circuit fault in Stator can be predicted by using this Model.

4 Results and Discussions

4.1 Dataset Used

A PMSM's current controller receives reference values exclusively from the left id-iq half-plane, in theory. Additionally, the current vector's length cannot be greater than a predetermined value. In the event that this is not the case, the motor could be destroyed by thermal overload. The physical quantities were recorded at 74 stationary operating points with a measuring time of 05 s for each operating point in order to cover the entire operating range in the current plane out of which Operating point 70 was considered for visualization. Even when the speed is constant and the current controller is in steady state, there is still a high-frequency phenomenon known as "current ripple." The inverter's switching operation principle provides an explanation for this.

The induced voltage of the flux linkage serves as the foundation for many torque estimation techniques. Because the induced voltage is directly correlated with the speed, it fails when the speed approaches zero. Because of this, the operating points were measured at a low speed of 120 RPM in order to examine how any potential estimating technique would depend on speed.

Table 1 demonstrates the dataset table.

Table 1 Dataset Table

Variable	Unit	Description
Time	S	Measured time
i_a	A	Measured a-current
i_b	A	Measured b-current
i_c	A	Measured c-current
U_a	V	Measured a-voltage
U_b	V	Measured b-voltage
U_c	V	Measured c-voltage
E_el	Rad	Measured electrical angle
U_dc	V	Measured supply voltage

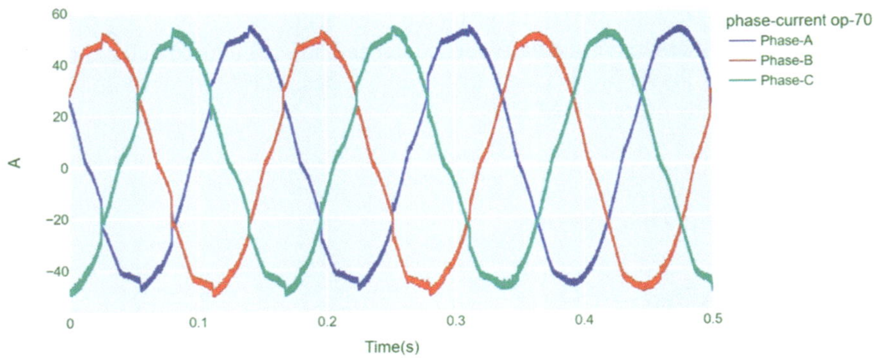

Fig. 3 Three Phase currents of the PMSM motor

4.2 Performance Evaluation

The presented neural network architecture aims at addressing the critical task of stator short circuit detection in Permanent Magnet Synchronous Motors (PMSMs). Leveraging recurrent neural networks (RNNs), specifically Long Short-Term Memory (LSTM) units, underscores a deliberate choice owing to the inherent sequentially characterizing motor operation data, typically manifested in time series format. This architectural design is adept at capturing temporal dependencies, thus facilitating the discernment of nuanced variations indicative of stator short circuits within PMSM systems.

The currents of three phase motor are plotted with some noise in the current signals from time series 0 s to 0.5 s in Fig. 3. Now, the torque is calculated by the given formula for real time data and Fast Fourier Transformation (FFT) is used to see any sudden anomalies in the torque as if there is any short circuit fault in stator, then Torque is the factor that is disturbed or affected first.

As we can see in Fig. 4, after ignoring some of the noise signals in the FFT, the plot is pretty accurate and has no sudden anomalies. So, we can conclude that no Fault has occurred. Comprising two LSTM layers supplemented by dropout regularization, the model structure is tailored to mitigate overfitting, a common challenge encountered in complex neural network configurations. Subsequent integration of fully connected (Dense) layers, coupled with LeakyReLU activation functions, underscores the model's capacity to assimilate intricate patterns inherent in the data. The specification of input shape, contingent upon the number of features and time steps, affords adaptability to diverse data configurations, enhancing the model's versatility.

As we can see in Fig. 5, the errors are less than 5 as soon as they reach the 5th epoch. So, the model is performing well on both the training and validation datasets and has learned the underlying patterns in the training data and generalizes effectively to new, unseen data. It is likely not overfitting or underfitting. During the training phase, model compilation entails the adoption of mean squared error (MSE) as the

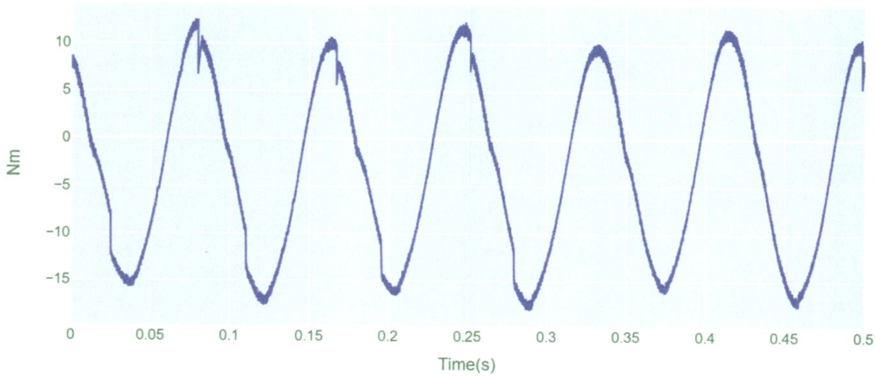

Fig. 4 Torque of the PMSM motor

loss function, paired with the Adam optimizer renowned for its adaptive learning rate properties. Mean Absolute Error (MAE) serves as the chosen evaluation metric, offering insights into the model's predictive accuracy. The training protocol spans 35 epochs, characterized by a batch size of 128 and a validation split of 20%, facilitating a comprehensive assessment of the model's learning dynamics.

As we can observe in Fig. 6, the model is learning well and generalizing effectively and both the training and validation loss are decreasing, indicating that the model is improving its ability to fit the training data and is likely to perform well on new, unseen data. Monitoring of training progression involves the tracking of loss and MAE metrics across epochs, both on the training and validation datasets. To ascertain training efficiency, the time in the module is utilized to quantify the duration of the

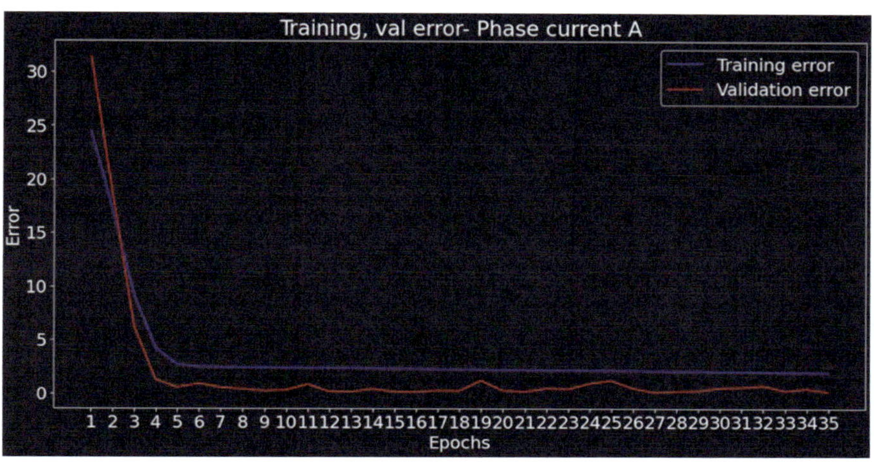

Fig. 5 Training error and validation error comparison plot

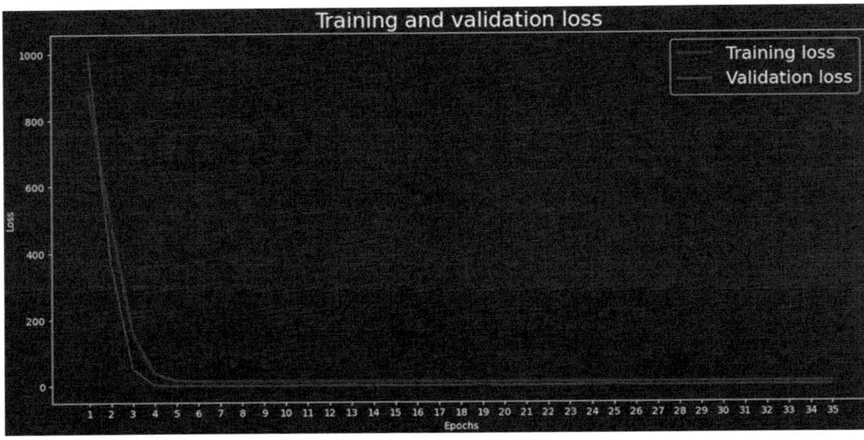

Fig. 6 Training loss and validation loss

training process, furnishing a quantitative benchmark for performance evaluation. Convergence behavior, manifested by a diminishing trend in loss and MAE over epochs, serves as a pivotal indicator of the model's efficacy in extracting salient features from the data.

Prudent considerations are warranted to preempt overfitting, particularly given the model's complexity and the inclusion of dropout regularization. Fine-tuning of hyperparameters, encompassing learning rate modulation, dropout rate adjustment, and optimization of layer dimensions, represents a viable avenue for augmenting model performance. Furthermore, validation on unseen data holds significance in validating the model's generalization capacity, thereby ensuring its efficacy in real-world scenarios.

As seen in Fig. 7, the blue line plot and red line plot overlap each other, so we can conclude that the model has predicted the data on a very accurate scale. Here the tested data includes the data provided in the dataset and the predicted data is the data predicted by the LSTM Model. Our findings contribute to the growing body of knowledge in fault detection methodologies, laying the foundation for future advancements in the quest for more reliable and resilient Artificial Intelligence and Machine learning systems.

5 Conclusions

In the proposed model, it has been embarked on a comprehensive exploration of stator short circuit fault detection and prediction in electrical machines, specifically focusing on Permanent Magnet Synchronous Motors (PMSM). Our endeavour

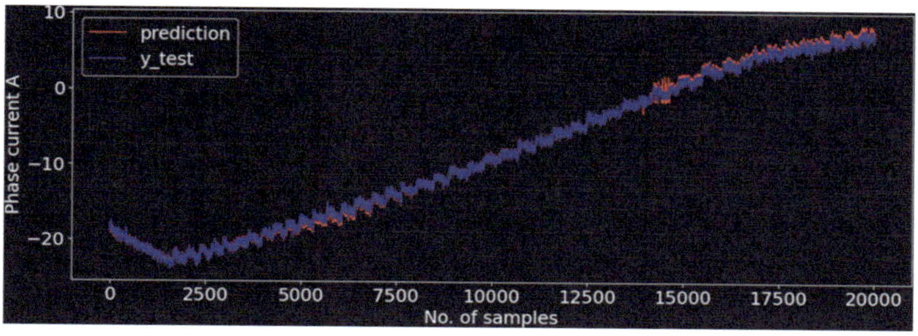

Fig. 7 Predicted data and testing data comparison plot

aimed to contribute to the enhancement of reliability, safety, and efficiency in industrial applications by developing advanced techniques for the early identification and forecasting of stator faults. Through an extensive literature review, we identified various fault detection techniques, including machine learning and artificial intelligence. Our focus was on leveraging machine learning algorithms and data-driven approaches for accurate fault detection. Short circuit faults in stator systems can cause dangerous situations that endanger both people and property. By reducing the possibility of catastrophic failures and accidents, early detection and prediction of these faults improve safety. Early defect identification and prediction reduces the likelihood of catastrophic failures and accidents, hence improving safety. Predictive maintenance can assist reduce total maintenance costs and optimize resource allocation. Unplanned maintenance and downtime caused by stator short circuit faults can result in large expenses for maintenance, repairs, and possible replacement of equipment. The people in charge of running industrial processes on a daily basis are called industrial plant operators. The results of the experiment can be used by maintenance staff who are in charge of maintaining electrical systems to better plan and carry out maintenance tasks, thereby preventing stator problems before they worsen. The project's insights on flaw detection technologies can be advantageous to engineers and researchers. They can use the knowledge gained for further advancements in the domain. By Using Fast Fourier transformation (FFT), Torque values were considered and observed and no sudden disturbance or anomalies were detected. So, we can conclude no short circuit fault occurred in Stator of PMSM Motor. Integration with broader condition monitoring systems allowed us to propose predictive maintenance strategies. The ability to predict the progression of stator faults contributes to proactive maintenance planning, minimizing downtime, and optimizing resource allocation. The LSTM Model is trained and tested for being used as a predictive model. The trained LSTM Model predicted the phase currents with very good accuracy. So, we can conclude that this model can be used to predict torque and also predict any short circuit anomalies that may occur in future. LSTM networks may require a lot of processing power, particularly when handling complex architectures and massive volumes of data. In order to meet latency requirements, real-time deployment might need efficient hardware resources. There may be difficulties in guaranteeing the availability of high-quality data from PMSM motors in industrial settings. The dependability of fault detection models can be impacted by noise, signal drift, and restricted sensor coverage. Since LSTM networks are frequently thought of as "black-box" models, it can be difficult to understand the decisions they make. Explainable AI methods might be required in industrial settings to promote decision-making and foster trust. The loads, temperatures, and operating conditions of PMSM motors are constantly changing in their dynamic environments. To provide accurate fault detection, LSTM models need to be resistant to these changes in the environment. For deployed models to remain effective over time and adjust to shifting operating conditions, they need to be calibrated and maintained on a regular basis. For long-term deployment, dependable maintenance procedures must be established. To guarantee smooth operation and the least amount of disturbance, careful planning and coordination are needed when integrating LSTM-based fault detection systems with current industrial control systems and maintenance activities.

References

1. Zhang C, Wang F, Wang Z, Yang J (2014) Analysis of stator winding inter-turn short circuit fault of PMSM for electric vehicle based on finite element simulation. 2014 IEEE conference and expo transportation electrification asia-pacific (ITEC Asia-Pacific), Beijing, pp 1–6
2. Wang Z, Yang J, Ye H, Zhou W (2014) A review of permanent magnet synchronous motor fault diagnosis. 2014 IEEE Conference and expo transportation electrification Asia-pacific (ITEC Asia-Pacific), Beijing, China, pp 1–5
3. Bouslimani S, Drid S, Chrifi-Alaoui L, Delahoche L (2022) On line inter-turn short-circuit fault diagnosis and nonlinear control of PMSM. 2022 19th international multi-conference on systems, signals & devices (SSD), Sétif, Algeria, pp 1139–1143
4. Lare P, Sarabi S, Delpha C, Nasr A, Diallo D (2021) Stator winding Inter-turn short-circuit and air gap eccentricity fault detection of a permanent magnet-assisted synchronous reluctance motor in electrified vehicle. 2021 24th international conference on electrical machines and systems (ICEMS), Gyeongju, Korea, pp 932–937
5. Zhao L, Bai Y, Wang X, Zhao P (2023) Inter-turn short circuit fault detection of a novel PMSM with fault tolerance. 2023 8th international conference on power and renewable energy (ICPRE), pp 209–213
6. Ebrahimi SH, Choux M, Huynh VK (2020) Modeling stator winding inter-turn short circuit faults in PMSMs including cross effects. 2020 International conference on electrical machines (ICEM), Gothenburg, Sweden, pp 1397–1403
7. Zhang C, Wang F, Wang Z, Yang J (2014) Analysis of stator winding inter-turn short circuit fault of PMSM for electric vehicle based on finite element simulation. 2014 IEEE conference and expo transportation electrification Asia-Pacific (ITEC Asia-Pacific), Beijing, pp 1–6
8. Ahn G, Lee J, Park CH, Youn M, Youn BD (2019) Inter-turn short circuit fault detection in permanent magnet synchronous motors based on reference voltage. 2019 IEEE 12th International symposium on diagnostics for electrical machines, power electronics and drives (SDEMPED), Toulouse, France, pp 245–250

Integrating Control and Computing in Vehicle Systems Through CAPL and CANoe User Interface Design

Aliya Rahman, Jayendra Kumar, Vani Ponugupati, Ayush Kumar Agrawal, Avanish Kumar, and Pratul Arvind

Abstract The automotive industry faces a significant challenge in the validation process, wherein engineers must physically visit vehicles to test and simulate faults from previous software iterations. These faults are then replicated in updated software to assess rectifications. The discussed process entails turning on the ignition, cranking the vehicle, performing brake presses, using the shifter for input commands, and scrutinizing Instrument Panel Cluster outputs while observing CAN signals graphically to evaluate Electronic Control Unit (ECU) functionality. This manual approach is time-consuming and often hindered by vehicle availability, necessitating a shift towards bench-level testing. This paper introduces a model that demonstrates vehicle functionality using panel design and CAPL (Communication Access Programming Language) within the CANoe environment. The model encompasses gear engagement and disengagement, vehicle speed variations, and various engine states. By utilizing bench-level testing and simulation techniques, engineers can efficiently evaluate software updates, identify faults, and verify ECU functionality without relying on physical vehicle access. This approach not only streamlines the validation process but also enhances accuracy and control in software modification assessments.

Keywords CANoe · Communication access programming language · Panel designer · Electronics control unit · Interface design

---

A. Rahman · J. Kumar (✉) · A. K. Agrawal
Department of ECE, National Institute of Technology Jamshedpur, Jamshedpur, India
e-mail: jkumar.ece@nitjsr.ac.in

V. Ponugupati
Transmission Control, E&E Department, TATA Motors, Pune, India

A. Kumar
Department of EE, DTU, New Delhi, India

P. Arvind
Department of ECE, ADGIPS, New Delhi, India

S. Kumar et al. (eds.), *Soft Computing Applications in Modern Power and Energy Systems*, Lecture Notes in Electrical Engineering 1298,
https://doi.org/10.1007/978-981-96-0058-8_22

1 Introduction

In the state of art, earliear the job of individuals in the automobile industry has become increasingly hectic and tiresome, particularly when it comes to vehicle testing. Automation is needed to reduce human work [1]. Currently, workers are required to physically go to vehicles, connect their respective ECUs, provide power to the ECUs, and conduct testing by repeatedly starting the vehicle and performing the same steps. In this paper the author will develop a model that accurately depicts the functioning of a car. The model comprises of a Graphical user interface using Panel designer in CANoe software, CAPL scripting in CANoe software and CANoe tool. The development of the model is achieved by utilizing CAPL and Panel design in CANoe. Users can input commands through the panel design, mimicking the actions performed in a practical car, and the model will replicate the behaviour of a real vehicle. The CAPL scripting in the Panel design is employed for input generation, and it also serves as the instrumentation panel cluster, displaying outputs such as gear changes and vehicle speed.

At the automotive industry, various communication protocols are utilized to facilitate data exchange and control within vehicles. These protocols cater to different communication needs and standards, ensuring efficient and reliable operation of vehicle systems. The prominent communication protocols used in cars include: CAN (Control Area Network), LIN (Local Interconnect Network), UART (Universal Asynchronous Receiver-Transmitter), Flex Ray, SPI, I2C (Inter-Integrated Circuit) and Ethernet. Among these protocols, CAN remains the dominant choice, with approximately 90% of vehicles using it due to its reliability, real-time capabilities, and established industry support. However, as automotive systems evolve to incorporate more complex functionalities and connectivity requirements, other protocols such as FlexRay, Ethernet, and wireless protocols like Wi-Fi and Bluetooth are gaining traction for specific applications [2–4].

2 Control and Computation Model of CANoe System

The focus is on writing scripts in CAPL to generate virtual signals that mimic real shifter inputs, acceleration pedal inputs and general inputs required in a vehicle. This implies that the CAPL scripts will define the behavior of virtual nodes, simulating the actions and responses expected from physical nodes associated with a shifter mechanism [2, 5] as shown in Fig. 1 shows the block diagram of the model.

The interaction between panels and CAPL relies on the utilization of system variables in the environment window. During the creation of a panel, it is essential to associate all dynamic control elements with a system variable. This linkage ensures that when a user interacts with a control, such as clicking or adjusting a value, the associated system variable undergoes a change [6].

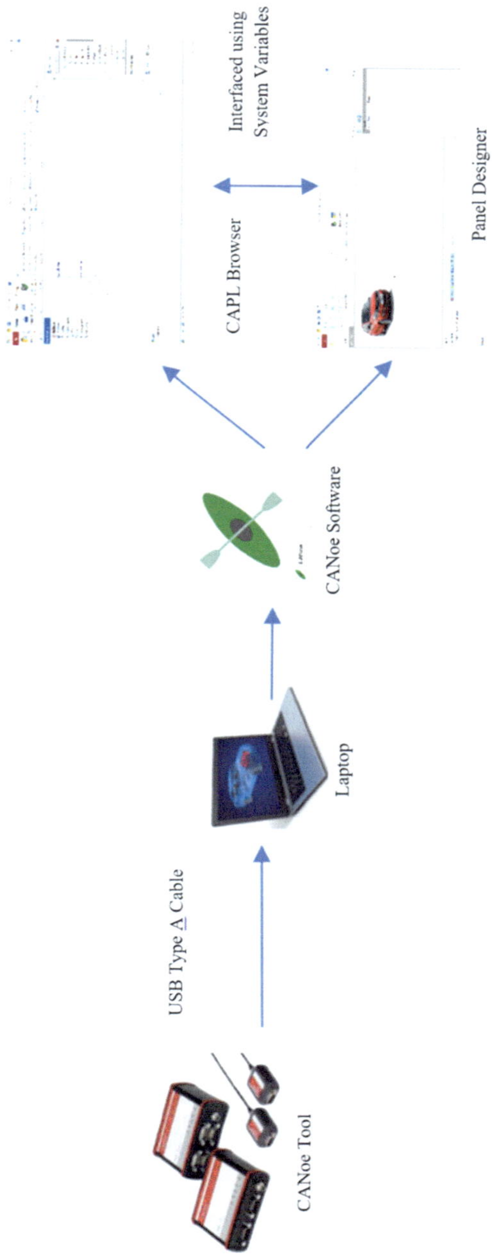

Fig. 1 Block diagram

This change triggers the execution of a system variable event procedure in CAPL. For a smooth integration, these variables should be pre-defined in the database alongside messages and signals. When these variables are part of the database definition, CAPL gains access to them, facilitating seamless communication.

3 Key Tools and Concepts in Automotive Development and Testing

The project relies on several key concepts and tools commonly used in the automotive industry. One crucial aspect is the Controller Area Network (CAN), which acts as the physical connection between different Electronic Control Units (ECUs) in a vehicle [7]. These ECUs communicate with each other using the CAN protocol, allowing them to work together seamlessly without needing a central computer [8]. Another important tool is CAPL (Communication Access Programming Language), which is like a special language used for testing ECUs in cars. It helps in setting up communication between different ECUs and is particularly useful with testing tools like CANalyzer and CANoe from Vector [4]. CAPL works in a similar way to programming languages like C and is used to create test cases and lists [6].

We also use Panel Designer, a tool for creating graphical panels used in the car's interface [9]. These panels have buttons and switches and can display different graphics, making it easy for users to interact with the car's systems. Panel Designer helps in making these panels, which are then linked with CANoe for practical use. CANoe software is another crucial tool. It's a versatile software used for developing and testing automotive systems [10–12]. With CANoe, we can write scripts in CAPL directly in a web browser, which helps automate test cases and generate reports automatically. This software is known for its features like network management and diagnostics, making it essential for testing and ensuring that automotive systems work reliably in real-world conditions [5]. Overall, these tools and concepts are essential for efficient development and testing of automotive systems. By using them effectively, we can improve productivity and ensure that cars are safe and reliable for users.

4 State of Art

To reduce testing time and enhance efficiency in automotive development, the author employs advanced methodologies and tools, with a spotlight on the critical role of CANoe software and CAPL scripting. The proposed automated testing of HVAC systems with CCUs achieves impressive accuracy in just 6.16 min through an emphasis on the CAN protocol [1]. Additionally, CANoe's integration for simulating ECU communication enhances testing efficiency, targeting key functionalities like engine start/stop, braking, and dashboard ECUs [2]. These results underscore CANoe's significance in automotive ECU development, facilitating efficient monitoring, troubleshooting, and validation processes. Furthermore, the author addresses a significant challenge in automotive lab testing by introducing the Virtual LABCAR, an IoT-based solution that reduces reliance on manual interventions and enables remote control [3]. The research also explores CANoe's role in simulating a vehicle body CAN network, providing valuable insights into efficient testing of electric vehicle control systems [4]. This discussion, coupled with an examination of the transition to Controller Area Network (CAN) systems, underscores Vector CANoe's pivotal role in designing, testing, and validating crucial functions, like ignition control, door locking, and window adjustment, emphasizing the critical importance of sophisticated simulation tools like CANoe in optimizing automotive electronics and control systems [5]. Additionally, the author discusses automated testing for Brake ECUs and ESP communication using the CANoe tool in [6], highlighting its accuracy, reusability, and scalability. The author emphasizes the necessity of automated testing to reduce failure rates and production costs in ECU development. Furthermore, the author investigates the evolution of embedded control systems in the automotive industry [7], focusing on BOSCH's Controller Area Network (CAN) protocol, and highlights CAN's efficiency in monitoring various parameters, emphasizing its reliability and practical viability demonstrated through PSoC and Vector CANoe/CANalyzer software. Safety and performance enhancement via On-Board Diagnostics (OBD) systems are emphasized by the author in [8]. Additionally, the author introduces the cost-effective "ECU Health Monitor," which leverages a CANUSB cable and Microsoft C# for comparable diagnostic validation, showcasing innovative approaches to enhancing diagnostic processes.

Upon reviewing existing papers, it's evident that many utilize CANoe software and CAPL scripting for automated testing, simulation, and diagnostics but overlook Panel Designer for ECU input/output and lack a model for vehicle functionality representation.

This paper contributes by introducing an innovative approach integrating Panel Designer and CAPL within CANoe, creating a model that accurately mimics vehicle functionalities like engine start/stop, gear shifting, and more. This method enhances ECU testing by enabling bench-level simulations, reducing the reliance on physical vehicles and streamlining development.

5 Configuring CANoe Software for Simulating, Controlling, and Monitoring Vehicle Model Behavior

In the development process outlined for our project, several key steps are crucial to ensure the successful integration and functionality of our vehicular model within the CANoe environment. Initially, the installation of the CANoe software is imperative, followed by meticulous configuration of settings to tailor the environment to our project requirements [13–15]. Upon establishing the CANoe tool's connection for licensing purposes, meticulous attention is directed towards channel mapping, where the Virtual CAN Bus is selected to facilitate seamless communication between system components. Subsequently, the incorporation of the necessary dbc file within the simulation window ensures access to essential data for system operation. In preparation for interfacing with the graphical user interface (GUI) and CAPL scripts, system variables are diligently added within the environment window. Leveraging the capabilities of Panel Designer, we craft a user-friendly interface 6 featuring buttons, LEDs, and other interactive elements crucial for system control and monitoring. Utilizing the capabilities of the CAPL Browser, essential scripts are authored to develop a robust vehicular model, facilitating simulation and analysis within the CANoe environment. Integrating these scripts within the network node ensures seamless communication and execution of functionalities. Navigating through various windows within the CANoe environment, including the Panel, Write, Analysis, Trace, and Graphics windows, enables comprehensive monitoring and analysis of system behavior [16]. Real-time observation of signal variations and graphical representation of critical parameters such as gear change, acceleration, and brake pedal status are facilitated, providing valuable insights into system performance. The dynamic display of changing signal values over time, coupled with the visual depiction of the brake pedal's activity, enhances our understanding of system behavior and aids in refining the vehicular model for optimal performance. In conclusion, the integration and visualization mechanisms implemented within the CANoe environment not only enable real-time monitoring but also provide a clear and comprehensive understanding of system behavior, essential for the development of robust vehicular models [17].

6 Simulation Results and Discussions

For each functionality, the user request provided from the panel was successfully executed and results were successfully observed. It demonstrates the efficacy of the controlling action facilitated by CAPL. This capability empowers users to not only execute commands but also monitor and visualize the corresponding signals effectively. By seamlessly integrating user control with signal monitoring and visualization, CAPL ensures a robust and user-friendly experience, enabling efficient interaction with the system's functionalities.

Starting a vehicle typically involves several sequential steps for ensuring safe and efficient operation. Initially, the driver inserts the key into the ignition and turns it to the "ON" position, activating the vehicle's electrical systems. Following this, the gear must be shifted into Neutral or Park mode as a safety precaution against unintended movement. Once the gear is appropriately set, turning the key further to initiate engine startup engages the starter motor, which cranks the engine until it ignites, enabling the vehicle to begin its operation. These steps, from key insertion to engine ignition, are fundamental procedures ensuring the proper functioning and safety of the vehicle during the startup process. Figure 2. illustrates the sequential states of the engine, ranging from Ignition ON to Engine ON. Figures 3 and 4. illustrates the positional states of a gear shifter denoted by P (Park), R (Reverse), N (Neutral), and D (Drive), corresponding to the respective shifter inputs. Corresponding LEDs light up on the dashboard to visually inform the driver of the selected mode. Shifting to the Reverse mode is allowed only when the car is completely stationary (speed is 0 km/hr). This is a safety measure to prevent accidents. The Reverse LED indicator lights up when in reverse mode, providing a visual indication.

Shifting to Drive mode enables forward motion. The Drive LED indicator lights up to confirm the mode selection. Shifting to Neutral or Park is done when the driver wants to bring the vehicle to a complete stop. This disengages the transmission and ensures the car is stationary. The Fig. 5 portrays the engagement and disengagement of the brake system. The brake being "released" signifies the absence of applied braking force, while "pressed" indicates the activation of the braking system. Overall, the figures provides a comprehensive overview of the gear shifter states and associated functionalities, incorporating technical terms to convey precise information about the system's operation.

7 Conclusion

In this paper, a model utilizing CANoe is presented. The CANoe software features a user-designed panel for controlling engine start/stop, gear shifting, acceleration pedal, and brake inputs, mimicking a practical car's Instrumentation Panel Cluster. The model, replicating real vehicle behavior, is implemented using a signal variable that serves as a bridge between CAPL language programming and user-defined panels. The design and testing of the model, encompassing gear engagement, disengagement, vehicle speed, and various engine states, were conducted using CAPL, Panel Designer, and CANoe.

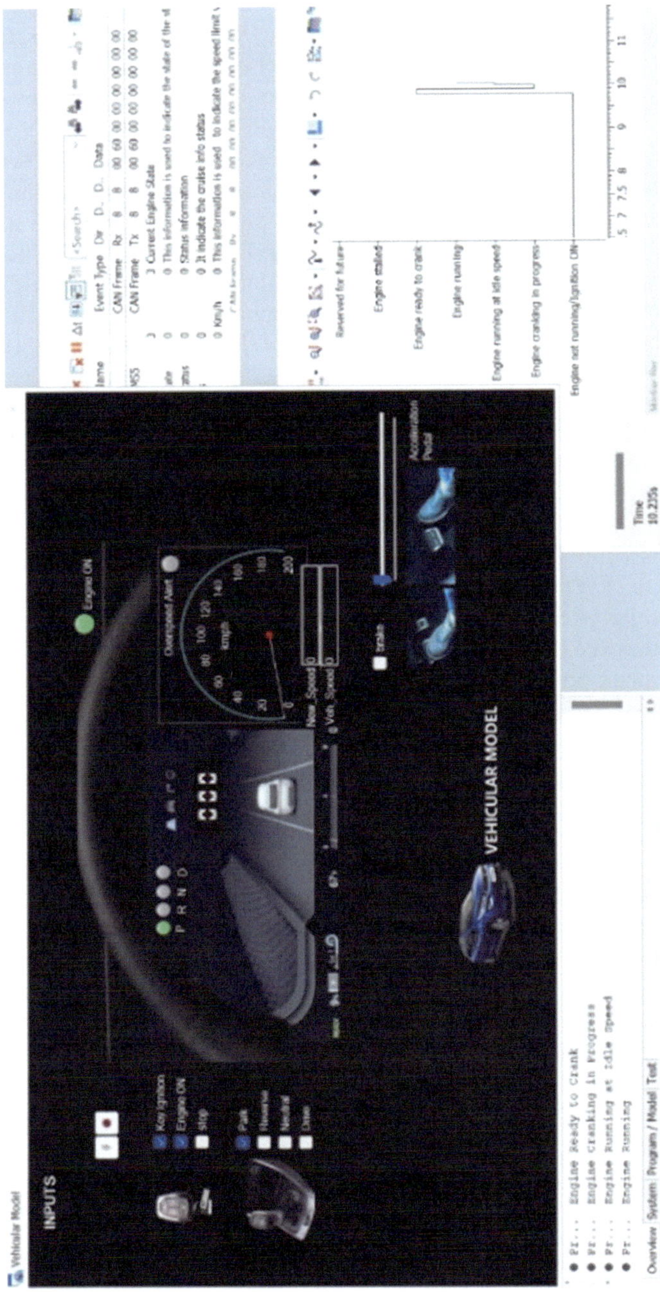

Fig. 2 Activation of vehicle systems in preparation for operation—ignition ON to engine on

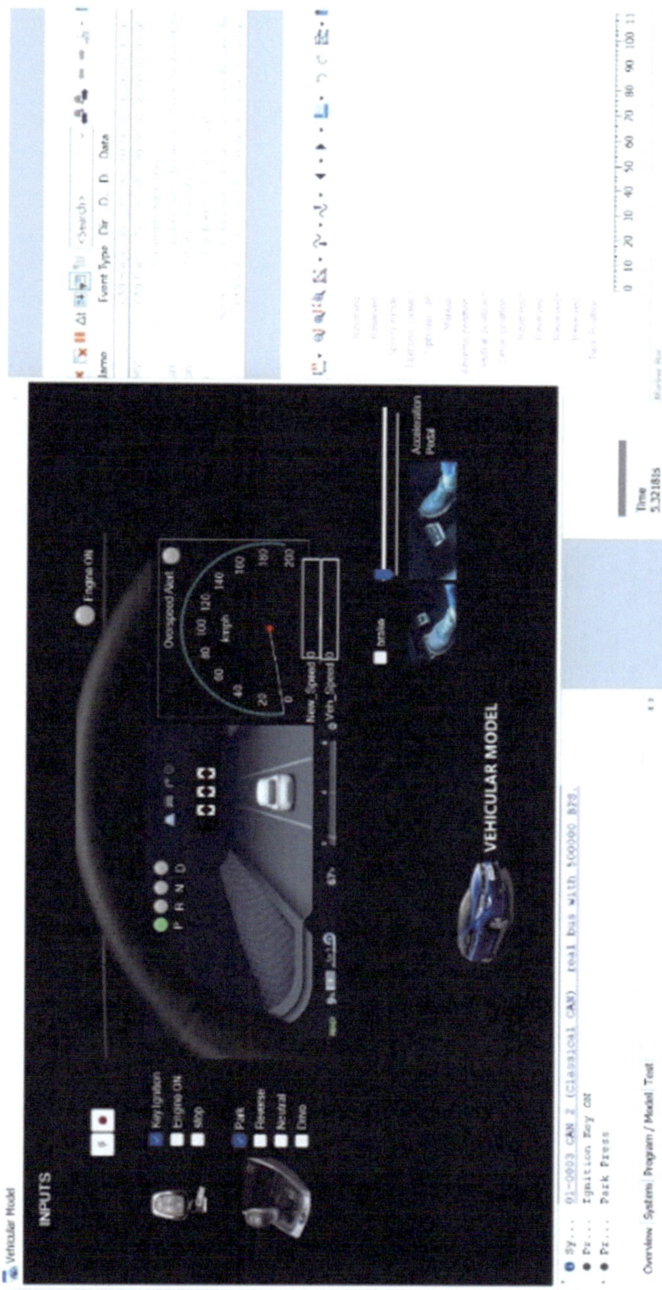

Fig. 3 Interactive park control: illumination of park LED and speed reduction to 0 upon activation of 'P' button

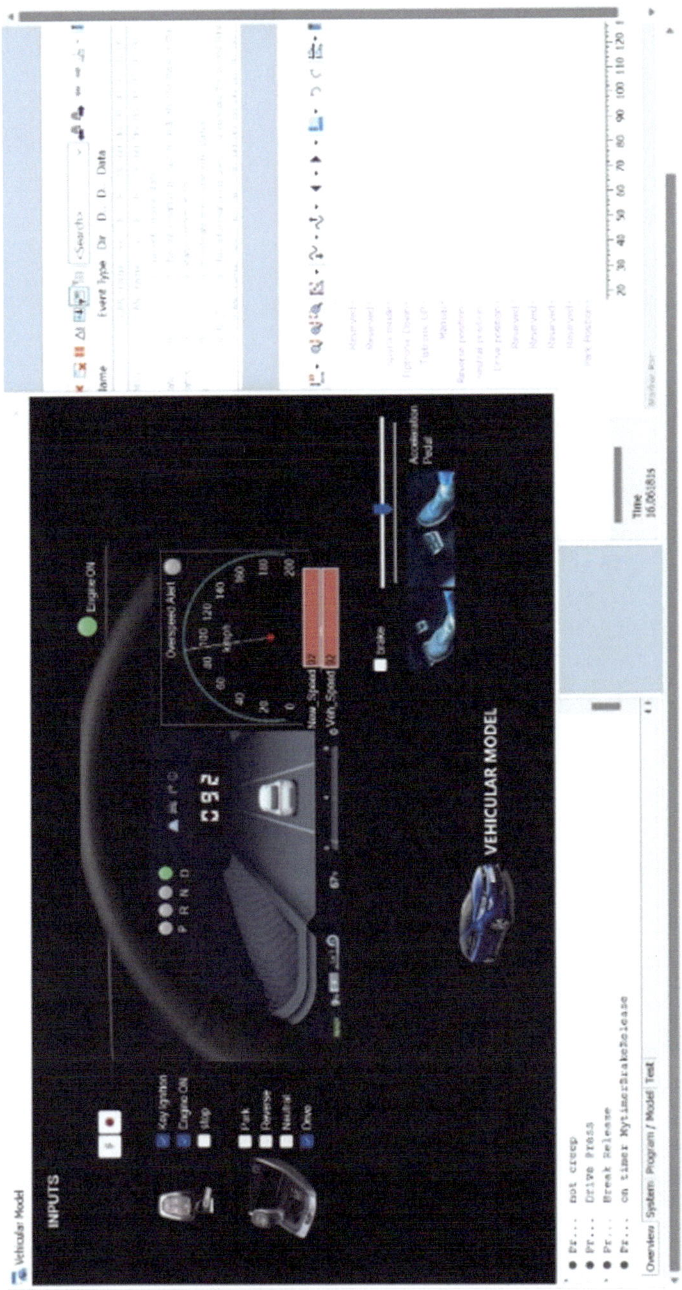

Fig. 4 Drive pressed with acceleration, speed at 92 km/hr

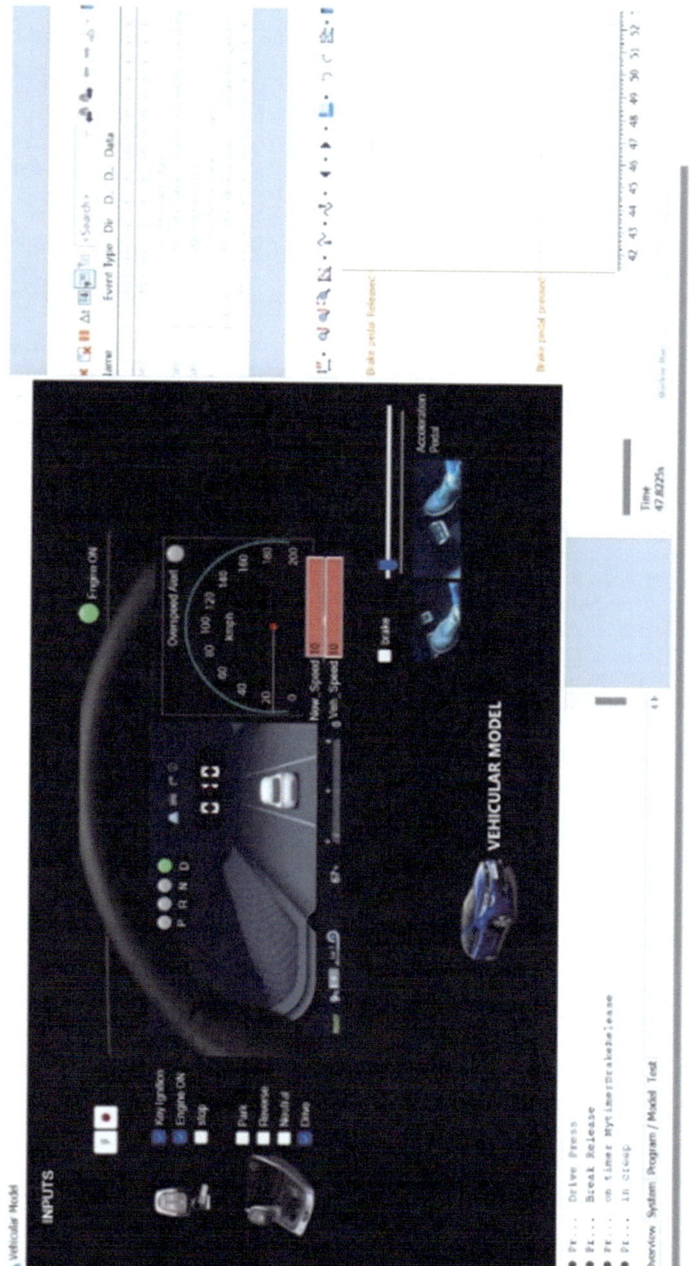

Fig. 5 Brake application graph: visualizing brake engagement and release dynamics

References

1. Kumbhar KV, Niture DV (2022) Development of test automation scripts for panel logic module using vector CANoe. In: 2022 International conference on recent trends in microelectronics, automation, computing and communications systems (ICMACC), pp 434–439. Hyderabad, India
2. Rudrappa G, Tigadi A, Kulkarni SB, Jadhav S, Chougula B (2022) CANoe tool usage for communication between automotive grade processors. In: 2022 3rd International conference for emerging technology (INCET), pp 1–5).Belgaum, India
3. Zhou F, Li S, Hou X (2008) Development method of simulation and test system for vehicle body CAN bus based on CANoe. In 2008 7th World congress on intelligent control and automation, pp 7515–7519. Chongqing, China
4. Vinay Kumar V, Raghava Charyulu M (2015) Implementing automotive features using CAN through vector CANoe
5. Karunendra DC, Prathik KM, Totappa RR, Sara G (2021) Method of configuring LABCAR based on IOT. In: 2021 IEEE International conference on electronics, computing and communication technologies (CONECCT), pp 01–04. Bangalore, India
6. Kavitha K (2018) Canoe tool for Ecu automated communication testing. Int J Innov Technol Exploring Eng 8(2S):390–392
7. Sharma D, Gupta M, Controller area network for automobile application using ASIC based on PSoC and analysing through vector CANoe
8. Sutar AD, Shinde SB (2018) ECU health monitor using CANUSB. In: 2018 Second international conference on inventive communication and computational technologies (ICICCT), pp 415–419. Coimbatore, India
9. Vector (n.d.) CANoe product information. Retrieved from https://cdn.vector.com/cms/content/products/canoe/canoe/docs/Product%20Informations/CANoe_ProductInformation_EN.pdf
10. VectorCAST (Year) CAPL quickstart. Retrieved from https://cdn.vector.com/cms/content/products/VectorCAST/Events/TechNights/CAPLQuickstart_Generic_2018_Final.pdf
11. Layton E (n.d.) Financial help. Retrieved from https://www.laytonecuremaps.co.uk/financial-help/
12. All About Circuits (n.d.) Automotive SoCs provide window to ADAS trends. Retrieved from https://www.allaboutcircuits.com/news/automotive-socs-provide-window-to-adas-trends/
13. Jawaharlal Nehru Technological University Anantapur, CAPL programming—controlled area network. Retrieved from https://www.studocu.com/in/document/jawaharlal-nehru-technological-university-anantapur/automation-and-robotics/hb-capl-programming-controlled-area-network/40669632
14. Patil SG, Ratnaparkhe VR (2020) CAN protocol–application in automation electronics. In: 2020 International conference on smart innovations in design, environment, management, planning and computing (ICSIDEMPC), pp 318–321. Aurangabad, India
15. Georgescu D, Stanciu L (2018) Designing and implementing a solution to manipulate signals in automated testing using CANoe. In: 2018 IEEE 12th international symposium on applied computational intelligence and informatics (SACI), pp 000309–000312. Timisoara, Romania
16. Gajul V, Mishra JKD, Tavhare S (2021) Automation solution for software testing of CAN based ECUs. In: 2021 Fourth international conference on electrical, computer and communication technologies (ICECCT), pp 1–6. Erode, India
17. Ranjan B, Gupta M (2013) Automobile control system using controller area network. Int J Comput Appl 67(18)

Optimizing Battery Energy Storage Systems for Cost-Efficient Energy Management at a University Campus: A Particle Swarm Optimization Approach

Komal D. Dhole and Atul Phadke

Abstract This study proposes a novel approach to optimizing the sizing of battery energy storage systems (BESS) tailored for university campus applications, employing Particle Swarm Optimization (PSO). The primary aim is to minimize electricity consumption charges while ensuring that the BESS capacity is optimized to avoid unnecessary costs. Through PSO, the solution space is effectively explored to identify the optimal BESS size, striking a balance between cost reduction and capacity requirements. Results demonstrate significant cost savings while maintaining sufficient energy storage capacity. Our research highlights the practical utility of PSO in real-world energy management, enhancing decision-making processes and resource allocation. Future research directions include investigating scalability for larger-scale energy systems, integrating dynamic factors such as fluctuating energy demand and renewable energy sources, and exploring additional applications for optimized battery systems. Overall, this study underscores the potential of PSO in designing tailored BESS solutions for diverse applications, addressing complex energy optimization challenges with precision and effectiveness.

Keywords BESS · Particle Swarm Optimization · OpenDSS · Energy Management

1 Introduction

The escalating demand for electricity, coupled with the increasing focus on sustainable energy solutions, has prompted institutions, like universities [1, 2], to explore innovative strategies for energy management. Among these strategies, battery energy

K. D. Dhole (✉) · A. Phadke
COEP Technological University, Pune, India
e-mail: dholekd21.elec@coeptech.ac.in

A. Phadke
e-mail: arp.elec@coeptech.ac.in

S. Kumar et al. (eds.), *Soft Computing Applications in Modern Power and Energy Systems*, Lecture Notes in Electrical Engineering 1298,
https://doi.org/10.1007/978-981-96-0058-8_23

storage systems (BESS) have emerged as promising tools for enhancing energy efficiency [3–6], reducing costs [7–12], and supporting renewable energy integration [13–15]. However, determining the optimal size of a BESS presents a complex optimization challenge, as it requires balancing cost considerations with energy storage requirements.

Existing literature highlights the multifaceted advantages of BESS implementation across various sectors, including residential [9, 16], commercial [17], and industrial [10, 17] settings. Previous research in battery energy storage optimization has demonstrated the effectiveness of BESS in various applications, including load leveling [18, 19], peak shaving [6, 10, 11, 20], frequency regulation [21–23], and grid stability enhancement [24–26]. In particular, the utilization of BESS in university campuses has emerged as a promising area of research, driven by the need for sustainable energy solutions and cost reduction measures. Studies have shown that BESS deployment can lead to enhanced energy efficiency, peak demand shaving, grid stabilization, and improved overall system resilience, making it an attractive option for institutions seeking to optimize their energy consumption.

However, determining the optimal size of a BESS remains a significant challenge [27]. Traditional methods often rely on heuristic approaches or simplistic cost-benefit analyses, which may overlook the dynamic nature of energy consumption patterns and the complex interactions within the grid environment. To address these challenges, researchers have increasingly turned to advanced optimization techniques such as Metaheuristic Algorithms, Machine Learning Algorithms, Hybrid Optimization Approaches, Multi-Objective Optimization, and Dynamic Programming [3–5, 7, 10, 12, 28–39]. In this context, optimization techniques play a crucial role in identifying the most cost-effective BESS configuration tailored to specific applications. Particle Swarm Optimization (PSO), a nature-inspired metaheuristic algorithm, has gained prominence for its effectiveness in solving complex optimization problems. By mimicking the social behavior of swarms, PSO efficiently navigates solution spaces to identify optimal solutions.

This paper proposes the application of PSO to optimize the sizing of BESS for a university campus setting. The primary objective is to minimize electricity consumption charges while ensuring that the BESS capacity in terms of power rating (kW) and energy capacity(kWh) to avoid unnecessary over sizing or under sizing of the BESS leading to cost savings. By employing PSO, we aim to strike a balance between cost reduction and capacity requirements, thereby maximizing the economic benefits of BESS deployment.

Efficient management of electricity consumption is a critical concern for educational institutions aiming to reduce costs and enhance sustainability. Among the various strategies available, the integration of battery energy storage systems (BESS) emerges as a promising solution for minimizing electricity expenses through strategic load management and peak shaving strategies.

The study focuses on exploring the optimal sizing of battery energy storage systems to minimize electricity charges in educational institutes. By leveraging load curve analysis and time-of-day (TOD) pricing schemes, the aim is to develop a methodology for determining the ideal BESS capacity to achieve significant cost savings while ensuring reliable power supply. The significance of this study lies in its potential to offer a practical and cost-effective approach to energy management in educational institutions. By optimizing the size of BESS and implementing tailored charging and discharging strategies, universities and colleges can reduce their reliance on grid electricity during peak hours, thereby lowering electricity expenses and contributing to overall sustainability goals.

In this research, we build upon previous research findings to develop a tailored approach for optimizing BESS size in educational institutes. By integrating real-world electricity consumption data and considering time-of-day (TOD) pricing dynamics, we aim to provide a practical and effective solution for reducing electricity charges while enhancing energy resilience in university campuses.

2 Formulation of Objective Function

The objective function for this research endeavor is structured to achieve two primary objectives:

1. Optimally determine the BESS optimal power rating (kW) and energy capacity (kWh),
2. Alongside the minimization of electricity charges across distinct Time-of-Day (TOD) zones.

The objective function of sizing the battery optimally can be formulated as below,

2.1 *Optimization of Power Rating and Energy Storage Capacity*

To minimize the power rating and energy storage capacity of the BESS, the final objective function is chosen and expressed as:

$$f_1 = min[x(1) + x(2)] \tag{1}$$

where,

$x(1) = P_{\text{BESS}}$ = Active power of BESS (kW)

$x(2) = E_{\text{BESS}}$ = Energy Storage capacity of BESS (in kWh) The minimization of f_1 corresponds to the optimization of the battery energy storage system, effectively managing peak power demand and overall energy consumption.

The inclusion of the constraint $P_{\text{BESS}}^{\min} \le P_{\text{BESS}} \le P_{\text{BESS}}^{\max}$ ensures that the battery size falls within a feasible range.

2.2 Minimization of Electricity Charges

The second component, f_2 given by Eq. (2) is designed to minimize the cumulative electricity charges incurred across the four TOD zones:

$$f_2 = min[r_1 kVAh_1 + r_2 kVAh_2 + r_3 kVAh_3 + r_4 kVAh_4] \tag{2}$$

where,

r_1, r_2, r_3 and r_4 = rates of electricity for the four TOD Zones

$kVAh_1$, $kVAh_2$, $kVAh_3$, and $kVAh_4$ = Power consumption during four TOD zones

2.3 Composite Objective Function

The overall objective function given in Eq. (3), denoted as f, represents a balanced combination of the P_{BESS} and E_{BESS} for the optimal reduction of f_1 and f_2:

$$f = \lambda_1 f_1 + \lambda_2 f_2 \tag{3}$$

Subject to the constraint,

$$P_{\text{BESS}}^{\min} \leq P_{\text{BESS}} \leq P_{\text{BESS}}^{\max}$$

where,

λ_1, λ_2 = weighting factors (allowing for the flexible adjustment of the importance assigned to each objective during the optimization process.)

In summary, the objective function f encapsulates the dual goals of optimizing both kW/kWh demands and minimizing electricity charges, providing a comprehensive metric for evaluating the efficacy of the battery energy storage system in achieving cost-efficient and sustainable energy management.

3 Methodology

3.1 Data Collection Through Fluke 435-II Power Quality Analyzer

The assessment of the university's power distribution systems involved the use of the Fluke 435-II Three Phase Energy and Power Quality Analyzer [40]. Equipped with a comprehensive array of measurements, this analyzer provided both an overarching perspective and intricate insights into system performance. Initial configurations encompassed language settings, nominal frequency, and voltage parameters, meticulously set at 250 V and 50 Hz. Additionally, phase identification, color coding, and time-stamping were meticulously configured to ensure accuracy (Figs. 1 and 2).

To verify precise connections, the Scope Waveform and Scope Phasor features were employed, ensuring proper alignment of voltage leads and current clamps, which were clearly marked for signal polarity. Utilizing the Fluke TLS430 Test

Fig. 1 Data collection with Fluke 435-II three phase energy and power quality analyzer

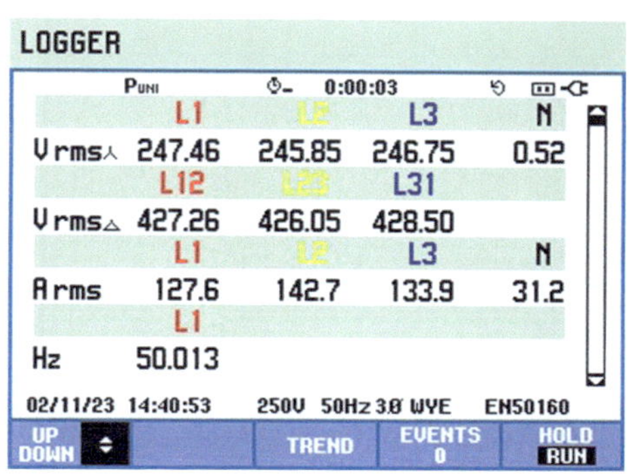

Fig. 2 Logger from analyzer

Leads and -Flexi-TF-II 6000 A AC Current Probes as voltage leads and current clamps, respectively, these instruments were connected to the live grid bus bars for data collection.

Data acquisition spanned 24 hours with a 5-minute interval, capturing comprehensive insights into system dynamics. The collected data, retrieved and analyzed using the Power Log 430-II software, facilitated the generation of a daily load curve. This curve served as the foundational dataset for subsequent modeling efforts aimed at optimizing the university's energy consumption patterns effectively.

3.2 *Load Curve Analysis*

Meticulous data collection was conducted using the Fluke 435-II Analyzer, capturing electricity consumption patterns over a 24-hour period with 5-minute intervals.

Following data collection, the recorded data underwent preprocessing and analysis. The system was modeled using OpenDSS, and the load curve was generated using MATLAB interfacing. This approach ensured precise data interpretation and visualization (Fig. 3).

The load curve analysis for the university revealed distinct consumption patterns. Lesser electricity consumption was observed during off-peak hours (22:00 to 06:00 Hrs), while peak demand was evident during peak hours (18:00 to 22:00 Hrs). This insight serves as a foundation for developing effective load-shifting strategies, where energy-intensive tasks are scheduled during off-peak periods to reduce electricity costs and optimize resource utilization.

Based on the analysis findings, an effective charging strategy was devised to leverage periods of low electricity demand for battery charging. Off-peak hours present an opportune time to charge the battery, optimizing energy utilization and reducing electricity charges.

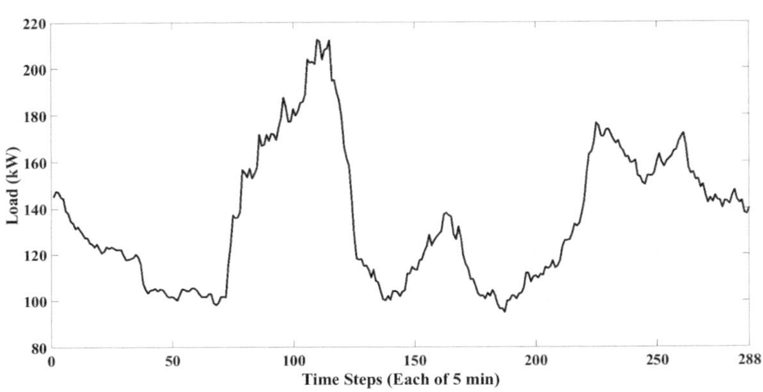

Fig. 3 Daily load curve

Table 1 Time of day (TOD) tarrif structure for university

Consumer category	Energy charges (Rs./kVAh)
HT-Public Services Government	9.53
Time-of-Day (TOD) Tariff (In addition to above base tariff)	–
00:00-6:00,22:00-24:00	−1.50
06:00-9:00,12:00-18:00	0.00
9:00-12:00	0.80
18:00-22:00	1.10

Table 1 illustrates the tariff structure applicable [41] to the university's electricity consumption. Additional wheeling charges are applicable, varying with the supply voltage level. For Extra High Voltage (EHV) supply, the wheeling charge is set at Rs. 0.60 per kVAh. Similarly, for High Tension (HT) supply, the wheeling charge is also Rs. 0.60 per kVAh. These charges are imposed to cover the costs associated with transmitting electricity at different voltage levels.

By aligning battery operations with off-peak hours (Zones 1 and 2) for charging and avoiding peak hours (Zone 4) for discharging, we aim to maximize cost savings while ensuring efficient utilization of the BESS.

Factors such as tariff structures, electricity demand profiles, and operational constraints influence the design and implementation of BESS strategies within each TOD zone.

3.3 Charging and Discharging Strategy

The charging and discharging strategy plays a crucial role in optimizing electricity consumption and reducing costs for electricity consumers. By leveraging time shifting, consumers can exploit off-peak hours when electricity rates are lower to charge battery energy storage systems (BESS) and discharge them during peak hours when rates are higher. This strategy aims to achieve optimal electricity consumption states, aligning with different Time-of-Day (TOD) tariff structures.

In the realm of university energy management, the charging and discharging protocol is custom-tailored following an in-depth examination of the daily load curve. This analysis facilitates the identification of optimal time frames for battery operations. For instance, the battery undergoes a four-hour charging cycle during off-peak hours, typically spanning from 01:00 to 05:00 Hrs. Conversely, a four-hour discharging phase occurs during peak demand periods, usually ranging from 18:00 to 22:00 Hrs. This strategy ensures efficient utilization of energy resources, aligning battery operations with the university's energy demands.

To implement and simulate this strategy effectively, the university utilizes OpenDSS, a robust electrical system simulation tool.

3.4 System Modeling and Simulation

In this section, we utilize the Open Distribution System Simulator (OpenDSS) [42] to model and simulate the electrical system of the university. OpenDSS is recognized for its capability to conduct steady-state analyses essential for utility distribution systems. It is available as open-source software, both as a standalone executable program and a Component Object Model (COM) DLL, allowing smooth integration with various software platforms.

Through the application of OpenDSS, our objective is to analyze the behavior and functionality of the university's electrical infrastructure, enabling strategic decisions regarding energy management optimization (Fig. 4).

Model Configuration The electrical network model is configured in OpenDSS software. Operating at a standard frequency of 50 Hz, the system is carefully designed to ensure uninterrupted power supply to the University premises.

The schematic diagram depicted in Fig. 1 serves as a visual representation of the distribution system. At the core of the model lies the source bus, functioning as the primary voltage source with a base voltage of 11 kV and a per unit value of 1.05. With the capability to withstand short-circuit currents ranging from 5 to 3000 amperes, the source bus ensures system stability under varying load conditions.

To connect the Source Bus to Bus 1, a Transformer is employed in a Delta-Wye configuration. Key parameters of the transformer include a primary voltage of 11 kV, a secondary voltage of 0.4 kV, and symmetric ratings of 750 kilovolt-amperes (kVAs) on both sides, accompanied by a reactance of 4 ohms. The proposed BESS would be connected to the same bus.

Additionally, a three-phase line spanning 1 meter is modeled to link Bus 1 and Bus 2. Each phase of the line is characterized by resistance and reactance per unit length, with capacitance values set to zero to streamline the analysis process.

To accurately represent variations in daily load profiles, three distinct load shapes are defined for each phase and connected to Bus 2. These load shapes are tailored with specific voltage, active power, and power factor ratings to align with operational requirements and reflect diverse load profiles.

Throughout the network, voltage base values are standardized at 11 kV and 0.4 kV to maintain consistency and coherence with the specified transformer and line ratings.

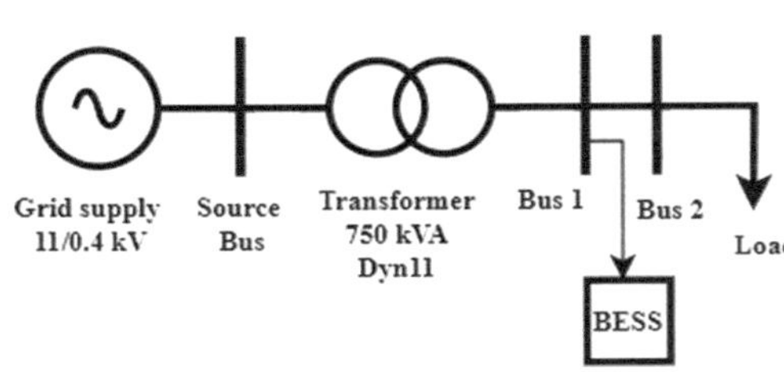

Fig. 4 Distribution system schematic

MATLAB OPENDSS Interafce The integration of MATLAB with OpenDSS via the COM interface facilitates seamless communication and data exchange between the two platforms. MATLAB scripts coordinate simulation and optimization processes, leveraging computational capabilities from both environments to enhance system efficiency and cost-effectiveness.

- Simulation Automation: MATLAB scripts automate OpenDSS simulations, streamlining the analysis process and ensuring efficient execution of simulation scenarios.
- Data Processing: MATLAB handles data processing tasks, extracting relevant insights from simulation results, and enabling comprehensive analysis of system performance.
- Optimization Framework: MATLAB serves as the platform for implementing Particle Swarm Optimization (PSO) algorithm, to fine-tune system parameters for optimal performance. This framework enables iterative refinement of system configurations to achieve desired objectives efficiently.

System Simulation Initially, the electrical network model in OpenDSS, where load profiles and other operational constraints are incorporated. MATLAB scripts then automate simulation runs within OpenDSS, ensuring efficient processing and accurate data capture. Following simulation, MATLAB conducts data analysis to extract insights on energy consumption patterns, focusing on optimizing electricity charges and maximizing battery utilization. The performance of various simulation scenarios is thoroughly evaluated to improve system parameters and optimization strategies iteratively.

4 Optimization Approach: PSO

Particle swarm algorithm operates on a population basis, resembling the genetic algorithm in this aspect. It is utilized for exploring optimal parameters within complex search spaces.The PSO algorithm has been implemented using [43, 44]. It involves a group of individuals known as particles traversing through a designated area in incremental steps. During each step, the algorithm assesses the objective function for each particle. PSO initiates with a set of random particles and iteratively refines these particles to identify optimal parameters. During each iteration, all particles undergo modifications influenced by two essential metrics: P_{best} and G_{best}. P_{best} signifies the most optimal solution reached by individual particles in preceding iterations, whereas G_{best} indicates the highest achievement attained by any particle across all iterations, recognized as the ultimate global solution [45].These updates are governed by equations (4) and (5), where each particle adjusts its position and velocity accordingly. Following the adjustment of velocities, the particles proceed to move, and the algorithm repeats this process.

$$v_{i+1} = c_1 r_1 (P_{best} - x_i) + c_2 r_2 (G_{best} - x_i) \quad (4)$$

$$x_{i+1} = v_{i+1} + x_i \quad (5)$$

In the context of particle swarm optimization (PSO):

i: Represents the current iteration number.

j: Denotes the index of a particular particle within the swarm.

v_i: Signifies the velocity of a particle at the ith iteration.

x_i: Indicates the position of a particle at the ith iteration.

P_{best}: Refers to the best solution found by a particle within its own history up to the ith iteration.

G_{best}: Represents the best global solution discovered by any particle within the entire swarm up to the ith iteration.

r_1 and r_2: Stand for random numbers generated between 0 and 1, used to introduce stochasticity into the algorithm.

c_1 and c_2: Represent the cognitive and social learning factors, respectively. These factors influence the balance between individual and social influences on the particles' movement. Adjusting these factors can have a significant impact on the convergence rate of the algorithm. Further insights into PSO can be explored in reference [44] (Fig. 5).

The PSO parameters chosen for this proposed algorithm are:

- The number of particles in the swarm (**NP**) is set to 100.
- The number of iterations (**NI**) is set to 100.
- The learning factors, denoted by **C1** and **C2**, are both set to 2.
- The inertia weight (**w**) is set to 0.9.
- The number of dimensions (**D**) is set to 1.

5 Results and Analysis

5.1 Load Consumption Variation with BESS

The graph in the Fig. 6 demonstrates the load consumption variations throughout the day, comparing scenarios with and without the installation of a battery energy storage system. The solid line represents load consumption without the battery, while the dotted line depicts load consumption after installing the optimal-sized battery (90.2891 kW and 448.1838 kWh). Charging and discharging the battery for 4 hours daily resulted in significant alterations in energy consumption patterns (Fig. 6).

To optimize battery performance, minimum and maximum charging capacities were set to be 50 kW and 215 kW respectively, ensuring efficient operation within the specified constraints.

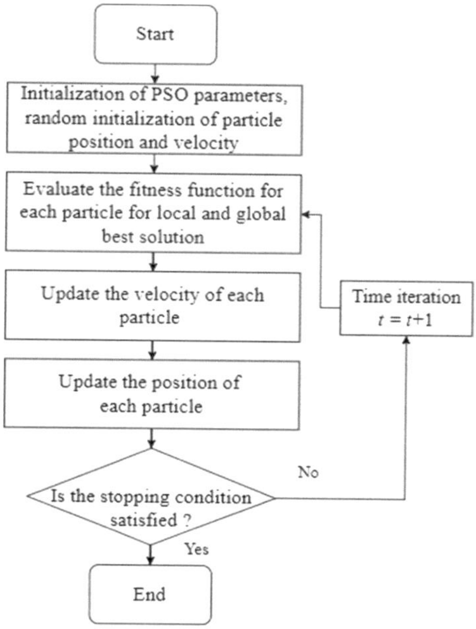

Fig. 5 flowchart for PSO algorithm

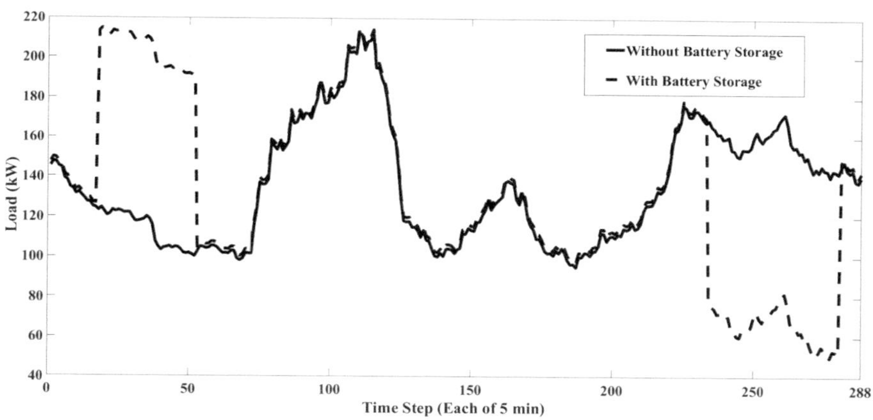

Fig. 6 Load consumption variation with BESS

Between 1:00 and 5:00 Hrs, marked by lower electricity rates (Rs 8.63 per unit), energy consumption rose from 123.3 kW to 213.5 kW post-battery installation. Conversely, from 18:00 to 22:00 Hrs, when rates peak (Rs 10.23 per unit), energy peak decreased from 172.14 kW to 80.39 kW. These adjustments culminated in reduced total energy charges, resulting in a decreased electricity bill.

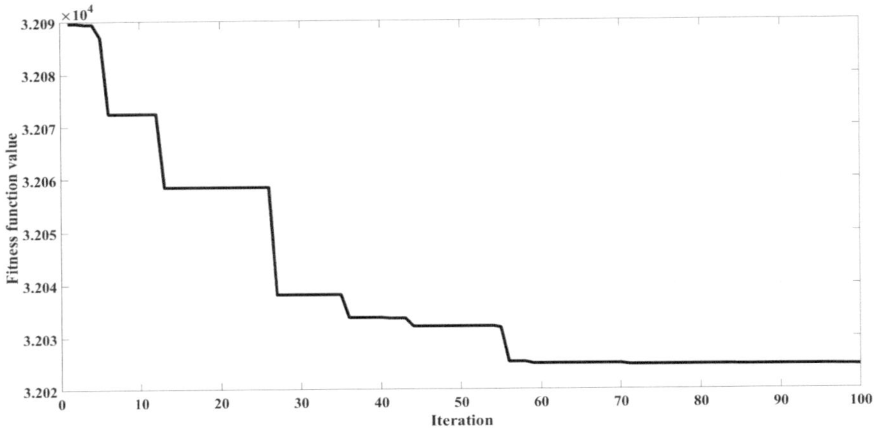

Fig. 7 PSO convergence characteristics

5.2 *Convergence Behavior of PSO Algorithm*

Figure 7 depicts the convergence rate of the PSO algorithm applied to optimize the sizing of Battery Energy Storage Systems (BESS).

The PSO algorithm iteratively refines fitness function values until convergence is reached. In this case, convergence occurred around the 60th iteration. Notably, the global best solution (G_{best}) was attained by this point, indicating optimal performance.

A slight improvement was observed in attainment after the 60th iteration, with further marginal gains up to the 70th iteration. However, no significant deviation from the optimized state was noticed beyond this point. This behavior can be attributed to the algorithm's symmetrical nature around the midpoint.

Ultimately, the fitness function converged to a value of Rs. 32025, representing the minimized electricity bill in rupees. This convergence signifies the achievement of an optimal balance between individual (personal best) and collective (global best) solution values.

In Table 2, the changes in electricity costs (in Rs.) under various scenarios are displayed. Following the utilization of the PSO algorithm in MATLAB, the optimal

Table 2 Electricity charges for the university in case of installed BESS and without BESS

Case	BESS capacity (kW)	Electricity charges (Rs.)
No BESS	0	32490
BESS analytical	25	32014
BESS based on PSO	90.28	31486

Table 3 Electricity charges (in Rs.) for different zones in case of installed BESS and without BESS

TOD	Electricity charges without BESS (in Rs.)	Electricity charges with BESS (in Rs.)
Zone 1	8481.5	9786.9
Zone 4	7223.4	4644.1

Table 4 Total kVAh consumption for different zones in case of installed BESS and without BESS

TOD	Total kVAh consumption without BESS	Total kVAh Consumption with BESS
Zone 1	982.8	1134.1
Zone 4	643.2	413.5

BESS size determined by PSO is 90.28 kW and 448.1838 kWh. This resulted in a decrease in electricity usage costs from Rs. 32,490 to Rs. 31,486 for the total kVAh consumption of the day.

The Table 3 presents the electricity usage costs (in Rs.) for both Zone 1 and Zone 2. Notably, there has been a substantial reduction in charges for Zone 4, amounting to Rs. 2579. This reduction is anticipated to lead to an overall decrease in the total electricity expenses.

Table 4 displays the comparative kVAh consumption for Zone 1 and Zone 4 prior to the installation of the BESS and subsequent to the installation of the optimally sized BESS obtained from the PSO algorithm.

The PSO algorithm effectively determined the optimal battery size, facilitating cost-effective energy management and enhancing the system's efficiency. By leveraging PSO convergence, the study achieved significant reductions in electricity costs.

The results demonstrate the tangible benefits of integrating a battery energy storage system into the university's electrical infrastructure. By strategically charging and discharging the battery during off-peak and peak hours, respectively, the study successfully mitigated electricity costs and optimized energy consumption patterns.

6 Conclusion

In this investigation, we scrutinized the efficacy of integrating a battery energy storage system (BESS) to optimize electricity expenses for educational institutions. Through meticulous data analysis, we discerned distinct consumption patterns, particularly during peak and off-peak periods. By tailoring a charging and discharging strategy to match the institution's load profile and tariff structure, we aimed to mitigate operational costs for consumesrs.

Our simulation experiments yielded promising results, showcasing a significant reduction in overall energy expenditure with BESS implementation. By strategically charging the battery during low-demand periods and discharging during peak hours, we observed a tangible decrease in the institution's electricity bills. Furthermore, leveraging Particle Swarm Optimization (PSO) allowed us to determine the optimal battery size, further enhancing cost-effectiveness.

By embracing innovative technologies like BESS and employing advanced optimization techniques such as PSO, consumers can not only trim expenses but also contribute to environmental sustainability.

Moving forward, future research endeavors could delve into additional factors affecting BESS efficacy, including diverse load profiles, renewable energy integration, and grid stability considerations. Real-world deployment and monitoring of BESS systems in educational settings would offer valuable insights into long-term performance and scalability.

References

1. Chedid R, Sawwas A, Fares D (2020) Optimal design of a university campus micro-grid operating under unreliable grid considering PV and battery storage. Energy 200:117510, ISSN: 0360-5442, https://doi.org/10.1016/j.energy.2020.117510
2. Muqeet HA, Munir HM, Javed H, Shahzad M, Jamil M, Guerrero JM (2021) An energy management system of campus microgrids: state-of-the-art and future challenges. Energies 14:6525. https://doi.org/10.3390/en14206525
3. Kumpatla M, Alam M (2023) Optimal placement and sizing of energy storage systems in networked microgrids, 1–6. https://doi.org/10.1109/SeFeT57834.2023.10245909
4. Wong LA, Ramachandaramurthy VK, Walker SL, Taylor P, Sanjari MJ (2019) Optimal placement and sizing of battery energy storage system for losses reduction using whale optimization algorithm. J Energy Storage 26:100892, ISSN 2352-152X, https://doi.org/10.1016/j.est.2019.100892
5. Vankudoth L, Badar AQH (2019) Distribution network optimization through siting and sizing of BESS. 2019 8th International conference on power systems (ICPS), Jaipur, India, pp 1–5, https://doi.org/10.1109/ICPS48983.2019.9067342
6. Manojkumar R, Kumar C, Ganguly S, Catalão J (2021) Optimal peak shaving control using dynamic demand and feed-in limits for grid-connected PV sources with batteries. IEEE Syst J 15(4):5560–5570. https://doi.org/10.1109/JSYST.2020.3045020
7. Arias NB, López JC, Hashemi S, Franco JF, Rider MJ (2021) Multi-objective sizing of battery energy storage systems for stackable grid applications. IEEE Trans Smart Grid 12(3):2708–2721. https://doi.org/10.1109/TSG.2020.3042186
8. Yang Z, Rabinowitz J, Li J (2022) Peak demand optimization of commercial buildings based on energy storage systems. GLOBECOM 2022-2022 IEEE global communications conference, Rio de Janeiro, Brazil, pp 5299–5304, https://doi.org/10.1109/GLOBECOM48099.2022.10001320
9. Shabbir N, Kütt L, Astapov V, Jawad M, Allik A, Husev O (2022) Battery size optimization with customer PV installations and domestic load profile. IEEE Access 10:13012–13025. https://doi.org/10.1109/ACCESS.2022.3147977
10. Saaravia-Guerrero JA, Espinosa-Juárez E (2021) Optimal sizing of BESS for industrial peak shaving applications considering different electricity billing rates. 2021 Fourth international conference on electrical, computer and communication technologies (ICECCT), Erode, India, pp 1–6, https://doi.org/10.1109/ICECCT52121.2021.9616916
11. Nigmatulina N, Belonogova N, Mendes G, Honkapuro S (2020) Peak-shaving potential of residential battery energy storage systems under power-based tariff pricing. CIRED, (2020)

Berlin workshop (CIRED 2020). Online conference, 278–281. https://doi.org/10.1049/oap-cired.2021.0329
12. Peña AA, Romero-Quete D, Cortes CA (2022) Sizing and siting of battery energy storage systems: a Colombian case. J Modern Power Syst Clean Energy 10(3):700–709. https://doi.org/10.35833/MPCE.2021.000237
13. Javadi M, Liang X, Gong Y, Chung CY (2022) Battery energy storage technology in renewable energy integration: a review. 2022 IEEE Canadian conference on electrical and computer engineering (CCECE), Halifax, NS, Canada, pp 435–440, https://doi.org/10.1109/CCECE49351.2022.9918504
14. Wali SB, Hannan MA, Reza MS, Ker PJ, Begum RA, Abd Rahman MS, Mansor M (2021) Battery storage systems integrated renewable energy sources: a biblio metric analysis towards future directions. J Energy Storage 35:102296, ISSN: 2352-152X, https://doi.org/10.1016/j.est.2021.102296
15. Abdalla AN, Nazir MS, Tao H, Cao S, Ji R, Jiang M, Yao L (2021) Integration of energy storage system and renewable energy sources based on artificial intelligence: An overview. J Energy Storage 40:102811, ISSN: 2352-152X, https://doi.org/10.1016/j.est.2021.102811
16. Rostamnezhad Z, Mary N, Dessaint L-A, Monfet D (2023) Electricity consumption optimization using thermal and battery energy storage systems in buildings. IEEE Trans Smart Grid 14(1):251–265. https://doi.org/10.1109/TSG.2022.3194815
17. Saini P, Gidwani L (2022) An investigation for battery energy storage system installation with renewable energy resources in distribution system by considering residential, commercial and industrial load models. J Energy Storage 45:103493, ISSN: 2352-152X, https://doi.org/10.1016/j.est.2021.103493
18. Gusev YP, Subbotin PV (2019) Using battery energy storage systems for load balancing and reactive power compensation in distribution grids. 2019 International conference on industrial engineering, applications and manufacturing (ICIEAM), Sochi, Russia, pp 1–5, https://doi.org/10.1109/ICIEAM.2019.8742909
19. Mehrjerdi H (2019) Simultaneous load leveling and voltage profile improvement in distribution networks by optimal battery storage planning. Energy 181:916–926, ISSN: 0360-5442, https://doi.org/10.1016/j.energy.2019.06.021
20. Argiolas L, Stecca M, Ramirez-Elizondo LM, Soeiro TB, Bauer P (2022) Optimal battery energy storage dispatch in energy and frequency regulation markets while peak shaving an EV fast charging station. IEEE Open Access J Power Energy 9:374–385. https://doi.org/10.1109/OAJPE.2022.3198553
21. Xiong L et al (2023) Voltage and frequency regulation with WT-PV-BESS in remote weak grids via fixed-time containment control. IEEE Trans Power Syst 38(3):2719–2735. https://doi.org/10.1109/TPWRS.2022.3190847
22. Cirio D et al (2023) Fast frequency regulation from a wind farm-BESS unit by model predictive control: method and hardware-in-the-loop validation. IEEE Trans Sustain Energy 14(4):2049–2061. https://doi.org/10.1109/TSTE.2023.3265187
23. Zhao H, Hong M, Lin W, Loparo KA (2019) Voltage and frequency regulation of microgrid with battery energy storage systems. IEEE Trans Smart Grid 10(1):414–424. https://doi.org/10.1109/TSG.2017.2741668
24. Wang J, Song Y, Hill DJ, Liu T (2021) Microgrid stability enhancement by incorporating BESS droop gain tuning. IEEE PES innovative smart grid technologies Europe (ISGT Europe). Espoo, Finland, 1–5. https://doi.org/10.1109/ISGTEurope52324.2021.9640181
25. Jannesar MR, Sadr S, Savaghebi M (2022) Dynamic stability enhancement in low-inertia power systems using battery energy storage. 12th Smart grid conference (SGC). Kerman, Iran, Islamic Republic of 2022, 1–5. https://doi.org/10.1109/SGC58052.2022.9998974
26. Zhu Y, Liu C, Dai R, Liu G, Xu Y (2019) Optimal battery energy storage placement for transient voltage stability enhancement. IEEE Power and energy society general meeting (PESGM). Atlanta, GA, USA 2019, 1–5. https://doi.org/10.1109/PESGM40551.2019.8973610
27. Hannan MA, Faisal M, Ker PJ, Begum RA, Dong ZY, Zhang C (2020) Review of optimal methods and algorithms for sizing energy storage systems to achieve decarbonization in microgrid

applications. Renew Sustain Energy Rev 131:110022, ISSN: 1364-0321, https://doi.org/10.1016/j.rser.2020.110022
28. Ying G, Jigeng L, Mengna H (2021) Machine learning based optimization model for energy management of energy storage system for large industrial park. Processes 9:825. https://doi.org/10.3390/pr9050825
29. Medghalchi Z, Taylan O (2023) A novel hybrid optimization framework for sizing renewable energy systems integrated with energy storage systems with solar photovoltaics, wind, battery and electrolyzer-fuel cell. Energy Conv Manag 294:117594, ISSN: 0196-8904, https://doi.org/10.1016/j.enconman.2023.117594
30. Mahesh A, Sushnigdha G (2022) Optimal sizing of photovoltaic/wind/battery hybrid renewable energy system including electric vehicles using improved search space reduction algorithm. J Energy Storage 56, Part A:105866, ISSN: 2352-152X, https://doi.org/10.1016/j.est.2022.105866
31. Lata P, Vadhera S (2020) Optimal placement and sizing of energy storage systems to improve the reliability of hybrid power distribution network with renewable energy sources. J Stat Manag Syst 23:17–31. https://doi.org/10.1080/09720510.2020.1714147
32. Gong K, Wang X, Jiang C, Shahidehpour M, Liu X, Zhu Z (2021) Security-constrained optimal sizing and siting of BESS in hybrid AC/DC microgrid considering post-contingency corrective rescheduling. IEEE Trans Sustain Energy 12(4):2110–2122. https://doi.org/10.1109/TSTE.2021.3080707
33. Kumar A, Rizwan M, Nangia U (2022) A new approach to design and optimize sizing of hybrid microgrids in deregulated electricity environment. CSEE J Power Energy Syst 8(2):569–579. https://doi.org/10.17775/CSEEJPES.2020.03200
34. Tjandra R, Wen S, Zhou D, Tang Y (2019) Optimal sizing of BESS for hybrid electric ship using multi-objective particle swarm optimization. 2019 10th international conference on power electronics and ECCE Asia (ICPE 2019 - ECCE Asia), Busan, Korea (South), pp 1460–1466, https://doi.org/10.23919/ICPE2019-ECCEAsia42246.2019.8797214
35. Cao Y, Wu Q, Zhang H, Li C (2023) Multi-objective optimal siting and sizing of BESS considering transient frequency deviation and post-disturbance line overload. Int J Electr Power Energy Syst 144:108575, ISSN: 0142-0615, https://doi.org/10.1016/j.ijepes.2022.108575
36. Arias-Cazco D (2023) Levelized cost of storage analysis with multi-objective optimization for BESS. IEEE Seventh Ecuador technical chapters meeting (ECTM). Ambato, Ecuador, pp 1–6. https://doi.org/10.1109/ETCM58927.2023.10309040
37. Brillianto Apribowo CH, Sarjiya, Hadi SP, Wijaya FD (2022) Optimal sizing and siting of battery energy storage systems with retired battery. 2022 International conference on technology and policy in energy and electric power (ICT-PEP), Jakarta, Indonesia, pp 327–332, https://doi.org/10.1109/ICT-PEP57242.2022.9988958
38. Naces RA, Tejada CJM, Ostia CF (2023) Application of earthworm algorithm in battery size optimization and load schedule of PV-BESS for residential houses. 2023 15th international conference on computer and automation engineering (ICCAE), Sydney, Australia, pp 498–502, https://doi.org/10.1109/ICCAE56788.2023.10111448
39. Wu D, Gui Q, Zhao W, Wang J, Shi S, Zhou Y (2020) Battery energy storage system (BESS) sizing analysis of bess-assisted fast-charge station based on double-layer optimization method. 2020 IEEE 3rd student conference on electrical machines and systems (SCEMS), Jinan, China, 2020, pp 658–662, https://doi.org/10.1109/SCEMS48876.2020.9352324
40. fluke 435 ii power quality analyzer manual, http://www.fluke-direct.ca/pdfs/cache/www.fluke-direct.ca/fluke/powerqualityanalyzer/435ii/manual/fluke435iipowerqualityanalyzermanual.pdf
41. Maharashtra state electricity distribution company Ltd. Commercial circular no. 341, Maharashtra state electricity distribution company Ltd. , 2023, https://www.mahadiscom.in/consumer/wp-content/uploads/2023/04/Order-226-of2022-1.pdf
42. Reference guide the open distribution system simulator (OpenDSS). Operation Manual V7.4.3. Electric Power Res Inst (EPRI) (2012)
43. MATLAB, MathWorks, 2022, R2022b, MATLAB R2022b. http://www.mathworks.com

44. Kennedy J, Eberhart R (1995) Particle swarm optimization. Proceedings of ICNN'95—International conference on neural networks, Perth, WA, Australia, pp 1942–1948, vol 4, https://doi.org/10.1109/ICNN.1995.488968
45. Zhou T, Sun W (2014) Optimization of battery-supercapacitor hybrid energy storage station in wind/solar generation system. IEEE Trans Sustain Energy 5(2):408–415. https://doi.org/10.1109/TSTE.2013.2288804

Optimizing Power Management of Wind PV Based Grid Connected Hybrid System

Pawan Kumar, Monika, and Umesh Kumar Sinha

Abstract This study represents the power management of the wind PV based grid connected hybrid system. The hybrid wind PV system is modelled and Maximum power point tracking (MPPT) approaches are discussed to maximize the grid power. Two renewable energy resources increase the power of the overall system and minimize the stress of the standalone system. The external weather conditions are uncertain due to which the single renewable system is not sufficient to generate the required power. The combination of two renewable resources gives better performance by extracting the maximum power. To extract the maximum power the system is necessary to operate on MPPT. In this study, Two MPPT techniques (Perturb and observe (P&O) and Cuckoo search optimisation (CSO)) are employed in the PV system and compared based on their power transfer capability to the grid and efficiency of the system. For the conversion of wind energy, a Doubly fed induction generator (DFIG) is used and vector control strategies is implemented to regulate the power transmission to the grid. To obtain the modelling and power management of the Hybrid system, the MATLAB/ Simulink environment is demonstrated for different stages of solar irradiance at constant low wind speed. The simulated results show the efficiency and effectiveness of the optimisation techniques for the system.

Keywords DFIG · P&O · CSO · Vector control strategies · Hybrid Wind PV system · MPPT

P. Kumar (✉) · U. K. Sinha
Department of Electrical Engineering, National Institute of Technology Jamshedpur, Jharkhand, India
e-mail: 2019rsee007@nitjsr.ac.in

U. K. Sinha
e-mail: uksinha.ee@nitjsr.ac.in

Monika
Department of Electrical Engineering, Birla Institute of Technology Mesra, Ranchi, Jharkhand, India
e-mail: monika.bitmesera.phd@gmail.com

S. Kumar et al. (eds.), *Soft Computing Applications in Modern Power and Energy Systems*, Lecture Notes in Electrical Engineering 1298,
https://doi.org/10.1007/978-981-96-0058-8_24

1 Introduction

Natural resources are now a significant and promising technological advancement. Because of the high cost of petroleum and the negative impact that conventional power sources have to the environment, they are used to produce electricity. In the past few decades, there has been an improvement in research on renewable energy [1]. In particular, hybrid systems are now more commonly used with photovoltaic and wind production sources. One of the most effective ways to supply electricity to isolated loads or grid-connected systems is through the utilisation of the combination under consideration. Numerous studies on the PV-wind hybrid system have been conducted recently [2–4]. The amount of electrical energy produced by this hybrid system is highly influenced by variations in the weather. Thus, the sharp variations in sun irradiance and wind speed are what producing the power oscillations. The control strategies need to be able to resolve the grid integration issues to enhance the injected power quality. Furthermore, this control strategy must operate at maximum power in all weather scenarios [5]. The photovoltaic system is covered in several works of literature. PV panels have nonlinearly evolving output characteristics. To implement the MPPT techniques various control optimisation techniques are recently used. The MPPT control has been the authors' main area of interest. To assess the efficiency and effectiveness of various MPPT techniques, a DC/DC chopper is required to carry out these operations and derive the highest possible power [6, 7]. The drive train, variable speed generator, AC/DC rectifier, and fixed-pitch wind turbine are all part of the wind power producing system. Wind energy is converted via a wind turbine from kinetic energy to mechanical energy and subsequently, via a generator to electrical energy.

Drive train models come in four different varieties. One mass model is typically employed to avoid the system becoming too complex [8]. The following three categories of AC generators are commonly seen in wind power conversion systems: (1) Squirrel cage induction generators (2) Doubly fed induction generator (DFIG) and (3) Permanent magnet synchronous generator (PMSG) [9, 10]. The hybrid wind-photovoltaic system integrates its generated power into the utility grid using a three-phase converter. The main aim of this paper is to model and simulate the proposed wind PV hybrid model in which the PV panel is integrated with a Wind based DFIG system and to examine how the Wind PV hybrid system behaves with P&O and CSO optimisation techniques in different solar irradiances. Also, the comparative investigation of both P&O and CSO in the perspective of their effectiveness and efficiency was analysed.

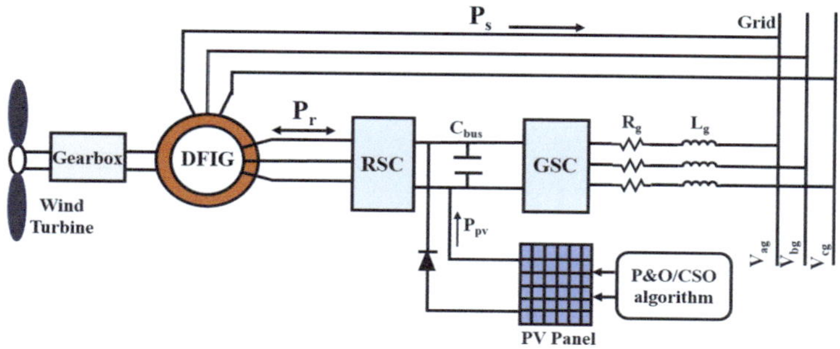

Fig. 1 Architecture of the proposed hybrid system

2 Wind PV Based Hybrid System

The architecture of the proposed Wind PV based hybrid system is shown in Fig. 1. The PV system is integrated with the wind system at the DC link between power electronics converters. The overall system is connected to the grid and delivers the power to the grid through the stator side of DFIG and the power electronics converter on the grid side. In a conventional wind conversion system, in case of low speed of wind, DFIG consumes the power from the grid. In the proposed system the requirement of power is fulfilled by the power generated by the PV system and the extra power is delivered to the grid. In the case of a lossless system, the grid power (Pg) can be represented by the equation as

$$P_g = P_s + (P_{pv} - P_r) \tag{1}$$

where P_s represents the stator power, P_{pv} represents the Photovoltaic power and P_r represents the rotor power.

A vector control strategy is employed to control and operate at MPPT of the DFIG system. P&O and CSO optimisation techniques are investigated in the PV system.

3 Photovoltaic System

A 400 KW PV panel system is implemented by connecting 37 parallel strings and 36 series connected strings. The curve of PV power vs PV voltage and PV current vs PV voltage at the three different solar irradiances (250 W/m^2, 500 W/m^2 and 1000 W/m^2) is shown in Figs. 2. and 3.

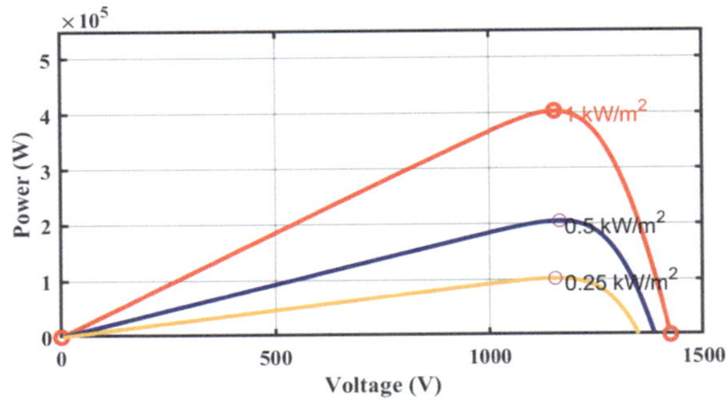

Fig. 2 Power vs voltage curve

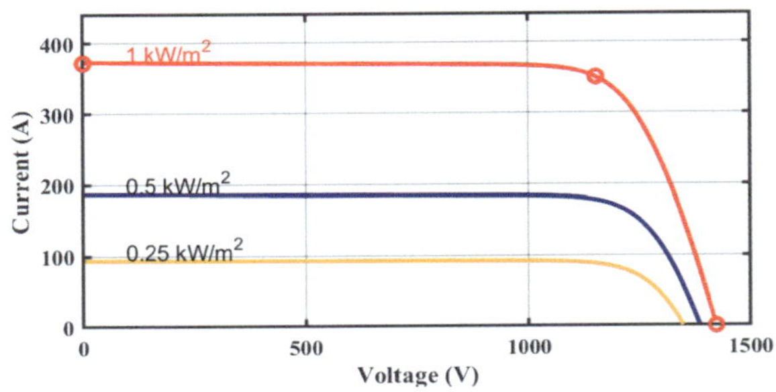

Fig. 3 Current vs voltage curve

3.1 Modelling of the Photovoltaic Generator

As seen in Fig. 4, the photoelectric phenomenon allows the solar cell to directly transform sunlight into electrical power. A single diode model represents the PV cell [11]. A current source with a diode, shunt resistance (Rsh), accompanied in parallel, and series resistance (Rs) constitute the solar cell model.

The modelling of the PV cell model can be represented by [12]

$$I_{pv} = -(I_{sh} + I_d - I_{psc}) \tag{2}$$

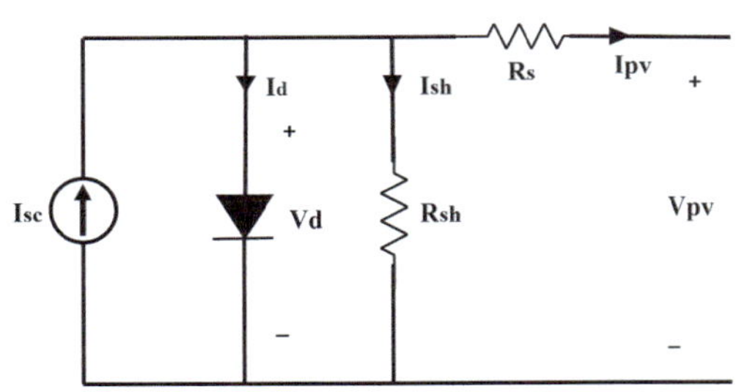

Fig. 4 Representation of PV cell

$$I_{pv} = -(\frac{V_d}{R_{sh}} + I_d - I_{psc}) \tag{3}$$

where I_d can be written as

$$I_d = I_s(e^{\frac{V_d}{V_t}} - 1) \tag{4}$$

Here, I_s represents the diode reverse saturation current.

$$V_t = \frac{nKT}{q} \tag{5}$$

and V_{pv} can be represented by

$$V_{pv} = V_d - I_{pv} * R_s \tag{6}$$

The PV power can be calculated by the expression as

$$\mathrm{P_{pv}} = \mathrm{V_{pv}} * \mathrm{I_{pv}} \tag{7}$$

3.2 *Modelling of DC-DC Boost Converter*

There are numerous chopper-based converters used to step up or down the DC voltage like buck converter, buck boost converter and boost converter. The most widely used DC-DC converter used in the PV system is the Boost converter which helps to increase the PV voltage at the desired level. In the proposed configuration the boost converter can be shown in Fig. 5. This converter helps to track the maximum power and ease to control the voltage [13].

For the boost converter to operate in step up mode, we can get the following relation of output voltage (Vo) and input voltage (Vin).

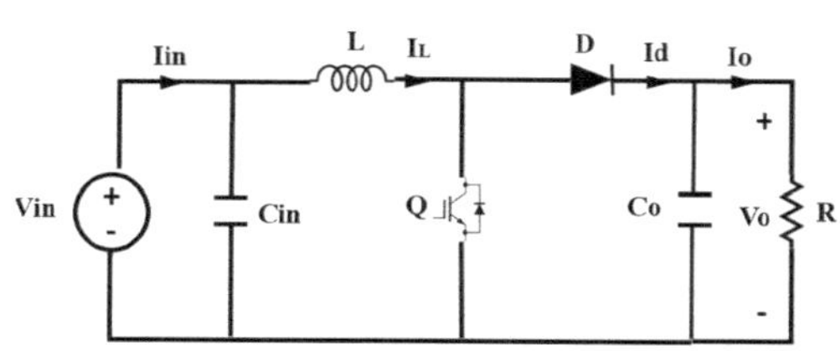

Fig. 5 The circuit configuration of DD-DC boost converter

$$\begin{Bmatrix} V_o = \frac{1}{1 - D_{boost}} V_{in} \\ I_o = (1 - D_{boost}) I_L \end{Bmatrix} \tag{8}$$

where $0 < D_{boost} < 1$

4 Wind Energy System

A 2 MW Wind based Doubly Fed Induction generator (DFIG) is implemented which operates on sub-synchronous speed i.e. at low wind speed. The modelling of the wind turbine and DFIG is employed in this section.

4.1 Modelling of Wind Turbine

The wind turbine is used to transform wind power into mechanical power. The power that a wind turbine (WT) develops is displayed by [14]:

$$P_t = \frac{1}{2} \rho \pi R^2 V^3 C_p(\beta, \lambda) \tag{9}$$

where ρ is representing air density (kg / m^3),

R represents the radius of the wind turbine; V represents wind speed (m/sec) and Cp represents the wind turbine power coefficient.

The Tip speed ratio can be represented as: -

$$\lambda = \frac{\Omega_t * R}{V} \tag{10}$$

Here Ω_t Turbine angular speed.

4.2 Modelling of Wind Turbine

According to Park reference system, the modelling of the DFIG is explained. The general modelling of the generator is described by Eqs. (11) and (12) [15, 16].

$$\begin{cases} V_s^q = -\omega_s \varphi_s^d + \dfrac{d\varphi_s^q}{dt} + R_s I_s^q \\ V_s^d = -\omega_s \varphi_s^q + \dfrac{d\varphi_s^d}{dt} R_s I_s^d \\ V_r^q = -(\omega_s - \omega_r)\varphi_r^d + \dfrac{d\varphi_r^q}{dt} + R_r I_r^q \\ V_r^d = -(\omega_s - \omega_r)\varphi_r^q + \dfrac{d\varphi_r^d}{dt} + R_r I_r^d \end{cases} \tag{11}$$

where flux linkages can represented as [16]

$$\begin{cases} \varphi_s^d = MI_r^d + L_s I_s^d \\ \varphi_s^q = MI_r^q + L_s I_s^q \\ \varphi_r^d = MI_s^d + L_r I_r^d \\ \varphi_r^q = MI_s^q + L_r I_r^q \end{cases} \tag{12}$$

Thus, for evidently simplified reasons, a stator flux along axis d and a reference d-q associated with the rotating field has been established [17].

$$\varphi = \varphi_s^d = \varphi_s \text{ and } \varphi_s^q = 0 \tag{13}$$

The flux linkages Eq. (10) becomes

$$\begin{cases} \varphi_s = L_s I_s^d + MI_r^d \\ 0 = L_s I_s^q + MI_r^q \end{cases} \tag{14}$$

We can presume that the electrical system is stable based on feasible assumptions about the wind turbine generators, which leads to a constant stator flux φs. Additionally, we may ignore the stator resistance. These results lead us to [18].

$$\begin{cases} V_s^d = 0 \\ V_s^q = V_s = \omega_s \varphi_s \end{cases} \tag{15}$$

The Real and reactive stator powers of a DFIG are expressed in the two-phase reference as follows:

$$\begin{cases} P_s = V_s^q I_s^q + V_s^d I_s^d \\ Q_s = -V_s^d I_s^q + V_s^q I_s^d \end{cases} \tag{16}$$

This Eq. (16) can be modified to simplify assumptions, obtaining:

$$
\begin{aligned}
P_s &= -V_s \frac{M}{L_s} i_r^q \\
Q_s &= -V_s \frac{M}{L_s} i_r^d + \frac{V_s^2}{L_s \omega_s}
\end{aligned} \tag{17}
$$

The above expression shows that both real and reactive can be regulated by the q and d-axis rotor currents.

5 MPPT Algorithms

5.1 *Perturb & Observe (P&O)*

PV applications require the use of MPPT methods. They are employed to maximise the photovoltaic output power of the array. Because it is the easiest to implement, the most used MPPT approach is Perturb and Observe (P&O). It entails the introduction of two voltage and current electrical sensors. The P&O algorithm is depicted in Fig. 6 [5, 19]. Using the current measured parameters, voltage Vpv(k) and current Ipv(k), the present power P(k) is computed and compared to the past power P(k-l). The subsequent duty cycle shift follows the same path if the output power rises. If not, there is an opposing change in the duty cycle.

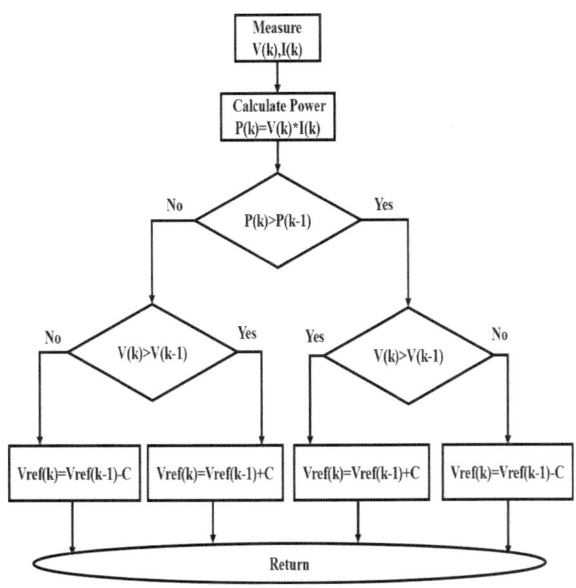

Fig. 6 Flowchart of P&O algorithm

5.2 *Cuckoo Search Optimization (CSO)*

The most well-known brood parasite bird is the Cuckoo Search Algorithm, a meta-heuristic algorithm that explains how to find the global solution for numerical optimisation problems involving Cuckoos. The CSO algorithm flowchart is shown in Fig. 7 [20]. The search process may become significantly slower when a set step is involved in the exploration and exploitation behaviour.

A random walk, or levy flight, has randomly chosen routes and steps that are defined in relation to step-lengths, which have a specific probability distribution. Both animals and insects exhibit this haphazard gait. What happens next depends on where you are right now [8, 21].

$$Z_i(t+1) = Z_i(t) + \alpha \otimes Levy(\lambda) \tag{18}$$

For any step size $\alpha > 0$. Assume $\alpha = 1$ in the majority of the situations. Enter-wise multiplication, or exclusive OR operation, is what the product symbolically represents. Following a levy distribution, a random walk with a random step size is known as a levy flight [8].

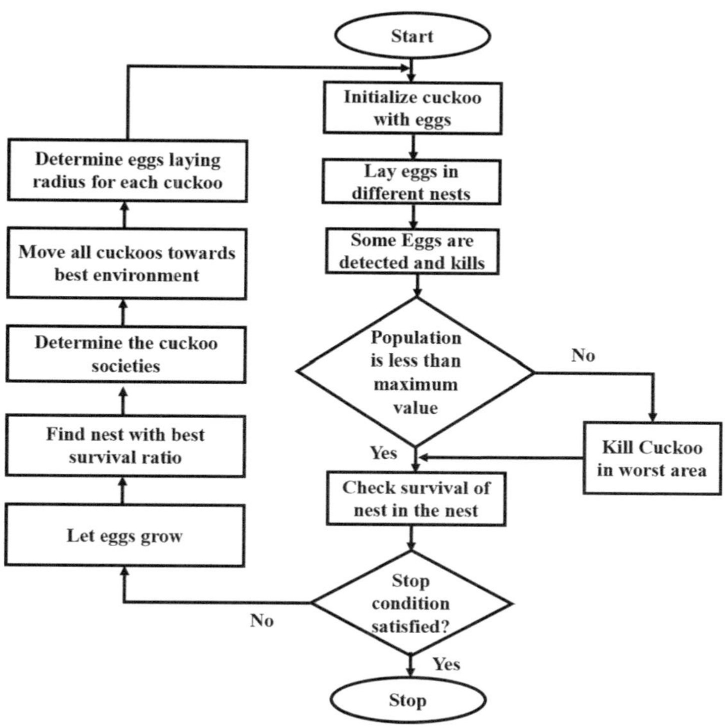

Fig. 7 Flowchart of CSO algorithm

$$Levy(\lambda) \sim u = t^{-\lambda}(1 < \lambda \leq 3) \quad (19)$$

where u represents the random variable and t represents the time taken by the iterations.

6 Results and Discussion

The simulation is run for the time t = 4 s with different solar irradiance as shown in Fig. 8 and constant wind speed. The rotor of DFIG rotates at a speed of 120 rad/sec which is lower than the synchronous speed (157 rad/sec) as shown in Fig. 9. Two MPPT optimisation algorithms are investigated in the overall system and it can be explained in the number of cases as given.

Case 1:- When solar irradiance is 0 W/m^2 (from t = 0 s to t = 0.5 s).

In this region, there is absence of solar irradiance (0 W/m^2). The PV voltage and PV current are zero as shown in Figs. 10 and 11. Similarly, the PV power also becomes zero during this region as shown in Fig. 12. The power transfers to the grid is only due to the power generated by the wind turbine as shown in Fig. 13.

Case 2:- When solar irradiance is 250 W/m^2 (from t = 0.5 s to t = 2 s).

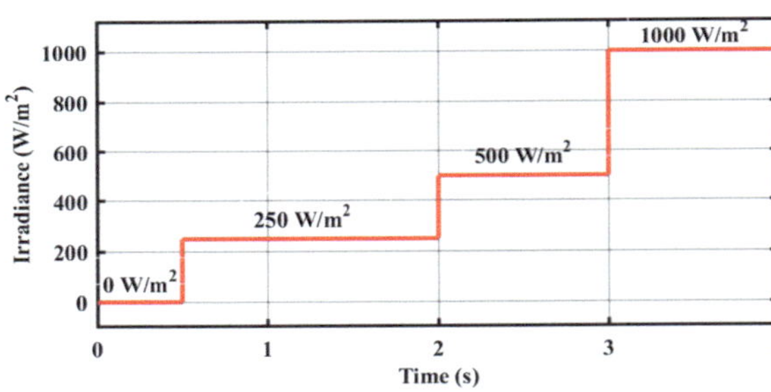

Fig. 8 Various stages of solar irradiance

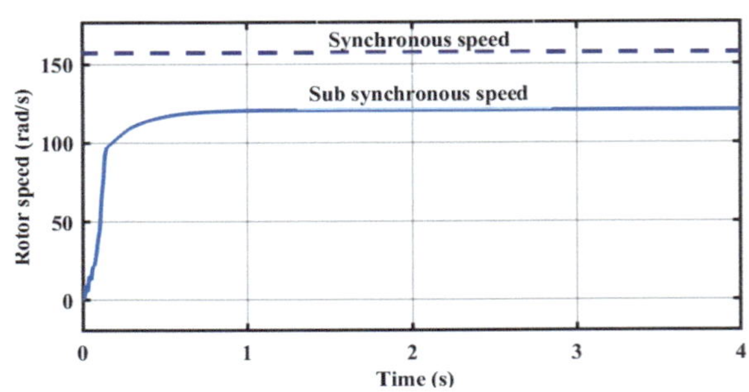

Fig. 9 Rotor speed of DFIG

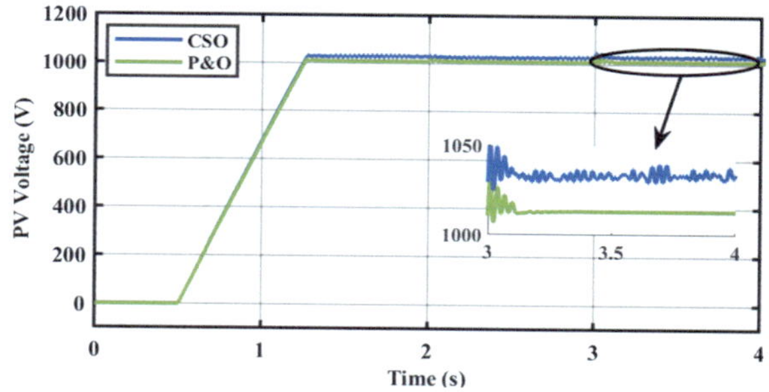

Fig. 10 PV voltage at different irradiances

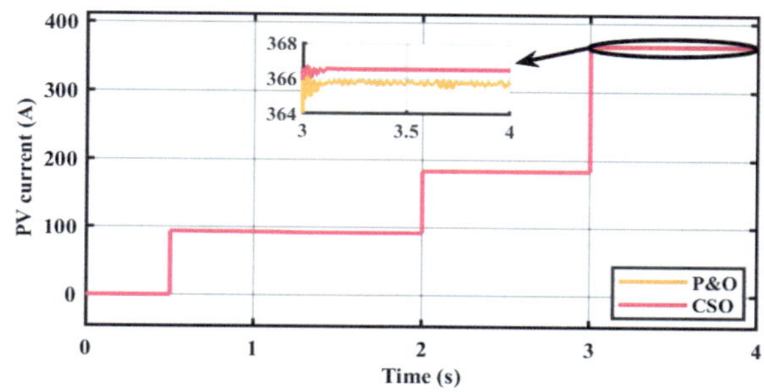

Fig. 11 PV current at different irradiances

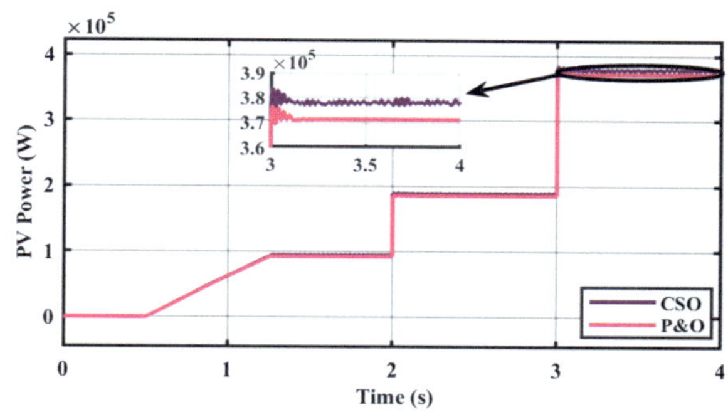

Fig. 12 PV power at different irradiances

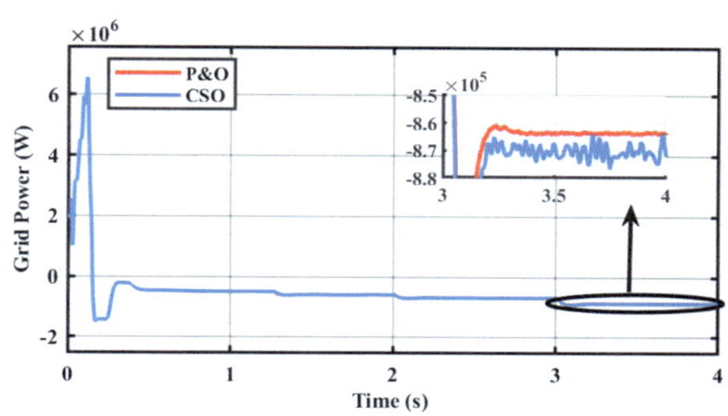

Fig. 13 Grid power at different irradiances

Table 1 Comparison of P&O and CSO at different solar irradiance

Irradiance (W/m^2)	Pg_{max}(KW)	P&O		CSO	
		Pg (KW)	Efficiency (%)	(Pg) (KW)	Efficiency (%)
250	−592.39	−583.75	98.54	−585.71	98.87
500	−695.42	−676.71	97.31	−680.30	97.82
1000	−894.33	−862.40	96.43	−868.91	97.15

In this region, PV voltage increases continuously at reach steady state at approximately time t = 1.2 s and constant current. Due to the effect of rising voltage, PV power also continuously increases and reaches a steady state at time t = 1.2 s as shown in Fig. 12. The grid power is transferred by the DFIG only till t = 1.2 s and after reaching the steady state there is an increment of grid power as shown in Fig. 13.

Case 3:- When solar irradiance is 500 W/m^2 (from t = 2 s to t = 3 s).

In this region solar irradiance increases to 500 W/m^2 the effect on PV voltage, PV current, PV power and Grid power can be shown in Figs. 10, 11, 12, and 13. The CSO algorithm gives more grid power in comparison to the P&O algorithm also the CSO optimization is more effective and gives higher efficiency.

Case 4:- When solar irradiance is 1000 W/m^2 (from t = 3 s to t = 4 s).

In this region, PV power increases to 371.3 KW for the P&O technique and 377.8 KW for the CSO optimization technique. The CSO can generate higher power than the P&O. The total grid power and the efficiency of the system is shown in Table 1.

Table 1 shows the power transferred to the grid through the wind energy system and PV system at different solar irradiances. Negative power shows the power received by the grid. Also, CSO techniques give a higher efficiency than the P&O technique at different solar irradiances.

7 Conclusion

This paper discusses the modelling and simulation study of a hybrid wind PV system. The system is investigated at the low wind speed and different solar irradiance, which transmit power to the grid. There are two techniques (P&O and CSO) employed to extract the maximum power from the PV panel and the performance of both techniques are compared. The comparative study shows that the CSO optimization technique gives higher efficiency and transfers more power to the grid. The power transfer by the grid to the rotor is reduced and the requirement of the rotor power is fulfilled by the PV system. This analysis gives a novel approach to designing a wind PV hybrid system at a location where wind speed is low but has a sufficient amount of solar radiation to operate the system effectively and efficiently.

References

1. Indragandhi V, Subramaniyaswamy V, Logesh R (2017) Resources, configurations, and soft computing techniques for power management and control of PV/wind hybrid system. Renew Sustain Energy Rev 69:129–143, https://doi.org/10.1016/j.rser.2016.11.209
2. Noureldeen O, Ibrahim AMA (2018) Modeling, implementation and performance analysis of a grid-connected photovoltaic/wind hybrid power system. Proceedings. 2018 International conference innovation trends computer engineering ITCE 2018, pp 296–304, 2018, https://doi.org/10.1109/ITCE.2018.8316641
3. Sanjari MJ, Gooi HB, Nair NKC (2020) Power generation forecast of hybrid PV-Wind system. IEEE Trans Sustain Energy 11(2):703–712. https://doi.org/10.1109/TSTE.2019.2903900
4. Kiran Kumar BM, Indira MS, Nagaraja Rao S (2022) Mathematical modeling and evaluation of performance characteristics of a hybrid solar PV and wind energy system. J Appl Sci Eng 25(4):685–697, https://doi.org/10.6180/jase.202208_25(4).0014
5. Slama A, Hamouda M, Khiat M (2020) Pv/wind hybrid energy system, modeling and simulation at variable weather conditions. Indones. J Electr Eng Informatics 8(4):637–647. https://doi.org/10.11591/ijeei.v8i4.2771
6. Nasef SA, Hassan AA, ElMadany HT, Zahran MB, El-Shaer MK, Abdelaziz AY (2022) Optimal power management and control of hybrid photovoltaic-battery for grid-connected doubly-fed induction generator based wind energy conversion system. Int J Renew Energy Res 12(1):408–421. https://doi.org/10.20508/ijrer.v12i1.12770.g8422
7. Díaz Martínez D, Trujillo Codorniu R, Giral R, Vázquez Seisdedos L (2021) Evaluation of particle swarm optimization techniques applied to maximum power point tracking in photovoltaic systems. Int J Circuit Theory Appl 49(7):1849–1867, https://doi.org/10.1002/cta.2978
8. N. K. N. K, Grid connected wind and photovoltaic hybrid energy system, no 1, pp 1–8
9. Hendrawati D, Soeprijanto A, Ashari M (2017) High performance maximum power point tracking on wind energy conversion system. Int J Power Electron Drive Syst 8(3):1359–1367. https://doi.org/10.11591/ijpeds.v8i3.pp1359-1367
10. Tang W, Hu J, Chang Y, Liu F (2018) Modeling of DFIG-based wind turbine for power system transient response analysis in rotor speed control timescale. IEEE Trans Power Syst 33(6):6795–6805. https://doi.org/10.1109/TPWRS.2018.2827402
11. Gude S, Jana KC (2020) Parameter extraction of photovoltaic cell using an improved cuckoo search optimization. Sol Energy 204(February):280–293. https://doi.org/10.1016/j.solener.2020.04.036
12. Vinod, Kumar R, Singh SK (2018) Solar photovoltaic modeling and simulation: As a renewable energy solution. Energy Reports 4:701–712, https://doi.org/10.1016/j.egyr.2018.09.008
13. Alhussain HMA, Yasin N (2020) Modeling and simulation of solar PV module for comparison of two MPPT algorithms (P&O & INC) in MATLAB/Simulink. Indones. J Electr Eng Comput Sci 18(2):666–677. https://doi.org/10.11591/ijeecs.v18.i2.pp666-677
14. Wadawa B, Errami Y, Obbadi A, Sahnoun S, Aoutoul M, Control of wind energy conversion system connected to the power grid and based on double fed induction generator
15. Bounifli F, Moudden AEL, Wahabi A, Hmidat A (2016) Vector control of a doubly-fed induction generator by using a classical PI and a fuzzy PI controllers, no 2, pp 18–24
16. Yaichi I, Semmah A, Wira P (2019) Control of doubly fed induction generator with maximum power point tracking for variable speed wind energy conversion systems, pp 1–10
17. Kumar A, Giribabu D (2016) Comparative study of the control strategies for the doubly fed induction, pp. 1–7
18. Ghosh S, Kamalasadan S, Senroy N, Enslin J (2016) Doubly fed induction generator (DFIG)-based wind farm control framework for primary frequency and inertial response application. IEEE Trans Power Syst 31(3):1861–1871. https://doi.org/10.1109/TPWRS.2015.2438861
19. Kumar GBA, Shivashankar (2022) Optimal power point tracking of solar and wind energy in a hybrid wind solar energy system. Int J Energy Environ Eng 13(1):77–103, https://doi.org/10.1007/s40095-021-00399-9

20. Pentapalli VG, Varma KR (2016) Cuckoo search optimization and its applications: a review. Int J Adv Res Comput Commun Eng ISO 5(11):556–562, https://doi.org/10.17148/IJARCCE.2016.511119
21. Eltamaly AM (2021) An improved cuckoo search algorithm for maximum power point tracking of photovoltaic systems under partial shading conditions. Energies 14(4), https://doi.org/10.3390/en14040953

Selection of Solar Panel Using Multi-criteria Decision Making Approach

Amit Patel, Rishi Dwivedi, and Kanika Prasad

Abstract In the changing geopolitical landscape, it's imperative that nations diversify their 'energy mix' with emphasis on self-reliance through sustainable and renewable energy. Photo Voltaic (PV) systems are capable of producing electricity through the sunlight gathered by solar panels. These play a crucial role in remote locations where people cannot get power from grids. The energy produced may be stored in batteries or directly used. The main component in a PV system is the solar panel. Selecting an appropriate solar panel is a complicated task as it involves evaluation of several available options on the basis of different criteria. The selection criteria are categorized into electrical, mechanical and economic. These criteria may sometimes be conflicting in nature, thus making the decision making process more complex. Therefore, in this paper, a multi-criteria decision making method is used to solve solar panel selection problem. In this hybrid approach, a method based on the removal effects of criteria (MEREC) is used for setting criteria weights, while an alternative ranking order method accounting for two-step normalization (AROMAN) is used for determining the best solar panel alternative for PV system.

Keywords Solar panel · MCDM · MEREC · AROMAN

A. Patel · K. Prasad (✉)
Production and Industrial Engineering, NIT Jamshedpur, Adityapur, India
e-mail: kprasad.prod@nitjsr.ac.in

A. Patel
e-mail: 2021ugpi030@nitjsr.ac.in

R. Dwivedi
Department of Finance, XISS, Ranchi, Jharkhand, India
e-mail: rishidwivedi@xiss.ac.in

S. Kumar et al. (eds.), *Soft Computing Applications in Modern Power and Energy Systems*, Lecture Notes in Electrical Engineering 1298,
https://doi.org/10.1007/978-981-96-0058-8_25

1 Introduction

The dynamic landscape of global geopolitics compels nations to fundamentally rethink their energy strategies. Prioritizing energy security and sustainability necessitates a strategic diversification of energy sources, with an urgent emphasis on harnessing domestic renewable resources. Photovoltaic (PV) systems stand at the forefront of this transition, offering a direct path to electricity generation from sunlight. Their decentralized nature, adaptability to various scales, and ability to integrate with energy storage systems make them particularly valuable for supplementing or replacing traditional power grids, especially in areas with limited infrastructure.

At the heart of any PV system lies the critical decision of selecting the appropriate solar panel [1]. This selection process is inherently complex, as the market presents a vast and often bewildering array of panels, each boasting a unique set of characteristics. Therefore, optimizing panel selection demands careful consideration of a wide range of technical, mechanical, economic, and societal factors that frequently present complex interdependencies [2].

Technical aspects, such as a panel's power output and solar cell conversion efficiency, have a direct bearing on the energy generation capacity of the system [3]. Mechanical durability is of paramount importance, as panels must be designed in such a way that they can endure severe environmental conditions which includes high temperatures, precipitation, wind, and potential physical damage–to ensure reliable performance over their expected lifespan. Economic considerations further complicate the decision-making process. Factors such as the initial capital investment, projected return on investment (ROI) timeframes, and recurring maintenance costs are crucial determinants of the financial viability of a PV installation. Additionally, subjective yet important aspects, such as a panel's visual appeal, brand reputation, and alignment with community values, can significantly influence decisions for individuals, businesses, and communities. Crucially, the environmental impact of solar panels, encompassing their entire lifecycle from the manufacturing process to their ultimate disposal, has emerged as a central concern in the pursuit of truly sustainable energy systems. Besides, minimizing the environmental footprint necessitates careful analysis of factors such as the energy intensity of production, material sourcing, and recyclability [4].

To navigate the multifaceted landscape of solar panel selection, this study advocates for the adoption of a multi-criteria decision-making (MCDM) framework. MCDM methodologies offer structured and transparent approaches for addressing challenges characterized by multiple, often conflicting objectives. This proposed framework integrates the Method based on the Removal Effects of Criteria (MEREC) for criteria weighting and the Alternative Ranking Order Method Accounting for Two-Step Normalization (AROMAN) for ranking solar panel options. Through this integration, the decision-making process becomes comprehensive and robust, aiming to facilitate well-informed decisions. The aim of current framework is to empower stakeholders with the tools necessary to make informed decisions, thereby catalysing the widespread adoption of PV systems. This, in turn, supports the broader transition toward sustainable and self-reliant energy solutions.

While various decision-making approaches have been applied to the solar panel selection problem in the past, the specific combination of MEREC and AROMAN represents a novel contribution to the field. MEREC offers a distinct advantage in determining the priority weights of the diverse evaluation criteria, while AROMAN provides a robust ranking system that considers the multi-layered nature of the problem. This synergistic utilization of MEREC and AROMAN enhances the efficacy of the decision-making process, promising more informed and nuanced outcomes in the realm of solar panel selection for PV systems.

2 Methodology

The MCDM problem is divided into two major steps. The first step is to determine the weights of factor which influence the selection process and the other is assessing the performance of feasible alternatives based on the evaluation criteria. The present study integrates two methods, namely MEREC and AROMAN method. MEREC is used to calculate criteria weights, and subsequently AROMAN method is applied to evaluate the available alternatives of solar panels for PV systems. The MEREC and AROMAN methods have been successfully applied for solving MCDM problems [5].

2.1 MEREC Method

In the first stage of decision-making, MEREC method is used to determine the criteria weights. This method was presented by Keshavarz-Ghorabaee et al. [6]. For this method, a performance measure for the alternatives is defined first. In this case, a simple logarithmic scale is utilized with equal weights to assess the performance of alternatives. Subsequently, priority weights for each criterion are calculated considering the effect on performance of feasible alternatives when one criterion is removed. The criterion having the most impact on the performance is assigned higher weights. Absolute deviation measure is utilized for understanding the effect of criterion removal. This measure reflects the change in the overall performance of an alternative when a criterion is removed. The following steps are used to calculate objective weights by MEREC method.

First Stage: Six steps followed in the MEREC method are as follows:

Step 1. Equation (1) is used to create the decision matrix (Z),where the matrix comprises n criteria (j = 1,2,3,……,n) and m alternatives (i = 1,2,3,…….n):

$$Z = \begin{matrix} Z_{11} & \ldots\ldots\ldots\ldots & Z_{1j} & \ldots\ldots\ldots\ldots & Z_{1n} \\ . & . & . & . & . \\ . & . & . & . & . \\ Z_{i1} & \ldots\ldots\ldots & Z_{ij} & \ldots\ldots\ldots & Z_{in} \\ . & . & . & . & . \\ . & . & . & . & . \\ Z_{m1} & \ldots\ldots\ldots & Z_{mj} & \ldots\ldots\ldots & Z_{mn} \end{matrix} \quad (1)$$

Step 2. Equation (2) is employed to obtain the normalized decision matrix (N_{ij}^{x}):

$$N_{ij}^{x} = \left\{ \frac{minZ_{ij}}{Z_{ij}}, \text{ } for\ benefecial\ criteria \right. \\ = \left\{ \frac{Z_{ij}}{maxZ_{ij}}, \text{ } for\ cost\ criteria \quad \forall \text{i, j.} \right. \quad (2)$$

Step 3. The performance score (S_i) of the alternatives are calculated while using Eq. (3):

$$S_i = ln\left(1 + \left(\frac{1}{m}\sum_j \left|ln\left(N_{lj}^{x}\right)\right|\right)\right), \forall \text{i.} \quad (3)$$

Step 4. Each criterion value is subtracted from performance score to obtain partial performance values ($S_{ij}^{;}$) of the alternatives, as defined in Eq. (4):

$$S_{ij}' = \ln\left(1 + \left(\frac{1}{m}\sum\nolimits_{j\|k} \left|\ln\left(N_{lj}^{x}\right)\right|\right)\right), \forall \text{i, j.} \quad (4)$$

Step 5. Equation (5) is applied to calculate the sum of absolute deviations (E_j):

$$E_j = \sum_i \left|S_{ij}^{;} - S_i\right|, \forall \text{j.} \quad (5)$$

Step 6. Finally, priority weight (W_j) for each criterion is calculated using Eq. (6):

$$W_j = \frac{E_j}{\sum_k E_k}, \forall \text{j.} \quad (6)$$

2.2 AROMAN Method

This method combines the normalized data from two-step normalization and obtains an average matrix from normalized data [7]. The AROMAN method has various advantages. While using this method for evaluation, both qualitative and quantitative evaluation can be done. Also, application of two different normalization techniques strengthens the process and permits sensitivity analysis. This methodology is capable of differentiating the criteria into beneficial and non-beneficial.

Second stage: There are six steps involved in the AROMAN method which are as follows:

Step 7. Equation (7) is employed to perform linear normalization (K_{ij}):

$$K_{ij} = \frac{Z_{ij} - minZ_{ij}}{maxZ_{ij} - minZ_{ij}}, \forall \text{i, j.} \tag{7}$$

Step 8. Equation (8) is utilized to perform vector normalization (K_{ij}^*) process:

$$K_{ij}^* = \frac{Z_{ij}}{\sqrt{\sum_{i=1}^{m} Z_{ij}^2}}, \forall \text{i, j.} \tag{8}$$

*Step 9***.** The aggregate normalization (K_{ij}^{norm}) process is done through Eq. (9):

$$K_{ij}^{norm} = \frac{\beta K_{ij} + (1 - \beta)K_{ij}^*}{2}, \forall \text{i, j.} \tag{9}$$

where, β is the weighting factor that has value between 0 and 1.

*Step 10***.** The values obtained in step 9 are multiplied with priority weights of the criteria as displayed in Eq. (10):

$$\widehat{K_{ij}} = W_{ij} * K_{ij}^{norm}, \forall \text{i, j.} \tag{10}$$

*Step 11***.** The criteria which are non-beneficial in nature need to be minimized. They are calculated while using Eq. (11):

$$L_i = \sum_{j=1}^{n} \widehat{K_{ij}^{(min)}}, \forall \text{i} \tag{11}$$

whereas, the values for beneficial criteria should be maximized. The values are calculated using Eq. (12:

$$P_i = \sum\nolimits_{j=1}^{n} \widehat{K_{ij}^{(max)}}, \forall \text{i}, \tag{12}$$

Step 12. Equation (13) is employed to calculate the R_i values. These values are arranged in ascending order to obtain final ranks of the alternatives:

$$R_i = L_i^{\lambda} + P_i^{(1-\lambda)}, \forall \text{i}, \quad (13)$$

The alternative having maximum R_i value is identified as the most appropriate option. The parameter λ denotes the coefficient of criterion variety. This parameter may be used with several variations. However, the value of λ is usually taken up as 0.5 to avoid any undefined results in an MCDM problem consisting of only benefit or cost criteria.

3 Illustrative Example

Balo and Şağbanşua [8] considered a problem of selecting the best solar panel alternative for a PV system and solved that using Analytic Hierarchy Process (AHP) method. Since the AHP method considers pairwise comparison, if the numbers of evaluation criteria are more the process becomes time consuming. Therefore, in this study, an integrated MCDM approach is used to determine the best solar panel for a PV system. The evaluation criteria for solar panel selection [9–12] which are often used are categorized into electrical, mechanical and economic. These criteria are further divided into 12 sub criteria. All the criteria for evaluation are considered as beneficial for which a higher value is desired. For cost also a higher value is acceptable as it is related to the quality of the solar panel. Six alternative solar panels are evaluated using the proposed technique. The decision matrix showing the performance of different solar panel alternatives on the selected criteria which are shown below in Table 1.

Table 1 Decision matrix for solar panel selection

	Power Rating	STC	Peak Efficiency	Number of Cells	Imp	Vmp	Isc	Voc	Energy Density	Weight	Cost/ Watt	Area
	(W)	(W/ m^2)	%		(A)	(V)	(A)	(V)	(W/ m^2)	(kg)	($)	(m^2)
	C1	C2	C3	C4	C5	C6	C7	C8	C9	C10	C11	C12
P1	175	142.1	14.21	50	8.16	24.5	8.7	30.8	11.52	15.4	105	1.4
P2	180	156.7	15.67	72	5.42	36.9	5.8	45.6	11.95	14.5	105	1.27
P3	179	136.5	14.2	54	7.89	25.38	8.24	33.53	11.82	35	175	1.46
P4	184.5	156.7	15.67	72	5.17	38.7	5.5	45.9	13.28	15.5	105	1.27
P5	177.5	135.1	13.5	54	7.6	26.3	8.22	33.3	11.26	18	105	1.48
P6	185.9	172.3	17.2	96	5.59	55.8	3.83	68.7	14.89	15	210	1.16

While solving a problem using MCDM method, the first step is to calculate priory weights of the criteria considered for evaluation. In this paper, MEREC method is followed to determine the criteria weights. The priority weights of the criteria obtained through MEREC method is displayed in Table 2.

Once the priority weights are determined, the next step is to evaluate the alternatives for which AROMAN approach is used. After following all the steps mentioned in Sect. 2.2 of Methodology section, performance score for solar panels is calculated. Finally, the alternatives are ranked as per increasing order of their performance score as shown in Table 3.

Balo and Şağbanşua [6] identified P6 as the best alternative. While applying the proposed method P6 is identified as the best alternative and P3 is identified as the second best alternative. Alternative P2 emerges out as the leas preferred option among all the six alternatives based on the considered criteria. Also, overall the obtained results may be justified by the performance of different solar panels shown in the decision matrix.

4 Conclusions

The world is constantly moving towards sustainable sources of energy. PV systems are an electric power system designed to supply usable solar power by means of solar panels. There are numerous solar panels available in the market and it is important to choose wisely among them. Since the selection process is governed by numerous factors, in this study, a hybrid MCDM approach combining MEREC and AROMAN methodology is used for solving the solar panel selection problem. This hybrid methodology is not subjected to biasness and performs evaluation based on the performance of alternatives. The study provides decision support to the academics and practitioners to deepen their understanding about solar panels as an alternative for sustainable energy and make better-informed decisions related to solar panel selection. The proposed methodology may be applied for various other kinds of decision-making problems. Future scope of this work may be to perform sensitivity analysis to study the effect of change in weights assigned to evaluation criteria on the performance of alternatives.

Table 2 Priority weights of criteria considered for solar panel selection

Power Rating	STC	Peak Efficiency	Number of Cells	Imp	Vmp	Isc	Voc	Energy Density	Weight	Cost/Watt	Area
(W)	(W/m^2)	%		(A)	(V)	(A)	(V)	(W/m^2)	(kg)	($)	(m^2)
C1	C2	C3	C4	C5	C6	C7	C8	C9	C10	C11	C12
0.0089	0.0298	0.0321	0.0770	0.0740	0.0900	0.1666	0.0882	0.0287	0.2069	0.1535	0.0443

Table 3 Performance score and ranking for solar panel selection

	Performance score	Rank
P1	0.189721	3
P2	0.172156	6
P3	0.235581	2
P4	0.174574	5
P5	0.18858	4
P6	0.249714	1

References

1. Elboshy B, Alwetaishi M, Aly RM, Zalhaf AS (2022) A suitability mapping for the PV solar farms in Egypt based on GIS-AHP to optimize multi-criteria feasibility. Ain Shams Eng J 13(3):101618
2. Kozlov V, Sałabun W (2021) Challenges in reliable solar panel selection using MCDA methods. Procedia Comput Sci 192:4913–4923
3. Hasan A, Dincer I (2020) A new performance assessment methodology of bifacial photovoltaic solar panels for offshore applications. Energy Convers Manage 220:112972
4. Krishankumar R et al (2021) Solving renewable energy source selection problems using a q-rung orthopair fuzzy-based integrated decision-making approach. J Cleaner Production 279:123329
5. Kara K et al (2024) The MEREC-AROMAN method for determining sustainable competitiveness levels: a case study for Turkey. Socioecon Plann Sci 91:101762
6. Keshavarz-Ghorabaee M et al (2021) Determination of objective weights using a new method based on the removal effects of criteria (MEREC). Symmetry 13(4):525
7. Boˇskoviˊc S et al (2023) An alternative ranking order method accounting for two-step normalization (AROMAN)–a case study of the electric vehicle selection problem. IEEE Access 11:39496–507
8. Baloa F, Şağbanşua (2016) The selection of the best solar panel for the photovoltaic system design by using AHP. In: 3rd International conference on power and energy systems engineering, energy Procedia, pp 50–53. Japan
9. Bruce T (2011) Investigation of cost and performance characteristic of photovoltaic panels. Research Dissertation, ENG 4111/4112 Project
10. IEA-PVPS (2011) Trends in photovoltaic applications a survey report of selected IEA Countries between 1992 and 2010. Report IEA-PVPS T1-20
11. Yılmaz S, Ozcalik HR (2015) Kesler, S,: The analysis of different PV power systems for the determination of optimal PV panel and system installation-a case study in Kahramanmaras. Renew Sustain Energy Rev 52:1015–1024
12. Haktanıra E, Kahramana C (2019) A novel interval-valued Pythagorean fuzzy QFD method and its application to solar photovoltaic technology development. Comput Ind Eng 132:361–372

MIX
Papier aus verantwortungsvollen Quellen
Paper from responsible sources
FSC® C105338

If you have any concerns about our products,
you can contact us on
ProductSafety@springernature.com

In case Publisher is established outside the EU,
the EU authorized representative is:
Springer Nature Customer Service Center GmbH
Europaplatz 3, 69115 Heidelberg, Germany

Printed by Libri Plureos GmbH
in Hamburg, Germany